INTRODUCTION TO MATHEMATICAL ANALYSIS

International Series in Pure and Applied Mathematics

G. Springer
Consulting Editor

Ahlfors: *Complex Analysis*
Bender and Orszag: *Advanced Mathematical Methods for Scientists and Engineers*
Buck: *Advanced Calculus*
Busacker and Saaty: *Finite Graphs and Networks*
Cheney: *Introduction to Approximation Theory*
Chester: *Techniques in Partial Differential Equations*
Coddington and Levinson: *Theory of Ordinary Differential Equations*
Conte and de Boor: *Elementary Numerical Analysis: An Algorithmic Approach*
Dennemeyer: *Introduction to Partial Differential Equations and Boundary Value Problems*
Dettman: *Mathematical Methods in Physics and Engineering*
Hamming: *Numerical Methods for Scientists and Engineers*
Hildebrand: *Introduction to Numerical Analysis*
Householder: *The Numerical Treatment of a Single Nonlinear Equation*
Kalman, Falb, and Arbib: *Topics in Mathematical Systems Theory*
McCarty: *Topology: An Introduction with Applications to Topological Groups*
Moore: *Elements of Linear Algebra and Matrix Theory*
Moursund and Duris: *Elementary Theory and Application of Numerical Analysis*
Parzynski and Zipse: *Introduction to Mathematical Analysis*
Pipes and Harvill: *Applied Mathematics for Engineers and Physicists*
Ralston and Rabinowitz: *A First Course in Numerical Analysis*
Ritger and Rose: *Differential Equations with Applications*
Rudin: *Principles of Mathematical Analysis*
Shapiro: *Introduction to Abstract Algebra*
Simmons: *Differential Equations with Applications and Historical Notes*
Simmons: *Introduction to Topology and Modern Analysis*
Struble: *Nonlinear Differential Equations*

INTRODUCTION TO MATHEMATICAL ANALYSIS

William R. Parzynski
Philip W. Zipse
Department of Mathematics and Computer Science
Montclair State College

McGraw-Hill Book Company

New York   St. Louis   San Francisco   Auckland   Bogotá   Hamburg
Johannesburg   London   Madrid   Mexico   Montreal   New Delhi
Panama   Paris   São Paulo   Singapore   Sydney   Tokyo   Toronto

This book was set in Times Roman by Santype-Byrd.
The editors were John J. Corrigan, James W. Bradley, and Claudia Tantillo;
the production supervisor was Phil Galea.
The drawings were done by J & R Services, Inc.
R. R. Donnelley & Sons Company was printer and binder.

INTRODUCTION TO MATHEMATICAL ANALYSIS

Copyright © 1982 by McGraw-Hill, Inc. All rights reserved.
Printed in the United States of America. Except as permitted under the United States Copyright Act of 1976, no part of this publication may be reproduced or distributed in any form or by any means, or stored in a data base or retrieval system, without the prior written permission of the publisher.

1234567890 DODO 898765432

ISBN 0-07-048845-2

Library of Congress Cataloging in Publication Data

Parzynski, William R.
    Introduction to mathematical analysis.

    (International series in pure and applied mathematics)
    Includes index.
    1. Mathematical analysis. I. Zipse, Philip W.
II. Title. III. Series.
QA300.P36   515   81-19304
ISBN 0-07-048845-2   AACR2

CONTENTS

	Preface	vii
Chapter 1	**Real Numbers and Functions**	1
1.1	Sets and Functions	1
*1.2	The Natural Numbers	6
*1.3	Development of Rational Numbers	11
*1.4	Construction of the Real Number System	19
1.5	Properties of Real Numbers	26
1.6	Functions and One-to-One Correspondences	29
Chapter 2	**Sequences and Sets of Real Numbers**	37
2.1	Limits of Sequences	37
2.2	Bounded Sequences	45
2.3	Sets of Real Numbers	53
Chapter 3	**Functions and Limits**	59
3.1	Bounded Functions	59
3.2	Limits of Functions	65
3.3	One-Sided Limits, Infinite Limits, and Limits at Infinity	77
3.4	Monotone Functions	89
Chapter 4	**Continuous Functions**	94
4.1	Continuity	94
4.2	Properties of Continuous Functions	102
4.3	Uniform Continuity	110
*4.4	Further Topics on Continuity	117

* Optional sections.

Chapter 5 Differentiable Functions 126

- 5.1 The Derivative 126
- 5.2 Properties of Differentiable Functions 131
- 5.3 L'Hospital's Rule 138
- *5.4 Further Topics on Differentiation 144

Chapter 6 The Riemann Integral 149

- 6.1 Definition of the Integral 149
- 6.2 Properties of the Integral 162
- 6.3 Fundamental Theorem of Calculus 172
- *6.4 Necessary and Sufficient Conditions for Riemann Integrability 178

Chapter 7 Sequences and Series of Functions 184

- 7.1 Infinite Series of Real Numbers 184
- 7.2 Pointwise and Uniform Convergence 195
- 7.3 Importance of Uniform Convergence 201
- 7.4 Power Series and Taylor Series 208

Chapter 8 Differentiable Functions of Several Variables 217

- 8.1 Sets and Functions in $\mathbf{R}^n$ 217
- 8.2 Differentiable Functions 232
- 8.3 Chain Rules and Taylor's Formula 239
- 8.4 Implicit- and Inverse-Function Theorems 246
- 8.5 Extrema of Functions of Several Variables 254

Chapter 9 Multiple Integrals 261

- 9.1 The Double Integral 261
- 9.2 Evaluation of Double Integrals 271
- 9.3 Change of Variables in an Integral 281
- 9.4 Improper Integrals 286

Chapter 10 Metric Spaces 297

- 10.1 Definition and Examples 297
- 10.2 Open and Closed Sets; Topology 302
- 10.3 Convergence and Completeness 308
- 10.4 Continuity and Compactness 315
- 10.5 Applications 324

Solutions and Hints to Selected Exercises 330

Index 353

PREFACE

This book has been written to serve as a text for students in a first course in mathematical analysis. Such a course would usually follow rather quickly the traditional freshman calculus and generally appears in college catalogs under the name of Introduction to Analysis, Mathematical Analysis, or Advanced Calculus. The body of material which has come to be known as freshman calculus serves as a prerequisite. This presumes that the reader is equipped with good skills in advanced high school or college algebra and in trigonometry.

Today calculus serves a much more diverse audience than in years past. Not only are mathematics and natural science majors and engineering students taking calculus but there are growing numbers of business, statistics, and computer science students in these courses. Traditionally, the purpose of freshman calculus has been to teach the student the facts and applications of calculus. This means that the student acquires skill in the mechanics of calculus and a certain level of proficiency at using calculus in those many areas where the methods of calculus prove useful. Often, however, little time in the freshman calculus course is spent on the theory which enhances one's understanding of the subject. This is especially true today, when the trend has been to expose the student as early as possible to a widening variety of calculus applications. It is the role of the sophomore or junior level mathematical analysis or advanced calculus course to provide the understanding that is so often lacking when the student leaves the freshman calculus course.

The authors believe that the purpose of advanced calculus is twofold: (1) to allow the student to become acquainted with, and develop a certain level of proficiency in, the techniques and methods of mathematical analysis (sometimes the proof is more important than the theorem) and (2) to be able to use these techniques and methods to reinforce and solidify an understanding of the learned calculus results. There is not an abundance of new material or facts to be learned; the emphasis instead is on looking at much the same calculus topics in greater depth and with a definite direction toward understanding. At the end of such a course the student should not only be better at doing and using calculus but should be well versed in the methods and techniques of mathematical analysis. This, after all, is the substance that provides an appreciation for the theory of calculus and lays the foundation for more advanced work in the mathematical sciences.

Since this book is intended for the reader who has not previously seen a theoretical and rigorous development of calculus, we have included numerous illustrative examples with detailed explanations. In the initial chapters some of the proofs tend to be expository in style, with a necessary sacrifice of elegance. This is done to provide more insight into the construction of mathematical proofs and to help develop the skills used in proving statements in mathematics. Many of the exercises call upon the student's ability to use the methods and techniques employed in the text, and, conversely, working the exercises provides a deeper and more thorough understanding of the theorems and their proofs. The exercises should be dealt with deliberately as they serve as an integral part of the text.

We have found that the first seven chapters provide a good arrangement of topics for a one-semester course in mathematical analysis. Chapter 1 contains a development of the real number system. If a developmental approach is desired then all the sections should be covered. For an axiomatic approach sections 1.2 to 1.4 can be skipped without loss of continuity. In either case the reader should strive for an overview of this material rather than a detailed exposition of each and every step in the construction process. Chapter 2 develops sequences and sets as tools to use in the calculus. The core of the material is in Chaps. 3 to 7, which should be covered in detail, except perhaps the optional sections (4.4, 5.4, and 6.4).

In a two-semester course the instructor may wish to include some or all of the optional sections, together with the last three chapters in the text. We have found that this can adequately be covered at a leisurely pace in such a one-year course. If it is desired to place more emphasis on multivariate calculus, Chaps. 8 and 9 should be supplemented with topics of the instructor's choosing, and Chap. 10 may become optional, depending on the audience and the direction of the particular course.

We would like to thank Professor George Springer for his many helpful suggestions during the various stages of development of the manuscript. We also appreciate the efforts of our typists, Elizabeth Bator, Betty Leszczak, and Amy Raskin.

William R. Parzynski
Philip W. Zipse

CHAPTER
ONE

REAL NUMBERS AND FUNCTIONS

1.1 SETS AND FUNCTIONS

In the study of analysis, as in many other areas of mathematics, the concepts of *set* and *function* are fundamental. Every language has certain words which are basic and remain undefined but of whose meaning there is universal acceptance. In mathematics the word "set" is such a term; a set is understood to be a *well-defined* collection of objects called *elements*. The term well-defined just means for us that some mechanism exists whereby one is able to determine whether or not a given element belongs to the set.

We denote sets by capital letters A, B, C, etc., and use lowercase letters a, b, c, etc., to represent elements. If an element x belongs to or (equivalently) is a member of the set S, we write $x \in S$; to designate that the element x does not belong to S we write $x \notin S$.

If each element in the set A is also a member of the set B, we say that A is a subset of B and write $A \subseteq B$ or equivalently $B \supseteq A$. We call two sets A and B *equal* and write $A = B$ provided $A \subseteq B$ and $B \subseteq A$. Two sets are equal, then, if the sets consist of precisely the same elements. If $A \subseteq B$ and $A \neq B$, we say that A is a proper subset of B and designate this by $A \subset B$ (equivalently $B \supset A$). If $A \subset B$, then every element in A is also in B but there is at least one element in B which fails to be in A.

In any discussion involving sets the letter U denotes the *universal set*, which is the set of all elements under discussion; the symbol $\emptyset$ denotes the

1

empty set, that set which contains no elements. Then for every set A we have

$$\emptyset \subseteq A \subseteq U$$

Specific sets can be defined by listing all the elements in the set or by stating a characteristic property which is unique to those elements belonging to the set. For example

$$G = \{\alpha, \beta, \gamma, \delta, \varepsilon\}$$

can be alternately defined by

$$G = \{x \mid x \text{ is one of the first five letters of the Greek alphabet}\}$$

which is read "G is the set of all elements x such that x is one of the first five letters of the Greek alphabet." When elements in a set are listed, each element should be listed exactly once; order is not important.

We use the above notation to define some natural ways of combining sets to construct new sets:

$$A \cup B = \{x \mid x \in A \text{ or } x \in B \text{ (or both)}\}$$

$A \cup B$ is called the *union* of A and B, read "A union B."

$$A \cap B = \{x \mid x \in A \text{ and } x \in B\}$$

$A \cap B$ is called the *intersection* of A and B, read "A intersection B."

$$A' = \{x \mid x \in U \text{ and } x \notin A\}$$

A' is called the *complement* of A (sometimes denoted $U - A$).

$$A \times B = \{(a, b) \mid a \in A \text{ and } b \in B\}$$

$A \times B$ is called the *cartesian product* of the sets A and B.

We can also consider the union and intersection of more than two sets. The union of a class of sets is just that set consisting of all those elements which belong to at least one set in the class. The intersection of a class of sets is the set consisting of all those elements which belong to each and every set in the class.

The cartesian product of A and B is the set of all ordered pairs (a, b), where a is any element in A and b is any element in B. Recall that the concept of ordered pairs of numbers was needed when graphing equations in the (x, y) plane. The plane is a geometric model of a cartesian product, and the graph of an equation is a subset of this cartesian product. The geometry allows us to visualize graphs and enhances our understanding of important concepts in algebra and calculus.

✱ **Definition 1.1** A *function* is a nonempty set X, a nonempty set Y, and a rule of correspondence f which associates with each element $x \in X$ a unique element $y \in Y$.

The element y associated with a given element $x \in X$ is denoted $f(x)$, and we write $y = f(x)$; y is called the *image of x under f*, and x is called a *preimage of y*. The function is often denoted by $f: X \to Y$ or sometimes just by f when the sets X, Y are clear from the context. The set X is called the *domain* of the function and the set $f(X) \subseteq Y$, defined by

$$f(X) = \{y \in Y \mid y = f(x) \text{ for some } x \in X\}$$

is called the *range* of the function. If $f(X) = Y$, we say that the function is *onto* Y. The graph of the function, written $gr(f)$, is a subset of the cartesian product $X \times Y$ and is defined by

$$gr(f) = \{(x, f(x)) \mid x \in X\}$$

As mentioned earlier, if X and Y are sets of numbers, the graph of f can be (and often is) visualized as a subset of the (x, y) plane. Some authors find it convenient to identify a function with its graph and to think of a function as a special kind of subset of the cartesian product $X \times Y$.

A function $f: X \to Y$ is called *one-to-one* if $f(x_1) = f(x_2)$ implies that $x_1 = x_2$, that is, if no two distinct elements in the domain of f are assigned to the same element in the range. Thus each range element has a unique preimage. One-to-one functions are useful because they allow us to define a new function called the *inverse function*. If $f: X \to Y$ is one-to-one, we define the function $f^{-1}: f(X) \to X$, read "f inverse," by the following: For each $y \in f(X)$ define $f^{-1}(y)$ to be the unique preimage of y under f. Then $f^{-1}(y) = x$ if and only if $f(x) = y$. It is clear that f^{-1} is onto X and that f^{-1} is one-to-one. If $f: X \to Y$ is one-to-one and onto Y then f is called a *one-to-one correspondence* between X and Y. In this case $f^{-1}: Y \to X$ is a one-to-one correspondence between Y and X.

Just as we can combine sets to construct new sets, we can combine given functions in the following way: If $f: X \to Y$ and $g: Y \to Z$, we define the *composite function* $g \circ f: X \to Z$ by $g \circ f(x) = g(f(x))$ for every $x \in X$. Thus composition of two functions is defined as the sequential action of the individual functions. For example, let f be the function which converts temperature in degrees Fahrenheit to temperature in degrees Celsius and let g be the function which converts temperature in degrees Celsius into absolute temperature in kelvins. Then $g \circ f$ is the function which converts temperature in degrees Fahrenheit into temperature in kelvins.

Two functions $f: X \to Y$ and $g: X \to Y$ are said to be *equal* and we write $f = g$ provided $f(x) = g(x)$ for every $x \in X$. If $f: X \to Y$, we shall sometimes want to consider f restricted to some nonempty subset $A \subseteq X$. We define $f_A: A \to Y$ by $f_A(x) = f(x)$ for every $x \in A$. If f is one-to-one then so is f_A for each nonempty $A \subseteq X$, and if f_A is onto Y for some nonempty $A \subseteq X$ then so is f onto Y. The function $i_X : X \to X$ defined by $i_X(x) = x$ for every $x \in X$ is called the *identity function on* X. i_X "maps" each element $x \in X$ onto itself. It is clear that i_X is one-to-one and onto X and that $i_X^{-1} = i_X$. If $f : X \to Y$ is one-to-one then $f^{-1} \circ f = i_X$ and $f \circ f^{-1} = i_{f(X)}$. In particular, if $f : X \to X$ is one-to-one and onto X then we have that $f^{-1} \circ f = f \circ f^{-1} = i_X$.

If X is a nonempty set, a *relation* R in X is a nonempty subset of $X \times X$; that is, $\emptyset \neq R \subseteq X \times X$. We have already encountered some relations in X; if $f : X \to X$ then $\text{gr}(f)$ is a relation in X. Of course, not every relation in X is the graph of some function. In order for the relation R in X to be the graph of a function it is necessary (and sufficient) that for each $x \in X$, R contains exactly one ordered pair with first component x. Graphs of functions are important relations, but there are other kinds of relations that are equally important. We investigate one type of relation, called an equivalence relation, in the following paragraph.

If R is a relation in X, we write $x \sim y$ provided $(x, y) \in R$.

Definition 1.2 A relation R in X is called an *equivalence relation* if:
(a) $x \sim x$ for every $x \in X$ (reflexive property).
(b) $x \sim y$ implies $y \sim x$ for all $x, y \in X$ (symmetric property).
(c) $x \sim y$ and $y \sim z$ implies $x \sim z$ for all $x, y, z \in X$ (transitive property).

Suppose $\mathcal{F} = \{f \mid f : X \to Y\}$; that is, $\mathcal{F}$ is the set of all functions f with domain X and range contained in Y. Let $x_0 \in X$ be fixed. We define the following relation R in $\mathcal{F} : (f, g) \in R$ (equivalently, $f \sim g$) provided $f(x_0) = g(x_0)$. It is clear that $f \sim f$ for every $f \in \mathcal{F}$, and if $f \sim g$ then $f(x_0) = g(x_0)$; hence $g(x_0) = f(x_0)$, and so $g \sim f$. Moreover, if $f \sim g$ and $g \sim h$ then $f(x_0) = g(x_0)$ and $g(x_0) = h(x_0)$. Therefore $f(x_0) = h(x_0)$ and $f \sim h$. It follows that R is an equivalence relation in $\mathcal{F}$.

A *partition* of a nonempty set X is a class of nonempty subsets of X which has the following property: Each element in X belongs to exactly one set in the class. One of the important facts about any equivalence relation in X is that it induces a partition of the set X. Suppose R is an equivalence relation in X. For each $x \in X$ we define the *equivalence class* of x by

$$[x] = \{y \in X \mid y \sim x\}$$

Theorem 1.1 The distinct equivalence classes of an equivalence relation in X form a partition of X.

PROOF Let R be an equivalence relation in X. Each equivalence class $[x]$ is a subset of X (by definition) and is nonempty since $x \in [x]$ (reflexive property). Moreover, each $x \in X$ is in at least one equivalence class, namely $x \in [x]$. It remains to show that an element $x \in X$ cannot be in two distinct equivalence classes. Suppose that $x \in [y]$. Then $x \sim y$ and $y \sim x$ by the symmetry property. Now, if $z \in [y]$ then $z \sim y$, and so $z \sim x$ by the transitive property. Thus $z \in [x]$, and so $[y] \subseteq [x]$. Similarly, if $z \in [x]$ then $z \sim x$ and by the transitive property $z \sim y$. Thus $z \in [y]$, and so $[x] \subseteq [y]$. It follows that $[y] = [x]$, and so $[x]$ is the only equivalence class containing x.

In our previous example of an equivalence relation in the set $\mathcal{F}$ of all functions f with domain X and range contained in Y, each equivalence class consists of all the functions in $\mathcal{F}$ which map x_0 to the same element in Y. Theorem 1.1 has a partial converse, which says that any partition of the nonempty set X induces an equivalence relation in X for which the distinct equivalence classes are precisely the sets in the partition. To see this define

$$R = \{(x, y) \mid y \text{ belongs to the same set in the partition as } x\}$$

The reflexive, symmetric, and transitive properties are easily verified. Therefore, in this sense, there is little difference between an equivalence relation in X and a partition of X.

The next example of an equivalence relation is useful in that it allows us to gauge the size of infinite sets. We define this equivalence relation below and return to a discussion of infinite sets in Sec. 1.6.

Let U be some fixed universal set and let $\mathcal{C}$ be a class of subsets of U ($\mathcal{C}$ is a set whose elements are subsets of U). We define a relation in $\mathcal{C}$ as

$$R = \{(A, B) \mid A, B \in \mathcal{C} \text{ and there exists a one-to-one correspondence } f : A \to B\}$$

Again we write $A \sim B$ when $(A, B) \in R$. For each $A \in \mathcal{C}$, $A \sim A$ since i_A is a one-to-one correspondence between A and itself. If $A \sim B$ then there is a one-to-one correspondence $f : A \to B$. We noted earlier (also see Exercise 1.4) that f^{-1} is a one-to-one correspondence between B and A and so $B \sim A$. If $A \sim B$ and $B \sim C$ then there exist one-to-one correspondences $f : A \to B$ and $g : B \to C$. The function $g \circ f : A \to C$ is one-to-one and onto C (see Exercises 1.5 and 1.6), and so $A \sim C$. Therefore the reflexive, symmetric, and transitive properties hold, and so R is an equivalence relation in $\mathcal{C}$.

In the next three sections we construct the set **R** of all real numbers. This material can be omitted if the reader chooses, and there is no loss in continuity in going directly to Sec. 1.5, where the important properties of **R** are summarized.

EXERCISES

1.1 Show that if $f : X \to Y$ is one-to-one then $(f^{-1})^{-1} = f$.
1.2 Show that if $f : X \to Y$ is onto Y then there exists a function $g : Y \to X$ such that $f \circ g = i_Y$.

1.3 Show that if $f: X \to Y$ is one-to-one then there exists a function $h: Y \to X$ such that $h \circ f = i_X$. Verify that $h_{f(x)} = f^{-1}$.

1.4 Show that if $f: X \to Y$ is a one-to-one correspondence between X and Y then $f^{-1}: Y \to X$ is a one-to-one correspondence between Y and X.

1.5 Show that if $f: X \to Y$ is one-to-one and $g: Y \to Z$ is one-to-one then $g \circ f: X \to Z$ is one-to-one.

1.6 Show that if $f: X \to Y$ is onto Y and $g: Y \to Z$ is onto Z then $g \circ f: X \to Z$ is onto Z.

1.7 Let X be the set of all residents of New Jersey. Determine which of the following are equivalence relations in X:

(a) $x \sim y$ provided y has the same natural parents as x.
(b) $x \sim y$ provided y lives within 5 miles of x.
(c) $x \sim y$ provided y has the same date of birth as x.
(d) $x \sim y$ provided y is a brother of x.

1.8 Find all functions $f: X \to X$ such that the graph of f, $\text{gr}(f)$, is an equivalence relation in X. Describe the equivalence classes.

1.9 Describe in what sense a function $f: X \to X$ can be considered as an example of a relation in X.

1.10 Show that if X and Y are nonempty sets then $X \times Y \sim Y \times X$, where $\sim$ is the equivalence relation from the final example of Sec. 1.1.

1.11 Let X be any set and let $P(X)$, called the *power set* of X, be the set of all subsets of X. Prove that there is no one-to-one correspondence $f: X \to P(X)$.

1.2 THE NATURAL NUMBERS

We begin our development of real numbers with the set **N** of natural numbers 1, 2, 3, ... All of us have had far more than merely a casual acquaintance with this set; indeed, our initial experiences in mathematics dealt mostly with counting and the arithmetic of natural numbers. We could certainly list many properties of natural numbers, properties which we first encountered in our beginning years in elementary school or even earlier. But if we are going to outline a development of the set of all real numbers, we must be quite specific about which properties of the set **N** will be assumed to be true. From these assumptions or axioms the other familiar properties will follow.

Let **N** be a set whose elements we shall call natural numbers. We take the statements P1 to P5 as our axioms:

P1. $1 \in \mathbf{N}$; that is, **N** is a nonempty set and contains an element we designate as 1.

P2. For each element $n \in \mathbf{N}$ there is a unique element $n^* \in \mathbf{N}$ called the *successor* of n.

P3. For each element $n \in \mathbf{N}$, $n^* \neq 1$; that is, 1 is not the successor of any element in **N**.

P4. For each pair $n, m \in \mathbf{N}$ with $n \neq m$, $n^* \neq m^*$; that is, distinct elements in **N** have distinct successors.

P5. If (a) $A \subseteq \mathbf{N}$, (b) $1 \in A$, and (c) $p \in A$ implies $p^* \in A$ then $A = \mathbf{N}$.

These five axioms are called the *Peano postulates*, and all the known properties of natural numbers can be shown to be consequences of them. P5, called the *principle of mathematical induction*, is an important tool in many mathematical proofs. It often appears in the following form.

If for each natural number n, $S(n)$ is a statement which depends on n then in order to prove that $S(n)$ is a true statement for each and every natural number n we define the set A to be the set of all those natural numbers n for which $S(n)$ is true ($A \subseteq \mathbf{N}$). If we can show that $S(1)$ is true ($1 \in A$) and if we can show that the truth of $S(p)$ implies that $S(p^*)$ is true ($p \in A$ implies $p^* \in A$) then it follows from the principle of mathematical induction that $S(n)$ is true for every natural number n ($A = \mathbf{N}$).

The axioms allow us to name the natural numbers in the conventional way.

(a) $1 \in \mathbf{N}$ by P1.
(b) $1^* \in \mathbf{N}$ by P2 and $1^* \neq 1$ by P3. Name $1^* = 2$; then 1, 2 are distinct natural numbers.
(c) $2^* \in \mathbf{N}$ by P2, and $2^* \neq 1$ by P3. $2^* \neq 2$ by P4 (since $2 \neq 1$). Name $2^* = 3$; then 1, 2, 3 are distinct natural numbers.
(d) $3^* \in \mathbf{N}$ by P2, and $3^* \neq 1$ by P3. $3^* \neq 2$ by P4 (since $3 \neq 1$). $3^* \neq 3$ by P4 (since $3 \neq 2$). Name $3^* = 4$; then 1, 2, 3, 4 are distinct natural numbers.

By continuing to name natural numbers in this way indefinitely, we get a set $A = \{1, 2, 3, 4, \ldots\}$ of distinct natural numbers. Since A satisfies the induction hypothesis of P5, it follows from Axiom P5 that $A = \mathbf{N}$ and so every natural number has been named. Therefore

$$\mathbf{N} = \{1, 2, 3, 4, 5, 6, \ldots\}$$

Next we develop an "arithmetic" in $\mathbf{N}$ by defining two binary operations called addition ($+$) and multiplication ($\cdot$).

Definition 1.3 Addition:
$$n + 1 = n^* \quad \text{for each } n \in \mathbf{N}$$
and
$$n + p^* = (n + p)^* \quad \text{for each } n \in \mathbf{N} \text{ and } p \in \mathbf{N}$$
Multiplication:
$$n \cdot 1 = n \quad \text{for each } n \in \mathbf{N}$$
and
$$n \cdot p^* = (n \cdot p) + n \quad \text{for each } n \in \mathbf{N} \text{ and } p \in \mathbf{N}$$

Notice that the operations of addition and multiplication are defined inductively: the definition of addition first gives the sum $n + 1$ and then the sums $n + 2 = (n + 1)^*$, $n + 3 = (n + 2)^*$, $n + 4 = (n + 3)^*$, etc. Similarly, the definition of multiplication first gives $n \cdot 1$ and then the products $n \cdot 2 = (n \cdot 1) + n$, $n \cdot 3 = (n \cdot 2) + n$, $n \cdot 4 = (n \cdot 3) + n$, etc. Thus, it should come as no surprise that the proofs of the various properties of the arithmetic of natural numbers are based on Axiom P5. As an example we prove the following.

Associative law for addition

$$(m + n) + p = m + (n + p) \quad \text{for all } m, n, p \in \mathbf{N}$$

PROOF Let $m, n \in \mathbf{N}$ be fixed but arbitrary and define

$$A = \{p \in \mathbf{N} \mid (m + n) + p = m + (n + p)\}$$

It is clear that $A \subseteq \mathbf{N}$, and since

$$(m + n) + 1 = (m + n)^* = m + n^* = m + (n + 1)$$

we have that $1 \in A$. Now suppose $p \in A$; then

$$(m + n) + p = m + (n + p)$$

Hence $\quad (m + n) + p^* = [(m + n) + p]^* = [m + (n + p)]^*$
$$= m + (n + p)^* = m + (n + p^*)$$

and so $p^* \in A$. It follows from Axiom P5 that $A = \mathbf{N}$. Thus $(m + n) + p = m + (n + p)$ for every natural number p. Since m and n were arbitrary, the associative law for addition is established.

The other properties of natural numbers can be proved in a similar fashion. They are listed below, and it is recommended that the reader prove them in the order given since some of the proofs will be simpler if previously established properties are used along with the five axioms in verifying a given property.

Commutative law for addition

$$m + n = n + m \quad \text{for all } m, n \in \mathbf{N}$$

Distributive laws

$$p \cdot (m + n) = (p \cdot m) + (p \cdot n) \quad \text{for all } m, n, p \in \mathbf{N}$$
$$(m + n) \cdot p = (m \cdot p) + (n \cdot p) \quad \text{for all } m, n, p \in \mathbf{N}$$

Associative law for multiplication

$$(m \cdot n) \cdot p = m \cdot (n \cdot p) \quad \text{for all } m, n, p \in \mathbf{N}$$

Commutative law for multiplication

$$m \cdot n = n \cdot m \quad \text{for all } m, n \in \mathbf{N}$$

Cancellation laws

$$p + m = p + n \text{ implies } m = n \quad \text{for all } m, n, p \in \mathbf{N}$$

and

$$p \cdot m = p \cdot n \text{ implies } m = n \quad \text{for all } m, n, p \in \mathbf{N}$$

We verify the first cancellation law and leave the second to the exercises.

PROOF Let $S(p)$ be the statement "$p + m = p + n$ implies $m = n$." If $1 + m = 1 + n$ then $m^* = n^*$, and so (by P4) $m = n$. Thus $S(1)$ is true. Suppose $S(p)$ is true; then $p + m = p + n$ implies $m = n$. Now if $p^* + m = p^* + n$ then $(1 + p) + m = (1 + p) + n$, and so $1 + (p + m) = 1 + (p + n)$. Since $S(1)$ is true, $p + m = p + n$. But $S(p)$ is true; hence $m = n$, and consequently $S(p^*)$ is true. It follows from the principle of mathematical induction that $S(p)$ is true for every natural number p; that is, $p + m = p + n$ implies $m = n$ for all $m, n, p \in \mathbf{N}$.

One of the first facts about natural numbers that we become familiar with is that some natural numbers are "larger" than others. This notion is called *order* and is introduced as follows. We define a relation in $\mathbf{N}$ called an order relation, symbolized by $<$ and read "less than." For any $m, n \in \mathbf{N}$ we write $m < n$ provided there is a natural number p such that $m + p = n$. This relation in $\mathbf{N}$ is clearly not an equivalence relation since it is neither reflexive nor symmetric. However, the transitive law is satisfied, for suppose $m < n$ and $n < p$. Then there are natural numbers q_1 and q_2 such that $m + q_1 = n$ and $n + q_2 = p$. Thus $p = n + q_2 = (m + q_1) + q_2 = m + (q_1 + q_2)$ and so $m < p$.

Now since $n + 1 = n^*$ for every natural number n, $n < n^*$; and so

$$1 < 2 < 3 < 4 < 5 < 6 < \ldots$$

The following law is fundamental and can be proved directly from the definition of the order relation in $\mathbf{N}$.

Law of trichotomy For every pair $m, n \in \mathbf{N}$ exactly one of the following holds:

(a) $m = n$.
(b) $m < n$.
(c) $n < m$.

We often write $m > n$ to mean $n < m$. Another relation in **N** that is frequently used is $\leq$ read "less than or equal to" and is defined by $m \leq n$ provided $m < n$ or $m = n$. This relation is both reflexive and transitive but not symmetric. It follows from the law of trichotomy that if $m \leq n$ and $n \leq m$ then $m = n$.

The next theorem, called the *well-ordering principle for* **N**, is an important property which is characteristic of the set of natural numbers. This principle is used frequently in the development of the real number system (Sec. 1.4) and, as we see in the exercises, is logically equivalent to the principle of mathematical induction.

Theorem 1.2 Every nonempty subset $A \subseteq \mathbf{N}$ has a first element; that is, there is a $p \in A$ such that $p \leq a$ for every $a \in A$.

PROOF We assume that A is a nonempty subset of **N** and that A has no first element and show that this leads to a contradiction. Define $M \subseteq \mathbf{N}$ by

$$M = \{x \in \mathbf{N} \mid x < a \text{ for each } a \in A\}$$

By the law of trichotomy $M \cap A = \emptyset$. Now $1 \notin A$; otherwise 1 would surely be the first element in A. Hence $1 < a$ for each $a \in A$, and so $1 \in M$. Assume $p \in M$; then $p < a$ for each $a \in A$. If $p + 1 \in A$ then $p + 1$, which is the first natural number larger than p, would be the first element in A, in contradiction to our assumption that A has no first element. Thus $p + 1 \notin A$, and so $p + 1 < a$ for each $a \in A$. Hence $p + 1 \in M$ and by induction $M = \mathbf{N}$. But $M \cap A = \emptyset$, and so $A = \emptyset$, which is a contradiction. Therefore A must have a first element.

EXERCISES

1.12 Use the law of trichotomy to prove the cancellation law:

$$p \cdot m = p \cdot n \text{ implies } m = n \qquad \text{for all } m, n, p \in \mathbf{N}$$

1.13 Prove

$$1 + 2 + 3 + \cdots + n = \frac{n(n+1)}{2} \qquad \text{for each } n \in \mathbf{N}$$

1.14 Prove

$$1^2 + 2^2 + 3^2 + \cdots + n^2 = \frac{n(n+1)(2n+1)}{6} \qquad \text{for each } n \in \mathbf{N}$$

1.15 Prove

$$1^3 + 2^3 + 3^3 + \cdots + n^3 = (1 + 2 + 3 + \cdots + n)^2 \qquad \text{for each } n \in \mathbf{N}$$

1.16 Prove

$$2^{n-1} \leq n! \quad \text{for each } n \in \mathbf{N}$$

1.17 Prove that the principle of mathematical induction follows from the well-ordering principle and thus the two are equivalent.

1.3 DEVELOPMENT OF RATIONAL NUMBERS

Integers are created to eliminate certain deficiencies which exist within the system of natural numbers. The definition of order and the law of trichotomy together imply that the equation

$$n + x = m$$

has a solution $x \in \mathbf{N}$ if and only if $n < m$. Thus it becomes desirable to extend our number system to include numbers which serve as solutions to this equation for the cases $n \geq m$. The extension will produce an identity for the operation of addition and an inverse for each element in the new system.

We begin by defining S to be the set of all ordered pairs of natural numbers

$$S = \{(m, n) \mid m, n \in \mathbf{N}\}$$

A relation is defined in S by

$$(m, n) \sim (a, b) \quad \text{provided } m + b = n + a$$

To see that this is an equivalence relation in S first observe that $m + n = n + m$ and so $(m, n) \sim (m, n)$ for every $(m, n) \in S$. Thus the relation is reflexive. If $(m, n) \sim (a, b)$ then $m + b = n + a$, and so $a + n = b + m$. Hence $(a, b) \sim (m, n)$, and the relation is symmetric. Finally, suppose that $(m, n) \sim (a, b)$ and $(a, b) \sim (x, y)$. Then $m + b = n + a$ and $a + y = b + x$. Now

$$(m + b) + (a + y) = (n + a) + (b + x)$$

and so

$$(a + b) + (m + y) = (a + b) + (n + x)$$

By the cancellation law $m + y = n + x$ and hence $(m, n) \sim (x, y)$. Therefore the relation is transitive.

We define integers to be the equivalence classes which result under the relation defined above.

Definition 1.4 An *integer* is an equivalence class of ordered pairs of natural numbers.

We designate the set of all integers by **J**, and, following the procedure with natural numbers, we set out to define an arithmetic in **J**.

Definition 1.5 Addition:

$$[(m, n)] + [(a, b)]$$
$$= [(m + a, n + b)] \quad \text{for all integers } [(m, n)], [(a, b)] \in \mathbf{J}$$

Multiplication:

$$[(m, n)] \cdot [(a, b)]$$
$$= [(ma + nb, mb + na)] \quad \text{for all integers } [(m, n)], [(a, b)] \in \mathbf{J}$$

At this stage it might be helpful for the reader to think of the equivalence class $[(m, n)]$ as the *difference* $m - n$. For example, the integer $[(1, 3)]$ should be thought of as the difference $1 - 3$ or -2. Addition and multiplication as defined above are then seen to be the usual definitions with which the reader is familiar. Actually, as we shall see below, the class $[(m, n)]$ consists of all ordered pairs (p, q) of natural numbers for which $p - q = m - n$. Of course, $m - n$ is a natural number if and only if $m > n$, where by the notation $m - n$ we mean the unique natural number which is a solution to the equation $n + x = m$.

To show that the operations are well defined it is necessary to verify that the sum and product are independent of the equivalence-class representatives used to define the operations. Toward this end suppose that $[(\hat{m}, \hat{n})] = [(m, n)]$ and $[(\hat{a}, \hat{b})] = [(a, b)]$. Then $(\hat{m}, \hat{n}) \sim (m, n)$ and $(\hat{a}, \hat{b}) \sim (a, b)$, and so $\hat{m} + n = \hat{n} + m$ and $\hat{a} + b = \hat{b} + a$. Now

$$(\hat{m} + \hat{a}) + (n + b) = (\hat{m} + n) + (\hat{a} + b) = (\hat{n} + m) + (\hat{b} + a)$$
$$= (\hat{n} + \hat{b}) + (m + a)$$

Hence $(\hat{m} + \hat{a}, \hat{n} + \hat{b}) \sim (m + a, n + b)$, and so $[(\hat{m} + \hat{a}, \hat{n} + \hat{b})] = [(m + a, n + b)]$. It follows that

$$[(\hat{m}, \hat{n})] + [(\hat{a}, \hat{b})] = [(m, n)] + [(a, b)]$$

The verification that multiplication is well defined is similar and is left to the reader (see Exercise 1.18).

To obtain a more familiar representation of the integers we take a closer look at the equivalence classes. Consider any class $[(m, n)]$; by the law of trichotomy either $m = n$, $m > n$, or $m < n$. We consider the three cases individually.

Case 1: $m = n$ We show that $[(m, m)] = \{(x, x) \mid x \in \mathbf{N}\}$. If $(a, b) \in [(m, m)]$ then $(a, b) \sim (m, m)$, and so $a + m = b + m$. Thus $a = b$ and so $(a, b) \in$

$\{(x, x) \mid x \in \mathbf{N}\}$. Conversely, if $(a, b) \in \{(x, x) \mid x \in \mathbf{N}\}$ then $a = b$ and so $a + m = b + m$. But then $(a, b) \sim (m, m)$, and we have $(a, b) \in [(m, m)]$.

Case 2: $m > n$ In this case there is a unique natural number p such that $m = n + p$, and hence $(m, n) = (n + p, n)$. We show that $[(m, n)] = \{(x + p, x) \mid x \in \mathbf{N}\}$. If $(a, b) \in [(m, n)]$ then $(a, b) \sim (m, n) = (n + p, n)$. Thus $a + n = b + (n + p) = (b + p) + n$, and so $a = b + p$. Therefore $(a, b) = (b + p, b) \in \{(x + p, x) \mid x \in \mathbf{N}\}$. Conversely, if $(a, b) \in \{(x + p, x) \mid x \in \mathbf{N}\}$ then $a = b + p$. Now $(b + p) + n = b + (n + p)$, and hence $(a, b) = (b + p, b) \sim (n + p, n) = (m, n)$. Hence $(a, b) \in [(m, n)]$.

Case 3: $m < n$ There is a unique natural number q such that $m + q = n$, and so $(m, n) = (m, m + q)$. We leave it to the reader to verify that $[(m, n)] = \{(x, x + q) \mid x \in \mathbf{N}\}$.

We have three types of integers: if $m = n$ then

$$[(m, n)] = \{(x, x) \mid x \in \mathbf{N}\} = [(1, 1)];$$

if $m > n$ with $m = n + p$ then

$$[(m, n)] = \{(x + p, x) \mid x \in \mathbf{N}\} = [(1 + p, 1)];$$

and if $m < n$ with $m + q = n$ then

$$[(m, n)] = \{(x, x + q) \mid x \in \mathbf{N}\} = [(1, 1 + q)].$$

The integer $[(1, 1)]$ is called zero and written $[(1, 1)] = 0$. Each integer of the form $[(1 + p, 1)]$ is called a *positive integer*, and we write $[(1 + p, 1)] = p$ for each $p \in \mathbf{N}$; each integer of the form $[(1, 1 + q)]$ is called a *negative integer* and we write $[(1, 1 + q)] = -q$ for each $q \in \mathbf{N}$. This yields

$$\mathbf{J} = \{0, 1, -1, 2, -2, 3, -3, \ldots\}$$

Observe that we have identified each positive integer $[(1 + p, 1)]$ with the natural number p, and so $\mathbf{N} \subset \mathbf{J}$. We justify this identification by showing that the operations defined in $\mathbf{J}$ agree with those defined for natural numbers in the previous section. If $m = [(1 + m, 1)]$ and $n = [(1 + n, 1)]$ are positive integers (natural numbers) then

$$[(1 + m, 1)] + [(1 + n, 1)] = [(1 + m + 1 + n, 1 + 1)]$$
$$= [(1 + m + n, 1)] = m + n$$

and

$$[(1 + m, 1)] \cdot [(1 + n, 1)] = [(1 + m + n + m \cdot n + 1, 1 + m + 1 + n)]$$
$$= [(1 + m \cdot n, 1)] = m \cdot n$$

Therefore $\mathbf{J}$ is an extension of the set $\mathbf{N}$ of natural numbers.

As with the system of natural numbers, the following laws are valid in **J**:

Associative law for addition
$$(j + k) + l = j + (k + l) \quad \text{for all } j, k, l \in \mathbf{J}$$

Cummutative law for addition
$$j + k = k + j \quad \text{for all } j, k \in \mathbf{J}$$

Associative law for multiplication
$$(j \cdot k) \cdot l = j \cdot (k \cdot l) \quad \text{for all } j, k, l \in \mathbf{J}$$

Commutative law for multiplication
$$j \cdot k = k \cdot j \quad \text{for all } j, k \in \mathbf{J}$$

Distributive law
$$j \cdot (k + l) = (j \cdot k) + (j \cdot l) \quad \text{for all } j, k, l \in \mathbf{J}$$

The system **J** has an additive identity and a multiplicative identity. If $j = [(m, n)]$ is an arbitrary integer then
$$j + 0 = [(m, n)] + [(1, 1)] = [(m + 1, n + 1)] = [(m, n)] = j$$
and
$$j \cdot 1 = [(m, n)] \cdot [(2, 1)] = [(2m + n, m + 2n)] = [(m, n)] = j$$
Therefore 0 is the identity for addition, and 1 is the identity for multiplication. Moreover
$$j \cdot 0 = [(m, n)] \cdot [(1, 1)] = [(m + n, m + n)] = [(1, 1)] = 0$$
and so multiplication by the integer 0 always yields the product 0. It is also true that if a product of two integers is 0 then at least one of the factors must be 0. This result is known as the *product law of arithmetic*, and we leave its proof to the reader (see Exercise 1.19).

We now show that the deficiency in the natural number system mentioned at the outset of this section has been removed; that is, the equation
$$k + x = j$$
has a solution $x \in \mathbf{J}$ for every pair of integers j, k. Suppose $k = [(m, n)]$ and $j = [(a, b)]$. Then
$$[(m, n)] + [(a + n, b + m)] = [(m + a + n, n + b + m)]$$
$$= [(a, b)]$$

and so $x = [(a + n, b + m)]$ is a solution to the equation $k + x = j$. In particular, there is a solution to the equation

$$k + x = 0$$

given by $x = [(1 + n, 1 + m)] = [(n, m)]$. $[(n, m)]$ is called the additive inverse of $[(m, n)]$. We note that the additive inverse of each positive integer $p = [(1 + p, 1)]$ is $[(1, 1 + p)] = -p$ and the additive inverse of each negative integer $-q = [(1, 1 + q)]$ is $[(1 + q, 1)] = q$. The additive inverse of $0 = [(1, 1)]$ is, of course, 0.

Subtraction is defined as follows: If $k = [(m, n)]$ and $j = [(a, b)]$ then the difference $j - k$ is defined as the sum of j and the additive inverse of k. That is,

$$j - k = [(a, b)] + [(n, m)] = [(a + n, b + m)]$$

Observe that $j - k$ is the solution x to the equation $k + x = j$. Also, $0 - k$ (written simply as $-k$) is $[(n, m)]$, the additive inverse of k.

Cancellation laws
(a) $j + k = j + l$ implies $k = l$ for all $j, k, l \in \mathbf{J}$.
(b) $j \cdot k = j \cdot l$ implies $k = l$ for all $j, k, l \in \mathbf{J}$ with $j \neq 0$.

An order relation $<$ is introduced into $\mathbf{J}$ by writing $j < k$ provided there is a natural number n with $j + n = k$. This relation is easily seen to be transitive and clearly extends the notion of order in $\mathbf{N}$. Moreover, if $j < k$ then $j + n = k$ for some $n \in \mathbf{N}$. Hence $-k + n = -j$, and so $-k < -j$. Therefore we have

$$\cdots < -3 < -2 < -1 < 0 < 1 < 2 < 3 < \cdots$$

Again, we write $k > j$ when $j < k$, and also $j \leq k$ means either $j < k$ or else $j = k$. Finally, we note that the law of trichotomy is valid in $\mathbf{J}$ but that $\mathbf{J}$ is not well-ordered in the sense that not every nonempty subset of $\mathbf{J}$ has a first element.

Having enriched our number system to include all solutions of the equation

$$k + x = j \qquad j, k \in \mathbf{J}$$

it is natural to ask about solutions to the equation

$$k \cdot x = j$$

If $k = 0$ and $j \neq 0$ then there can be no solution since $0 \cdot x = 0$ for every integer x. Also, if $k = j = 0$ then every x is a solution. Thus we consider the equation $k \cdot x = j$ only for the case where $k \neq 0$. Now $2 \cdot x = 1$ does not have a solution in $\mathbf{J}$, or, equivalently, there is no integer which serves as a multiplicative inverse for the integer 2. This motivates us to enlarge the number system again.

Let
$$T = \{j/k \mid j, k \in \mathbf{J}, k \neq 0\}$$
For now, j/k is simply an ordered pair of integers with $k \neq 0$. Later we shall interpret j/k as a quotient of two integers. A relation is defined in T by
$$j/k \sim r/s \quad \text{provided } js = rk$$
We leave to the reader verification that $\sim$ is an equivalence relation (see Exercise 1.22), and we define rational numbers to be the equivalence classes resulting rom this equivalence relation.

Definition 1.6 *A rational number* is an equivalence class of ordered pairs of integers.

We designate the set of all rational numbers by $\mathbf{Q}$. For example, 1/2 is in the same class as 2/4 and $(-3)/(-6)$, but 2/3 is in a different equivalence class. It is convenient to refer to the rational number [1/2] by simply 1/2 (or 2/4 or $-3/(-6)$) or any other ordered pair of integers in the class. The rational number [2/3] will be referred to by 2/3 or 4/6 or $-10/(-15)$, etc. We shall be free to designate an equivalence class by any member of that class. Next we define an arithmetic in the set $\mathbf{Q}$ of all rational numbers.

Definition 1.7 Addition:
$$j/k + r/s = (js + rk)/ks \quad \text{for all rational numbers } j/k, r/s \in \mathbf{Q}$$
Multiplication:
$$j/k \cdot r/s = jr/ks \quad \text{for all rational numbers } j/k, r/s \in \mathbf{Q}$$

The operations of addition and multiplication of rational numbers are well defined; we leave the details to the reader (see Exercise 1.23).

We identify rational numbers of the form $j/1$ by j (where $j \in \mathbf{J}$) and call such rational numbers *whole numbers*. Note that the ordered pair $12/4 \sim 3/1$, and so 12/4 is the whole number which we designate by 3. This representation identifies a certain subset of $\mathbf{Q}$, namely the whole numbers, with the set $\mathbf{J}$ of all integers. Thus $\mathbf{J} \subset \mathbf{Q}$. We ask the reader to justify this identification (see Exercise 1.24) by showing that the arithmetic in $\mathbf{Q}$ agrees with the arithmetic previously defined in $\mathbf{J}$. Therefore the number system $\mathbf{Q}$ is an extension of the system of all integers.

The deficiency in $\mathbf{J}$ has been eliminated; that is, the equation
$$v \cdot x = u$$
always has a solution $x \in \mathbf{Q}$, where $u, v \in \mathbf{Q}$ and $v \neq 0$. In fact, if $v = j/k$

with $j \neq 0$ and $u = r/s$ then $x = rk/js$ is a solution to the equation $v \cdot x = u$. In particular, $x = k/j$ is a solution to the equation

$$v \cdot x = 1 \quad \text{where } v = j/k$$

$x = k/j$ is called the *multiplicative inverse* of the nonzero rational number j/k.

Division is defined as follows. If $j/k = v \neq 0$ and $r/s = u$ then the quotient $\frac{u}{v}$ is defined as the product of u and the *multiplicative* inverse of v. That is,

$$\frac{u}{v} = (r/s) \cdot (k/j) = rk/js$$

Observe that $\frac{u}{v}$ is the solution to the equation $v \cdot x = u$. Also, $\frac{1}{v} = k/j$, the multiplicative inverse of v.

We summarize the properties of $\mathbf{Q}$ in the following definition and theorem.

Definition 1.8 A *field* F is a nonempty set together with two binary operations addition ($+$) and multiplication ($\cdot$) satisfying:
(a) The associative law for addition

$$(u + v) + w = u + (v + w) \quad \text{for all } u, v, w \in F$$

(b) The commutative law for addition

$$u + v = v + u \quad \text{for all } u, v \in F$$

(c) Identity for addition

There is an element $0 \in F$ such that $u + 0 = u$ for every $u \in F$

(d) Inverse for addition

For each $u \in F$ there is an element $-u \in F$ such that $u + (-u) = 0$

(e) The associative law for multiplication

$$(u \cdot v) \cdot w = u \cdot (v \cdot w) \quad \text{for all } u, v, w \in F$$

(f) The cummutative law for multiplication

$$u \cdot v = v \cdot u \quad \text{for all } u, v \in F$$

(g) Identity for multiplication

There is an element $1 \in F$ such that $u \cdot 1 = u$ for every $u \in F$

(h) Inverses for multiplication

For each $u \in F$ except $u = 0$ there is an element $1/u \in F$ such that $u \cdot 1/u = 1$

(i) The distributive law

$$u \cdot (v + w) = (u \cdot v) + (u \cdot w) \qquad \text{for all } u, v, w \in F$$

Theorem 1.3 **Q** is a field.

A rational number $u = j/k$ is called *positive* if j and k are either both positive integers or both negatives integers. If one of the two integers j and k is a positive integer and the other is a negative integer then $u = j/k$ is called a *negative rational number*. The rational number 0 is neither positive nor negative. We note that these definitions agree with those defined in **J**; that is, a positive whole number is a positive integer, and a negative whole number is a negative integer. It follows immediately from Definition 1.7 that the sum and product of two positive rationals is again a positive rational.

An order relation $<$ is introduced into **Q** by writing $u < v$ provided there is a positive rational number w with $u + w = v$. Clearly $0 < u$ if and only if u is a positive rational number. The law of trichotomy holds, as does each of the following two order properties:

1. If $u < v$ then $u + w < v + w$ for every $w \in \mathbf{Q}$.
2. If $u < v$ and $w > 0$ then $u \cdot w < v \cdot w$.

Any field which has an order relation defined in it satisfying the law of trichotomy and properties 1 and 2 above is called an *ordered field*. The rational number system is an ordered field.

It may appear that we must surely be finished with extending our number systems, but there still remains a deficiency in the system of rational numbers. We explore this deficiency in the next section, where we conclude our development of the real number system.

EXERCISES

1.18 Verify that multiplication of integers is a well-defined operation. *Hint*: Assume that $[(\hat{m}, \hat{n})] = [(m, n)]$ and $[(\hat{a}, \hat{b})] = [(a, b)]$ and then show that

$$(\hat{m}\hat{a} + \hat{n}\hat{b}) + (mb + na) + (m + n)(\hat{a} + \hat{b}) = (\hat{m}\hat{b} + \hat{n}\hat{a}) + (ma + nb) + (m + n)(\hat{a} + \hat{b})$$

1.19 Prove the product law of arithmetic: If $j \cdot k = 0$, where $j, k \in \mathbf{J}$ and $j \neq 0$, then $k = 0$.

1.20 Prove the cancellation laws in **J**.

1.21 An integer j is called *even* if there is an integer k with $j = 2k$; an integer is called *odd* if it fails to be even. Prove that if j and k are both odd integers, the product $j \cdot k$ is odd.

1.22 Prove that the relation $\sim$ defined in T is an equivalence relation.

1.23 Verify that the operations of addition and multiplication of rational numbers (Definition 1.7) are well-defined operations.

1.24 Show that addition and multiplication of whole numbers agree with the previously defined addition and multiplication in **J**.

1.25 Prove the order properties in **Q**:
(a) $u < v$ implies $u + w < v + w$ for every $w \in \mathbf{Q}$.
(b) $u < v$ and $w > 0$ implies $u \cdot w < v \cdot w$.

1.26 Show that for any rational number r there is a natural number n with $n > r$. Hint: If $r = p/q > 0$, consider $n = p + 1$.

1.27 Show that if r and s are two rationals of the same sign (r and s are either both positive or both negative) and $r < s$ then $1/s < 1/r$.

1.4 CONSTRUCTION OF THE REAL NUMBER SYSTEM

In this section we develop the system of all real numbers. To justify the need for extending the system of rational numbers, we again explore a deficiency which exists in the rationals. This is the kind of motivation that prompted us to develop the integers from the natural numbers and then the rational numbers from the integers.

What kind of deficiency can exist in the system of all rational numbers? After all, the rational numbers satisfy those many properties which make them an ordered field. What more could we want? The deficiency in the system of rational numbers is a bit more subtle; we call it *completeness*. Before exhibiting this property we record the following lemma.

Lemma There is no rational number r such that $r^2 = 2$.

PROOF Suppose $r^2 = 2$. r can be written in the form $r = p/q$, where $q \in \mathbf{N}$, in infinitely many distinct ways. Let $A = \{q \in \mathbf{N} \mid \text{there is a } p \in \mathbf{J}$ with $p/q = r\}$. By the well-ordering property of **N**, A has a first element, which we designate by q_0. Then $r = p_0/q_0$ for some $p_0 \in \mathbf{J}$. Since $p_0^2/q_0^2 = r^2 = 2$, $p_0^2 = 2q_0^2$, and so p_0^2 is even. It follows that p_0 is even (see Exercise 1.21). Thus $p_0 = 2k$ for some $k \in \mathbf{J}$, and so $2q_0^2 = p_0^2 = 4k^2$ or $q_0^2 = 2k^2$. Thus q_0^2 is even, and so q_0 must be even. Let $q_0 = 2n$, where $n \in \mathbf{N}$. Then $r = p_0/q_0 = 2k/2n = k/n$. Clearly, then, $n \in A$ and $n < q_0$, a contradiction to the hypothesis that q_0 is the first element in A. Therefore there can be no rational number r with $r^2 = 2$.

The following is a variation of the *Dedekind cut method* for defining real numbers: the cut method was presented by the German mathematician Richard Dedekind more than a century ago.

Definition 1.9 A *ray* in the set **Q** of all rational numbers is a nonempty proper subset $U \subset \mathbf{Q}$ satisfying the following properties:
(a) If $x \in U$ and $y > x$ then $y \in U$.
(b) U has no first element.

The reader should recognize U as an *interval of rationals* on the real line, that is, a set consisting of all the rational points to the right of some fixed real point P. Later the ray U will be identified with the determining point P. There are two cases depending on whether or not the point P is rational:

1. The set $U' = \mathbf{Q} - U$ of all rationals not in U has a largest element r_0.
2. U' has no largest element.

If U' has a largest element r_0 then the ray U in $\mathbf{Q}$ can be defined by

$$U = \{x \in \mathbf{Q} \mid x > r_0\}$$

It is easy to demonstrate that any set of the form

$$\{x \in \mathbf{Q} \mid x > r\}$$

for some fixed $r \in \mathbf{Q}$ is a ray in $\mathbf{Q}$ of type 1. We call such a set a *rational ray* and identify the ray $\{x \in \mathbf{Q} \mid x > r\}$ with the rational number r. An ordered field F is called *complete* if each ray in the field can be identified (in the above sense) with some number in the field. In other words, every ray U must be of type 1; that is, there must exist a $t \in F$ such that $U = \{x \in F \mid x > t\}$.

Theorem 1.4 $\mathbf{Q}$ is not complete.

PROOF Let $V = \{x \in \mathbf{Q} \mid x > 0 \text{ and } x^2 > 2\}$. We first demonstrate that V is a ray in $\mathbf{Q}$. Clearly $\emptyset \neq V \subset \mathbf{Q}$; in fact each natural number except 1 is in V. Suppose $x \in V$ and $y > x$. Then $y > 0$ and $y^2 > xy > x^2 > 2$, and so $y \in V$. Now if x_0 is the first element in V then $x_0 > 0$ and $x_0^2 > 2$. By Exercises 1.26 and 1.27 there is a natural number n with

$$n > \frac{2x_0}{x_0^2 - 2} \quad \text{and} \quad \frac{1}{n} < \frac{x_0^2 - 2}{2x_0}$$

Then

$$x_0^2 - 2 > \frac{2x_0}{n}$$

and so

$$x_0^2 - \frac{2x_0}{n} > 2$$

It follows that

$$x_0^2 - \frac{2x_0}{n} + \frac{1}{n^2} > 2$$

and hence
$$\left(x_0 - \frac{1}{n}\right)^2 > 2$$

Now
$$x_0 - \frac{1}{n} > x_0 - \frac{x_0^2 - 2}{2x_0} = \frac{x_0^2 + 2}{2x_0} > 0$$

and so $x_0 - 1/n \in V$. But $x_0 - 1/n < x_0$, contradicting the fact that x_0 is the first element in V. Therefore V has no first element, and this proves that V is a ray in $\mathbf{Q}$.

We show that V' has no largest element and so V is not a rational ray. Suppose r_0 is the largest element in V'. Since $1 \in V'$, we must have $r_0 \geq 1$. Since $r_0 > 0$ and $r_0 \notin V$, $r_0^2 \leq 2$. But $r_0^2 \neq 2$ by the preceding lemma, and so $r_0^2 < 2$. Again using Exercises 1.26 and 1.27, we find that there is a natural number m with

$$m > \frac{2r_0 + 1}{2 - r_0^2} \quad \text{and} \quad \frac{1}{m} < \frac{2 - r_0^2}{2r_0 + 1}$$

Now
$$\frac{2r_0}{m} + \frac{1}{m} < 2 - r_0^2$$

and since $1/k^2 \leq 1/k$ for every $k \in \mathbf{N}$,

$$\frac{2r_0}{m} + \frac{1}{m^2} < 2 - r_0^2$$

Hence
$$r_0^2 + \frac{2r_0}{m} + \frac{1}{m^2} < 2$$

and so
$$\left(r_0 + \frac{1}{m}\right)^2 < 2$$

It follows that $r_0 + 1/m \in V'$ and yet $r_0 + 1/m > r_0$, where r_0 is the assumed largest element in V'. This contradiction guarantees that V' has no largest element.

Since we have produced a ray in $\mathbf{Q}$ of type 2, the ordered field of rational numbers is not complete.

Any ray in $\mathbf{Q}$ of type 2 is called an *irrational ray*. We are now prepared to define what we mean by a real number.

Definition 1.10 A real number is a ray in $\mathbf{Q}$.

We designate the set of all real numbers by $\mathbf{R}$ and point out again that there are two kinds of real numbers: the rational rays in $\mathbf{Q}$, called the *rational real numbers*, and the irrational rays in $\mathbf{Q}$, called *irrational numbers*.

Again, following the pattern of development established in the preceding section, we define an arithmetic in **R**.

Definition 1.11 Addition:
$$U + V = \{u + v \mid u \in U \text{ and } v \in V\} \quad \text{for all } U, V \in \mathbf{R}$$

Multiplication:
(a) $0 \notin U, 0 \notin V$:
$$U \cdot V = \{uv \mid u \in U \text{ and } v \in V\}$$

(b) $0 \in U, 0 \notin V$:
$$U \cdot V = \{r \in \mathbf{Q} \mid r > uy \text{ for some } u \in U \text{ and some } y \in V' \text{ with } y \geq 0\}$$

(c) $0 \notin U, 0 \in V$:
$$U \cdot V = \{r \in \mathbf{Q} \mid r > xv \text{ for some } x \in U' \text{ with } x \geq 0 \text{ and some } v \in V\}$$

(d) $0 \in U, 0 \in V$:
$$U \cdot V = \{r \in \mathbf{Q} \mid r > xy \text{ for some } x \in U' \text{ and some } y \in V'\}$$

In order to show that addition and multiplication of real numbers are well-defined operations it is necessary to establish that in each case $U + V$ and $U \cdot V$ is a ray in **Q**. We ask the reader to make this verification in Exercise 1.29. Now, as noted above, each rational ray in **Q**, $\{x \in \mathbf{Q} \mid x > r\}$ for some fixed $r \in \mathbf{Q}$, is identified with the rational number r. Therefore $\mathbf{Q} \subset \mathbf{R}$; again, the reader is asked to justify this identification by showing that the operations defined above agree with the addition and multiplication of rational numbers defined in the preceding section (see Exercise 1.30). Hence **R** is an extension of the number system **Q**.

The real number $\{x \in \mathbf{Q} \mid x > 0\} = 0$ is the additive identity in **R**, and if U is any ray in **Q** (U is any real number) then
$$-U = \{x \in \mathbf{Q} \mid x > -r \text{ for some } r \in U'\}$$
is the additive inverse of U. Moreover, $\{x \in \mathbf{Q} \mid x > 1\} = 1$ is the multiplicative identity in **R**, and if U is any nonzero real number then

$$U^{-1} = \begin{cases} \left\{x \in \mathbf{Q} \mid x > \dfrac{1}{r} \text{ for some } r \in U'\right\} & \text{if } 0 \in U \\ \left\{x \in \mathbf{Q} \mid x > \dfrac{1}{r} \text{ for some } r \in U' \text{ with } r > 0\right\} & \text{if } 0 \notin U \end{cases}$$

is the multiplicative inverse of U. We summarize the properties of **R** in the following theorem.

Theorem 1.5 **R** is a field.

A real number U is called *positive* if U' contains some positive rational number. It is easy to demonstrate that the sum and product of two positive real numbers is again positive and that for any real number U either U is positive or else $-U$ is positive (such a real number is called negative) or else $U = 0$. We introduce an order relation into $\mathbf{R}$ by defining

$$U < V \quad \text{provided that } V \subset U$$

Equivalently, $U < V$ if $U' \subset V'$, where $U' = \mathbf{Q} - U$ and $V' = \mathbf{Q} - V$. Then U is positive if and only if $U > 0$. The order relation is clearly transitive. Moreover, the law of trichotomy holds, as do the order properties:

1. If $U < V$ then $U + W < V + W$ for every $W \in \mathbf{R}$.
2. If $U < V$ and $W > 0$ then $U \cdot W < V \cdot W$.

To summarize, $\mathbf{R}$ is an ordered field.

The following lemma will be useful in establishing the completeness of $\mathbf{R}$.

Lemma If $r \in Q$ and $V \in \mathbf{R}$ then $r \in V$ if and only if $r > V$.

PROOF Let $r \in V$. Then $\{x \in \mathbf{Q} \mid x > r\} \subset V$, and so $V < \{x \in \mathbf{Q} \mid x > r\} = r$. Conversely, suppose $r > V$. Then by the definition of order $\{x \in \mathbf{Q} \mid x > r\} \subset V$. Hence there is an $r_0 \in V$ with $r_0 \leq r$. Thus $r \in V$.

The following theorem establishes that the deficiency in the rational numbers system has been removed. A ray in $\mathbf{R}$ is a nonempty proper subset $A \subset \mathbf{R}$ satisfying: (a) If $x \in A$ and $y > x$ then $y \in A$; and (b) A has no first element.

Theorem 1.6: Dedekind's theorem $\mathbf{R}$ is complete.

PROOF Let A be any ray in $\mathbf{R}$; we must show that A' has a largest element. Define the set

$$U = \{x \in \mathbf{Q} \mid x > W \text{ for every } W \in A'\}$$

(Here, of course, the set $A' = \mathbf{R} - A$.) We first establish that U is a ray in $\mathbf{Q}$.

Let $V \in A$ be arbitrary. Then $V > W$ for every $W \in A'$, since $V \leq W$ implies $W \in A$. Now, $r \in V$ implies $r > V$ (by the lemma), and so $r > W$ for every $W \in A'$. Thus $r \in U$. It follows that each $V \in A$ satisfies $V \subseteq U$; in particular, $U \neq \emptyset$. Now let $W \in A'$ be arbitrary. Then $r \in W'$ implies $r \leq W$ (again, by the lemma), and so $r \in U'$.

Consequently, each $W \in A'$ satisfies $W' \subseteq U'$; in particular $U' \neq \emptyset$. It has thus been established that U is a nonempty proper subset of $\mathbf{Q}$ and moreover, for every $W \in A'$ and $V \in A$, $V \subseteq U \subseteq W$.

Next, properties (a) and (b) of Definition 1.9 are shown to hold for the set U.

(a) Suppose that $x \in U$ and $y > x$. Then $x > W$ for every $W \in A'$, and so $y > W$ for every $W \in A'$. It follows that $y \in U$.

(b) Suppose that r_0 is the first element in U. Then $r_0 \in U$, and so $r_0 > W$ for every $W \in A'$. Hence $r_0 \in A$. Let $V \in A$ be arbitrary. Now $k \in V$ implies that $k \in U$ (since each $V \in A$ satisfies $V \subseteq U$), and so $k \geq r_0$ (since r_0 is the first element in U). Thus $r_0 \notin V$ (since V has no first element). By the lemma, $r_0 \leq V$. Thus r_0 must be the first element in A, but this is a contradiction (recall that A has no first element). Therefore, U has no first element.

It follows from Definition 1.9 that U is a ray in $\mathbf{Q}$; that is, $U \in \mathbf{R}$.

Now if $V \in A$ then $V \subseteq U$, and so $U \leq V$. Since A has no first element, we must have $U \in A'$ (otherwise U would be the first element in A). If $W \in A'$ then $U \subseteq W$, and so $W \leq U$. Thus U is the largest element in A'. It follows that $\mathbf{R}$ is complete.

The system $\mathbf{R}$ is a complete ordered field. In the following paragraphs a more useful, but totally equivalent, form of the completeness axiom is developed.

A real number u is called an *upper bound* of the nonempty set $S \subset \mathbf{R}$ provided $u \geq x$ for every $x \in S$; a real number l is called a *lower bound* of S provided $l \leq x$ for every $x \in S$. S is said to be *bounded above* if S has an upper bound; similarly, S is said to be *bounded below* if S has a lower bound. The set S is called *bounded* if it is bounded above and bounded below.

If u is an upper bound of S and $u' > u$ then clearly u' is an upper bound of S; analogously if l is a lower bound of S and $l' < l$ then l' is a lower bound of S. We designate the set of all upper bounds of S by U_S and the set of all lower bounds of S by L_S. Of course, if S is not bounded above then $U_S = \emptyset$; and if S is not bounded below, $L_S = \emptyset$.

Definition 1.12 $M \in \mathbf{R}$ is called the *least upper bound of S* (or *supremum of S*), and we write $M = \sup S$ provided:
(a) $M \in U_S$.
(b) $M \leq u$ for every $u \in U_S$.

$m \in \mathbf{R}$ is called the *greatest lower bound of S* (or *infimum of S*), and we write $m = \inf S$ provided:
(a) $m \in L_S$.
(b) $m \geq l$ for every $l \in L_S$.

The following property of **R**, known as the *least-upper-bound property*, is equivalent to Dedekind's theorem.

Theorem 1.7 Every nonempty set $S \subset \mathbf{R}$ which is bounded above has a least upper bound.

PROOF Let $\emptyset \neq S \subset \mathbf{R}$ and suppose that S is bounded above. Define the set A by

$$A = \{x \in \mathbf{R} \mid x > u \text{ for some } u \in U_S\}$$

We first establish that A is a ray in **R**. $A \neq \emptyset$ since S is bounded above. Also, $A' \neq \emptyset$ since $S \subseteq A'$. Therefore A is a nonempty proper subset of **R**. We show that A is a ray in **R**:
(a) If $x \in A$ and $y > x$ then $x > u$ for some $u \in U_S$. Clearly $y > u$, and so $y \in A$.
(b) Suppose a_0 is the first element in A. Then $a_0 > u$ for some $u \in U_S$. Now $u + u < a_0 + u < a_0 + a_0$ and so $u < \frac{1}{2}(a_0 + u) < a_0$. Hence $\frac{1}{2}(a_0 + u) \in A$, and this contradicts our assumption that a_0 is the first element in A. Therefore A has no first element.

Since A is a ray in **R**, it follows from Dedekind's theorem that A' has a largest element M. We demonstrate that M is the least upper bound of S.
(a) If $M < x$ for some $x \in S$ then $M < \frac{1}{2}(M + x) < x$ implies that $\frac{1}{2}(M + x) \notin U_S$, and so $\frac{1}{2}(M + x) \notin A$. But then $\frac{1}{2}(M + x) \in A'$, and this contradicts the fact that M is the largest element in A'. Therefore $M \geq x$ for every $x \in S$ and so $M \in U_S$.
(b) Since $M \notin A$, $M \leq u$ for every $u \in U_S$.
Consequently, $M = \sup S$.

The equivalence of Dedekind's theorem and the least-upper-bound property is established by proving that Dedekind's theorem is a consequence of the least-upper-bound property. We ask the reader to supply the details in Exercise 1.35.

EXERCISES

1.28 Prove that for any $r \in \mathbf{Q}$ the set $\{x \in \mathbf{Q} \mid x > r\}$ is a ray in **Q**.
1.29 Verify that addition and multiplication of real numbers are well-defined operations.
1.30 Show that addition and multiplication of rational real numbers (Definition 1.11) agrees with the addition and multiplication of rationals in Sec. 1.3 (Definition 1.7).
1.31 Verify that $\{x \in \mathbf{Q} \mid x > 0\}$ is the additive identity in **R** and that

$$-U = \{x \in \mathbf{Q} \mid x > -r \text{ for some } r \in U'\}$$

is a ray in **Q** satisfying $U + (-U) = 0 = \{x \in \mathbf{Q} \mid x > 0\}$.

1.32 Verify that **R** is an ordered field.

1.33 Prove that a nonempty set S of real numbers is bounded if and only if there is a nonnegative real number K such that $-K \leq x \leq K$ for every $x \in S$.

1.34 Use the least-upper-bound property of **R** to establish that every nonempty set $S \subset \mathbf{R}$ which is bounded below has a greatest lower bound.

1.35 Use the least-upper-bound property of **R** to prove Dedekind's theorem.

1.36 Verify that the least-upper-bound property does not hold in the ordered field of rational numbers.

1.37 Prove that, if they exist, the least upper bound and the greatest lower bound of a nonempty set $S \subset \mathbf{R}$ are unique.

1.38 Prove Bernoulli's inequality: $(1 + x)^n \geq 1 + nx$ for every real number $x \geq -1$ and every $n \in \mathbf{N}$. *Hint*: Use mathematical induction.

1.39 Prove the binomial theorem: For every pair of real numbers x and y and every natural number n

$$(x + y)^n = x^n + nx^{n-1}y + \ldots + \frac{n(n-1)\ldots(n-k+1)}{k!} x^{n-k}y^k + \ldots + nxy^{n-1} + y^n$$

1.5 PROPERTIES OF REAL NUMBERS

The set of all real numbers **R** is a nonempty set together with two binary operations, ($+$) addition and ($\cdot$) multiplication, which satisfy the following axioms:

1. $(a + b) + c = a + (b + c)$ for all $a, b, c \in \mathbf{R}$.
2. $a + b = b + a$ for all $a, b \in \mathbf{R}$.
3. There is a $0 \in \mathbf{R}$ with $a + 0 = a$ for every $a \in \mathbf{R}$.
4. For each $a \in \mathbf{R}$ there is an element $-a \in \mathbf{R}$ with $a + (-a) = 0$.
5. $(a \cdot b) \cdot c = a \cdot (b \cdot c)$ for all $a, b, c \in \mathbf{R}$.
6. $a \cdot b = b \cdot a$ for all $a, b \in \mathbf{R}$.
7. There is a $1 \in \mathbf{R}$ with $a \cdot 1 = a$ for every $a \in \mathbf{R}$.
8. For each $a \in \mathbf{R}$ such that $a \neq 0$ there is an element $1/a \in \mathbf{R}$ with $a \cdot (1/a) = 1$.
9. $a \cdot (b + c) = (a \cdot b) + (a \cdot c)$ for all $a, b, c \in \mathbf{R}$.

The above axioms tell us that **R** is a field. There is a subset P of **R** which satisfies the following three properties:

1. If $a, b \in P$, then $a + b \in P$.
2. If $a, b \in P$, then $a \cdot b \in P$.
3. For every $a \in \mathbf{R}$ exactly one of the following holds: $a \in P$ or $-a \in P$ or $a = 0$.

Any field which has a subset P satisfying properties 1 to 3 above is called an *ordered field*. The order relation $<$ is defined by $a < b$ provided $b - a \in P$. P is the set of all positive real numbers, and it follows from the above that:

1. If $a < b$ and $b < c$ then $a < c$ (transitive law).
2. For all $a, b, \in \mathbf{R}$ exactly one of the following holds: $a < b$ or $b < a$ or $a = b$ (law of trichotomy).
3. If $a < b$ then $a + c < b + c$ for all $c \in \mathbf{R}$.
4. If $a < b$ and $c > 0$ then $a \cdot c < b \cdot c$.

In addition, one further axiom will be needed in order to guarantee that we have all the usual and familiar properties of the real number system.

Completeness axiom (or Dedekind's axiom) Every nonempty subset of $\mathbf{R}$ which is bounded above has a least upper bound.

The reader should recognize the completeness axiom, given here, as the least upper bound property of $\mathbf{R}$ which was proved in Sec. 1.4. We refer to $\mathbf{R}$ as a *complete ordered field*.

Our first result in this section is a property of $\mathbf{R}$ we shall find repeated use for in the chapters ahead. It is known as the *archimedean property*, and the fact that the real number system satisfies this property leads us to refer to $\mathbf{R}$ as a *complete archimedean ordered field*. Before stating the archimedean property we need the following lemma

Lemma If $M = \sup S$ and $y < M$ then there is an $x_0 \in S$ with $x_0 > y$.

PROOF For suppose $x \leq y$ for every $x \in S$. Then $y \in U_S$, and $y < M$ contradicts that $M = \sup S$. Thus there must be some $x_0 \in S$ with $x_0 > y$.

Theorem 1.8: Archimedean property If $a, b \in \mathbf{R}$ with $a > 0$ and $b > 0$ then there is an $n \in \mathbf{N}$ with $na > b$.

PROOF Suppose that $na \leq b$ for every $n \in \mathbf{N}$. Let

$$S = \{na \mid n \in \mathbf{N}\}$$

S is bounded above since b is an upper bound of S. By the completeness axiom, $M = \sup S$ exists. Now $M - a < M$, and so by the preceding lemma there is an $n_0 \in \mathbf{N}$ such that $n_0 a > M - a$. But then $n_0 a + a > M$, and so $(n_0 + 1)a > M$. Since $n_0 + 1 \in \mathbf{N}$, we are led to the contradiction that $(n_0 + 1)a \in S$ and $(n_0 + 1)a > \sup S$. Thus $na \leq b$ cannot hold for every $n \in \mathbf{N}$.

The following corollary is an especially useful form of the archimedean property.

Corollary Given any $\varepsilon > 0$ there is a natural number n such that $1/n < \varepsilon$.

We adopt the following notation for *intervals* of real numbers. It is assumed that a and b are real numbers with $a < b$. A bounded interval is a set of the form

$$[a, b] = \{x \in \mathbf{R} \mid a \leq x \leq b\} \qquad (a,b) = \{x \in \mathbf{R} \mid a < x < b\}$$
$$[a, b) = \{x \in \mathbf{R} \mid a \leq x < b\} \qquad (a,b] = \{x \in \mathbf{R} \mid a < x \leq b\}$$

$[a, b]$ is called a *closed interval*, and (a, b) is called an *open interval*. Closed sets and open sets will be discussed more fully in the next chapter. The bounded intervals $[a, b)$ and $(a, b]$ are neither open nor closed. Note that each of the above four types of intervals is a bounded subset of $\mathbf{R}$ with least upper bound b and greatest lower bound a. An unbounded interval is a set of the form

$$[a, \infty) = \{x \in \mathbf{R} \mid x \geq a\} \qquad (-\infty, b] = \{x \in \mathbf{R} \mid x \leq b\}$$
$$(a, \infty) = \{x \in \mathbf{R} \mid x > a\} \qquad (-\infty, b) = \{x \in \mathbf{R} \mid x < b\}$$

Using this notation, we can write $\mathbf{R} = (-\infty, \infty)$.

Theorem 1.9 Every open interval (a, b) contains a rational number.

PROOF
Case 1: $0 < a < b$ By the archimedean property there is a $k \in \mathbf{N}$ such that $1/k < b - a$. Let $A = \{n \in N \mid n/k > a\}$. Again by the archimedean property $A \neq \emptyset$. Now, by the well-ordering principle for N, A has a first element, say n_0. Then $n_0/k > a$ and $(n_0 - 1)/k \leq a$. Moreover,

$$\frac{n_0}{k} \leq a + \frac{1}{k} < a + (b - a) = b$$

Therefore the rational number n_0/k is contained in the interval (a, b).

Case 2: $a \leq 0 < b$ By the archimedean property there is a $k \in \mathbf{N}$ with $1/k < b$. Clearly $1/k \in (a, b)$.

Case 3: $a < b \leq 0$ Then $0 \leq -b < -a$. By the previous cases there is a rational $r \in (-b, -a)$, and so the rational number $-r \in (a, b)$.

Corollary Every open interval (a, b) contains infinitely many rational numbers.

It should be clear that we need not be restricted to open intervals in the above theorem. In fact, every interval contains an open interval as a subset, and so each interval on the real line contains infinitely many rational numbers.

In Section 1.4 we saw that there is no rational number r such that $r^2 = 2$. We denote the positive real number r which satisfies $r^2 = 2$ by $\sqrt{2}$ ($\sqrt{2}$ represents the ray V in Q defined by $V = \{x \in Q \mid x > 0$ and $x^2 > 2\}$). $\sqrt{2}$ is an *irrational number*.

Theorem 1.10 Every interval I of real numbers contains an irrational number.

PROOF Choose $a, b \in I$ with $a < b$ and choose the nonzero rational number r such that $a/\sqrt{2} < r < b/\sqrt{2}$. Then $a < r\sqrt{2} < b$. Clearly $r\sqrt{2}$ is irrational (since $r\sqrt{2}$ rational implies $(1/r)(r\sqrt{2}) = \sqrt{2}$ rational), and $r\sqrt{2} \in I$.

Corollary Every interval I of real numbers contains infinitely many irrational numbers.

We call a set $D \subseteq \mathbf{R}$ *dense* in $\mathbf{R}$ provided $D \cap I \neq \emptyset$ for every interval I. The preceding theorems say that $\mathbf{Q}$ is dense in $\mathbf{R}$ and also $\mathbf{R} - \mathbf{Q}$ (the set of all irrational numbers) is dense in $\mathbf{R}$.

EXERCISES

1.40 Prove the transitive law.
1.41 Prove the law of trichotomy.
1.42 Prove the corollary to the archimedean property.
1.43 Prove the corollaries to Theorems 1.9 and 1.10.
1.44 Is the system $\mathbf{Q}$ of all rational numbers an archimedean ordered field? Explain.

1.6 FUNCTIONS AND ONE-TO-ONE CORRESPONDENCES

We begin by giving some examples of real-valued functions defined on intervals $I \subseteq \mathbf{R}$, which we shall have occasion to discuss and use in the chapters ahead.

Example 1.1 The absolute-value function is defined by

$$|x| = \begin{cases} x & \text{if } x \geq 0 \\ -x & \text{if } x < 0 \end{cases}$$

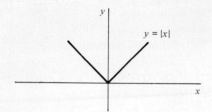

Figure 1.1

The domain of $f(x) = |x|$ is the set of all real numbers **R** and the range is the set of all nonnegative real numbers $\{x \in \mathbf{R} \mid x \geq 0\}$. The graph is given in Fig. 1.1.

Some properties of the absolute-value function which can easily be established are

1. $|-a| = |a|$ for every $a \in \mathbf{R}$.
2. $|ab| = |a| \cdot |b|$ for every pair $a, b \in \mathbf{R}$.
3. $||a| - |b|| \leq |a + b| \leq |a| + |b|$ for every pair $a, b \in \mathbf{R}$.

We define the distance between any two real numbers a and b to be $|a - b|$. We ask the reader to verify (see Exercise 1.46) that distance satisfies the following properties:

$$|a - b| \begin{cases} = 0 & \text{if and only if } a = b \\ = |b - a| & \text{for every pair } a, b \in \mathbf{R} \\ \leq |a - c| + |c - b| & \text{for all } a, b, c \in \mathbf{R} \end{cases}$$

Example 1.2 The signum function is defined by

$$\operatorname{sgn}(x) = \begin{cases} 1 & \text{if } x > 0 \\ 0 & \text{if } x = 0 \\ -1 & \text{if } x < 0 \end{cases}$$

The domain of $f(x) = \operatorname{sgn}(x)$ is the set of all real numbers **R** and the range is the finite set $\{-1, 0, 1\}$. The graph is given in Fig. 1.2. Observe that $|x| = x \operatorname{sgn}(x)$ for every real number $x \in \mathbf{R}$.

Example 1.3 The greatest-integer function is defined by

$[\![x]\!] = $ greatest integer which is less than or equal to x

That is, $[\![x]\!] = k$ for every $x \in [k, k + 1)$ holds for each and every integer k. The domain of $f(x) = [\![x]\!]$ is the set of all real numbers **R**, and the range is the set of integers **J**. The graph of the greatest integer function is given in Fig. 1.3. It is clear from the graph of $f(x) = [\![x]\!]$ why this kind of function is sometimes referred to as a *step function*.

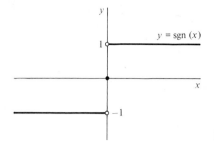

Figure 1.2

The functions illustrated in this section may seem somewhat contrived or artificial, but such functions do arise in the applications of mathematics. For example, one might consider postage as a function of weight for first class mail delivery. This is an example of a step function whose defining rule could well use the greatest-integer function in its formulation. In Chap. 3 we shall encounter a function requiring four separate formulas for its defining rule but arising in a quite natural way.

Functions can also be constructed from previously defined functions by using the algebraic properties of the real number system. For example, if f and g are two functions each of which has as its range a subset of $\mathbf{R}$ then we can define the functions

$$(f + g)(x) = f(x) + g(x)$$
$$(f - g)(x) = f(x) - g(x)$$
$$(f \cdot g)(x) = f(x) \cdot g(x)$$

Each of these functions has as its domain the set of all x which are both in the domain of f and in the domain of g. The function

$$(kf)(x) = k \cdot f(x)$$

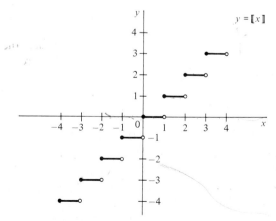

Figure 1.3

where k is any real number. The domain of kf is identically the domain of f. The function

$$\frac{f}{g}(x) = \frac{f(x)}{g(x)}$$

has as its domain all those values x which are both in the domain of f and in the domain of g and for which $g(x) \neq 0$.

As we saw in Sec. 1.1, we can also construct a composition function from two functions f and g, where

$$g \circ f(x) = g(f(x))$$

$g \circ f$ has as its domain the set of all x in the domain of f for which $f(x)$ is in the domain of g.

Recall in Section 1.1 we defined two sets A and B to be equivalent, written $A \sim B$, provided there is a one-to-one correspondence $f: A \to B$. Moreover, the equivalence relation holds in the set of all subsets of $\mathbf{R}$. A set $A \subset \mathbf{R}$ is called *finite* if there is a natural number n such that $A \sim \{1, 2, \ldots, n\}$. In addition, it is conventional to call the empty set a finite set. A set which is not finite is called *infinite*. A set $A \subseteq \mathbf{R}$ is called *countably infinite* (or *denumerable*) provided $A \sim \mathbf{N}$, and A is *countable* if it is either finite or countably infinite. If a set is not countable, it is called *uncountable*; as we shall soon see, every interval in $\mathbf{R}$ is an uncountable set.

As a first example consider the set E of even natural numbers. The function $f: E \to \mathbf{N}$ defined by $f(x) = x/2$ for each $x \in E$ is easily seen to be one-to-one and onto $\mathbf{N}$. Therefore $E \sim \mathbf{N}$, and so E is countably infinite. Of course, E is a proper subset of $\mathbf{N}$, and this property is characteristic of infinite sets. In fact, some authors define an infinite set to be any set which is equivalent to a proper subset of itself. We ask the reader to prove (see Exercise 1.49) that a finite set is never equivalent to a proper subset of itself.

Countably infinite sets are in some sense the "smallest" infinite sets. In fact it can be shown that every infinite set has a proper subset which is countably infinite. The following theorem shows that an infinite subset of a countably infinite set is necessarily countably infinite.

Theorem 1.11 If $f: A \to B$ is one-to-one and B is countable then A is countable.

PROOF There is nothing to prove if A is finite; so suppose that A is infinite. Now $A \sim f(A)$, where $f(A)$ is the range of f, and so $f(A)$ is infinite. Since $f(A) \subseteq B$, it follows that B is infinite (a finite set has no infinite subsets). By hypothesis B is countable, and so B must be countably infinite. Let $\varphi: \mathbf{N} \to B$ be one-to-one and onto and denote $b_n = \varphi(n)$ for each $n \in \mathbf{N}$. Then $B = \{b_1, b_2, b_3, \ldots\}$. Now let k_1 be the first natural number such that $b_{k_1} \in f(A)$; let k_2 be the first natural number greater than k_1 such that $b_{k_2} \in f(A)$; let k_3 be the first natural number greater than k_2 such that $b_{k_3} \in f(A)$, etc. Then $f(A) = \{b_{k_1}, b_{k_2}, b_{k_3}, \ldots\}$. Define $g: f(A) \to \mathbf{N}$ by $g(b_{k_j}) = j$ for $j = 1, 2, 3, \ldots$. Clearly g is one-to-one and onto. But f if one-to-one from A onto $f(A)$. Therefore $g \circ f$ is one-to-one from A onto $\mathbf{N}$. It follows that $A \sim \mathbf{N}$. Consequently A is countable.

If B is any countable set and $S \subseteq B$ then the identity function $i_S : S \to B$, which is clearly one-to-one, proves the following corollary.

Corollary Every subset of a countable set is countable.

The cartesian product of two sets would seen to be "larger" than either of the individual sets. Of course, if either A or B is empty, then $A \times B$ is empty. If A is finite with n elements and B is finite with m elements then $A \times B$ is finite with nm elements (see Exercise 1.50). If $A \sim \mathbf{N}$ and B is finite with m elements ($m \in N$) then $A \times B$ is countably infinite (see Exercise 1.51). The same result holds if $A \neq \emptyset$ is finite and B is countably infinite since $A \times B \sim B \times A$ for all sets A and B by Exercise 1.10. The next result shows that if A and B are each countably infinite then so too is the cartesian product of A and B.

Theorem 1.12 If A and B are countably infinite then $A \times B$ is countably infinite.

PROOF Since A and B are each countably infinite, we can write $A = \{a_1, a_2, a_3, \ldots\}$ and $B = \{b_1, b_2, b_3, \ldots\}$. To display a one-to-one correspondence between $A \times B$ and N we set up the array

$$(a_1, b_1) \quad (a_1, b_2) \quad (a_1, b_3) \quad \ldots$$
$$(a_2, b_1) \quad (a_2, b_2) \quad (a_2, b_3) \quad \ldots$$
$$(a_3, b_1) \quad (a_3, b_2) \quad (a_3, b_3) \quad \ldots$$
$$\ldots\ldots\ldots\ldots\ldots\ldots\ldots\ldots\ldots\ldots\ldots$$

of elements in $A \times B$. We define the function $\varphi: A \times B \to \mathbf{N}$ by "counting" in a diagonal zigzag through the array, obtaining

$$\varphi(a_1, b_1) = 1$$
$$\varphi(a_1, b_2) = 2 \quad \varphi(a_2, b_1) = 3$$
$$\varphi(a_1, b_3) = 4 \quad \varphi(a_2, b_2) = 5 \quad \varphi(a_3, b_1) = 6$$
$$\varphi(a_1, b_4) = 7 \quad \varphi(a_2, b_3) = 8 \quad \varphi(a_3, b_1) = 9 \quad \varphi(a_4, b_1) = 10$$

and so on. φ is a one-to-one function from $A \times B$ onto $\mathbf{N}$. Therefore, $A \times B$ is countably infinite.

If X is a nonempty set and $\sim$ is any equivalence relation in X, let $\mathcal{C}$ be the set of all equivalence classes arising from the equivalence relation. We define $\varphi: \mathcal{C} \to X$ by choosing a single element x from each equivalence class $E \in \mathcal{C}$ and defining $\varphi(E) = x$. φ is one-to-one since the equivalence classes partition X; that is, they are pairwise disjoint, with union X (see Theorem 1.1). It follows from Theorem 1.11 that if X is countable then so is $\mathcal{C}$. Recall that the set of all integers $\mathbf{J}$ was defined to be the set of all equivalence classes of a suitably chosen equivalence relation defined in $S = \mathbf{N} \times \mathbf{N}$ (see Definition 1.4). $\mathbf{N} \times \mathbf{N}$ is countable by Theorem 1.12, and so $\mathbf{J}$ is countable. Since $\mathbf{J}$ is clearly not finite, it follows that $\mathbf{J}$ is countably infinite. Similarly, we defined a rational number to be an equivalence class resulting from an equivalence relation defined in $T = \mathbf{J} \times (\mathbf{J} - \{0\})$ (see Definition 1.6). T is countably infinite (again by Theorem 1.12), and so Theorem 1.11 yields the following result.

Theorem 1.13 The set $\mathbf{Q}$ of all rational numbers is countably infinite.

The set $\mathbf{Q}$ of all rational numbers has the interesting property of being a countably infinite set which is dense in $\mathbf{R}$. In other words, $\mathbf{Q}$ is countable and has elements (infinitely many, in fact) in each and every interval I on the real line. Next, we demonstrate that an interval is never a countable set.

Theorem 1.14 For any $a, b \in \mathbf{R}$, where $a < b$, the open interval (a, b) is uncountable.

PROOF (a, b) is clearly infinite; suppose that it is countably infinite. Then there is a one-to-one correspondence $\varphi: \mathbf{N} \to (a, b)$, and so we can list the elements in (a, b) by $(a, b) = \{x_1, x_2, x_3, \ldots\}$, where, as before, $x_n = \varphi(n)$ for each $n \in \mathbf{N}$. We arrive at a contradiction by exhibiting an element $x \in (a, b)$ which is not in the list $x_1, x_2, x_3, \ldots$; that is, x is not in the range of φ, and so we have contradicted that φ is onto (a, b).

First define $y_1 = x_1$; let $y_2 = x_j$, where j is the first natural number greater that 1 such that $x_j < y_1$. If there is no such j then each number $x \in (a, y_1)$ is missing from the list. Next, let $y_3 = x_k$, where k is the first natural number greater than j such that $x_k \in (y_2, y_1)$. If there is no such k then each real number $x \in (y_2, y_1)$ is missing from the list. Next, let $y_4 = x_l$, where l is the first natural number greater than k such that $x_l \in (y_2, y_3)$. If there is no such l then each real number $x \in (y_2, y_3)$ is missing from the list. Next, let $y_5 = x_m$, where m is first natural number greater than l such that $x_m \in (y_4, y_3)$. If there is no such m then each real number $x \in (y_4, y_3)$ is missing from the list (see Fig. 1.4). Continue the process, always choosing y_{n+1} to be the next appropriate x value in the list $x_1, x_2, x_3, \ldots$ which is between y_n and y_{n-1} if such a value exists. If the process terminates at some stage, we have an interval of x's

Figure 1.4

in the interval (a, b) which fail to be in the list $x_1, x_2, x_3, \ldots$ and the proof is complete If the process does not terminate, note that the even-numbered y's satisfy $y_2 < y_4 < y_6 < \ldots$, the odd-numbered y's satisfy $y_1 > y_3 > y_5 > \ldots$, and each even-numbered y value is less than each odd-numbered y value. Then the set $\{y_2, y_4, y_6, \ldots\}$ is bounded above (each odd y value is an upper bound), and so by the least-upper-bound property of $\mathbf{R}$, $x = \sup\{y_2, y_4, y_6, \ldots\}$ exists. Now, if $x = y_{2n}$ for some $n \in \mathbf{N}$ then $x < y_{2n+2}$, a contradiction of the fact that x is an upper bound of the set $\{y_2, y_4, y_6, \ldots\}$. Hence $x > y_{2n}$ for every $n \in \mathbf{N}$. If $x = y_{2n-1}$ for some $n \in \mathbf{N}$ then $x > y_{2n+1}$, a contradiction of the fact that x is the least upper bound of the set $\{y_2, y_4, y_6, \ldots\}$. Thus $x < y_{2n-1}$ for every $n \in \mathbf{N}$. Also, x cannot be one of the x values in the list $x_1, x_2, x_3, \ldots$ which were skipped, that is, not taken to be a y value, since each skipped x value is either less than some even-numbered y value or greater than some odd-numbered y value. Consequently, x does not appear in the list $x_1, x_2, x_3, \ldots$. But clearly $x \in (y_2, y_1) \subset (a, b)$, and so we have produced an x in (a, b) which is not in the range of φ. Therefore (a, b) is uncountable.

Every interval I contains some open interval (a, b), and hence each I is uncountable.

EXERCISES

1.45 Prove the properties of the absolute-value function:
$|-a| = |a|$ for every $a \in \mathbf{R}$.
$|ab| = |a| \cdot |b|$ for every pair $a, b \in \mathbf{R}$.
$||a| - |b|| \leq |a + b| \leq |a| + |b|$ for every pair $a, b, \in \mathbf{R}$.

1.46 Verify the following properties for the distance $d(a, b) = |a - b|$ between a and b:
$d(a, b) = 0$ if and only if $a = b$.
$d(a, b) = d(b, a)$ for every pair $a, b \in \mathbf{R}$.
$d(a, b) \leq d(a, c) + d(c, b)$ for all $a, b, c \in \mathbf{R}$.

1.47 Find the domain and range of each of the following functions and graph each function.
(a) $[\![2x]\!] - [\![x]\!]$ (b) $\dfrac{|x|}{x}$ (c) $|[\![x]\!]|$
(d) $[\![\operatorname{sgn}(x)]\!]$ (e) $\dfrac{2}{x + |x|}$ (f) $\dfrac{|x| + \operatorname{sgn}(x)}{[\![x]\!]}$

1.48 Prove that if $f: A \to B$ is a function defined on the countable set A then the range of f is countable.

1.49 Prove that a finite set cannot be equivalent to any of its proper subsets.

1.50 Let $f: A \to \{1, 2, \ldots, n\}$ and $g: B \to \{1, 2, \ldots, m\}$ be one-to-one onto functions. Show that $\varphi: A \times B \to \{1, 2, \ldots, nm\}$ defined by $\varphi(a, b) = [f(a) - 1]m + g(b)$ is one-to-one and onto.

1.51 Let $f: A \to \mathbf{N}$ and $g: B \to \{1, 2, \ldots, m\}$ be one-to-one onto functions. Show that $\varphi: A \times B \to \mathbf{N}$ defined as in Exercise 1.50 is a one-to-one function onto $\mathbf{N}$.

1.52 (a) Prove that the union of two countable sets is countable.

(b) Use part (a) and the principle of mathematical induction to prove that the union of n countable sets (where n is any natural number) is again a countable set.

1.53 Prove that the cartesian product of n countable sets ($n \in \mathbf{N}$) is again a countable set.

1.54 Determine which of the following sets are countable and which are uncountable:

(a) Set of positive rationals.

(b) Set of all irrationals in (0, 1).

(c) Set of all terminating decimals. *Note*: A decimal terminates if all digits are 0 from some point on.

(d) $\{r + \sqrt{n} \mid r \in \mathbf{Q} \text{ and } n \in \mathbf{N}\}$

(e) $\{x \in \mathbf{R} - \mathbf{Q} \mid x \text{ cannot be written as the square root of a nonnegative rational}\}$.

(f) $\{x \in \mathbf{R} \mid x \text{ is a solution to } ax^2 + bx + c = 0 \text{ for some } a, b, c \in \mathbf{Q}\}$.

1.55 Give an example of a countably infinite subset of the set of irrational numbers which is dense in $\mathbf{R}$.

CHAPTER TWO

SEQUENCES AND SETS OF REAL NUMBERS

2.1 LIMITS OF SEQUENCES

In this chapter sequences and sets of real numbers are studied. Sequences are used throughout the text as a tool, and the idea of limit of a sequence will prepare the reader for the more general notion of the limit of a function, which is introduced in the next chapter. As we begin the study of limits, we shall be especially concerned with the methods and techniques used in the proofs of our results.

A *sequence* is a function whose domain is the set of natural numbers **N**. For the most part, we shall be concerned with sequences of real numbers, that is, functions defined on **N** whose range is a subset of **R**. To represent the values of a sequence we use the notation $a_1, a_2, a_3, \ldots$ instead of the usual notation for a function $f(1), f(2), f(3), \ldots$.

Let r be any real number. An important type of sequence, called a *geometric progression*, is given by

$$a_1 = 1 + r, \ a_2 = 1 + r + r^2, \ a_3 = 1 + r + r^2 + r^3, \ldots$$

The general term in this sequence is given by

$$a_n = 1 + r + r^2 + r^3 + \ldots + r^n \quad \text{for } n \in \mathbf{N}$$

To derive an alternate expression for a_n multiply each side of this equation by r

$$ra_n = r + r^2 + r^3 + r^4 + \ldots + r^{n+1}$$

Subtracting this equation from the first gives $(1-r)a_n = 1 - r^{n+1}$, and so

$$a_n = \frac{1 - r^{n+1}}{1 - r} \quad \text{provided } r \neq 1$$

The reader may find it instructive to supply an inductive argument to verify the above expression for a_n. Intuitively, if $-1 < r < 1$, a_n tends to the value $1/(1-r)$ as n becomes large. Furthermore, if $|r| \geq 1$, a_n does not tend to any real number value. We formalize this idea of limit in the following definition.

Definition 2.1 The limit of the sequence $\{a_n\}$ is A, and is written $\lim_{n\to\infty} a_n = A$, if given any $\varepsilon > 0$ there is a natural number N such that $|a_n - A| < \varepsilon$ whenever $n > N$.

If $\{a_n\}$ has a limit A, we say that the sequence *converges*; otherwise we say that $\{a_n\}$ *diverges*. It is understood that A is a real number.

Example 2.1 We prove that $\lim_{n\to\infty} (1/n) = 0$.

SOLUTION By the archimedean property, given any $\varepsilon > 0$ there is a natural number N such that $1/N < \varepsilon$. Now if $n > N$, then $|1/n - 0| = 1/n < 1/N < \varepsilon$.

The next example is a special case of a geometric progression with $r = 1/2$.

Example 2.2 Let $a_n = 2 - 1/2^n$, and let $\varepsilon > 0$ be given. We prove that $\lim_{n\to\infty} a_n = 2$.

SOLUTION

$$|a_n - 2| = \frac{1}{2^n} = \frac{1}{(1+1)^n} \leq \frac{1}{1+n}$$

by Bernoulli's inequality (see Exercise 1.38). Hence $|a_n - 2| < 1/n$ for every natural number n. As in the previous example, we choose $N \in \mathbf{N}$ so large that $1/N < \varepsilon$ (archimedean property); then if $n > N$, we have $|a_n - 2| < 1/n < 1/N < \varepsilon$, and so $\lim_{n\to\infty} a_n = 2$.

Example 2.3 We prove that $\lim_{n\to\infty} [(2n^2 + 1)/(n^2 + 3n)] = 2$.

SOLUTION Let $\varepsilon > 0$ be given. Then

$$\left|\frac{2n^2 + 1}{n^2 + 3n} - 2\right| = \left|\frac{2n^2 + 1 - 2(n^2 + 3n)}{n^2 + 3n}\right|$$

$$= \left|\frac{1 - 6n}{n^2 + 3n}\right| = \frac{6n - 1}{n^2 + 3n} < \frac{6n}{n^2} = \frac{6}{n}$$

Now choose $N \in \mathbf{N}$ so large that $1/N < \varepsilon/6$; that is, choose $N > 6/\varepsilon$. Then for $n > N$ we have

$$\left| \frac{2n^2 + 1}{n^2 + 3n} - 2\varepsilon \right| < \frac{6}{n} < \frac{6}{N} < \varepsilon$$

As with any mathematical operation, it is important to know (and show) that the limit, if it exists, is unique.

Theorem 2.1 If $\lim_{n \to \infty} a_n$ exists, then it is unique.

PROOF Suppose that $\lim_{n \to \infty} a_n = A$ and $\lim_{n \to \infty} a_n = A'$. Given any $\varepsilon > 0$ there is a natural number N_1 such that $|a_n - A| < \varepsilon/2$ whenever $n > N_1$ and there is a natural number N_2 such that $|a_n - A'| < \varepsilon/2$ whenever $n > N_2$. If $n > \max(N_1, N_2)$, where $\max(N_1, N_2)$ just means the larger of the two natural numbers N_1 and N_2, then $|a_n - A| < \varepsilon/2$ and $|a_n - A'| < \varepsilon/2$. Consequently,

$$|A - A'| = |A - a_n + a_n - A'| < |A - a_n| + |a_n - A'|$$

$$= |a_n - A| + |a_n - A'| < \frac{\varepsilon}{2} + \frac{\varepsilon}{2} = \varepsilon$$

But A and A' are fixed real numbers, and ε is an arbitrary positive real number. Therefore, the above inequality says that $|A - A'|$ is less than every positive real number. But by definition $|A - A'| \geq 0$. The only way out is for $|A - A'|$ to equal 0, and so $A = A'$.

As the sequences get more complex, the techniques used in Examples 2.1 to 2.3 become less practical. Then, too, if all limits are established by using the definition, these techniques are used over and over again. To avoid such repetition and to facilitate limit proofs in general, we develop the following useful results.

Theorem 2.2 If $a_n \leq b_n \leq c_n$ for every $n \in \mathbf{N}$ and $\lim_{n \to \infty} a_n = A = \lim_{n \to \infty} c_n$ then $\lim_{n \to \infty} b_n = A$.

PROOF Suppose $a_n \leq b_n \leq c_n$ for every natural number n. Then

$$a_n - A \leq b_n - A \leq c_n - A$$

It follows that

$$b_n - A \leq |c_n - A|$$

and

$$-(b_n - A) \leq -(a_n - A) \leq |a_n - A|$$

Now given any $\varepsilon > 0$, since $\lim_{n \to \infty} a_n = A$ and $\lim_{n \to \infty} c_n = A$ there exists a natural number N_1 such that $|a_n - A| < \varepsilon$ whenever $n > N_1$

and there exists a natural number N_2 such that $|c_n - A| < \varepsilon$ whenever $n > N_2$. If $n > \max(N_1, N_2)$ then $b_n - A \leq |c_n - A| < \varepsilon$ and also $-(b_n - A) \leq |a_n - A| < \varepsilon$. Therefore $|b_n - A| < \varepsilon$ for all natural numbers $n > \max(N_1, N_2)$. This proves that $\lim_{n \to \infty} b_n = A$.

The above theorem is sometimes referred to as "the old squeeze play." It should be clear to the reader that the conclusion to the theorem is obtained even if $a_n \leq b_n \leq c_n$ holds for every natural number from some point on; that is, there is an $n_0 \in \mathbf{N}$ such that $a_n \leq b_n \leq c_n$ is true for all natural numbers $n \geq n_0$.

Recall that a set S is bounded if there are real numbers m and M such that $m \leq x \leq M$ for all $x \in S$. This is equivalent to there being a nonnegative real number K such that $|x| \leq K$ for every $x \in S$. Just choose K to be the maximum of $|m|$ and $|M|$. A sequence $\{a_n\}$ is called bounded if there is a nonnegative real number K such that $|a_n| \leq K$ for every natural number n. The following lemma is necessary to prove our next theorem.

Lemma If $\lim_{n \to \infty} a_n = A$ exists then the sequence $\{a_n\}$ is bounded.

PROOF If $\varepsilon = 1$, there is a natural number N such that $|a_n - A| < 1$ whenever $n > N$. It follows that $|a_n| - |A| < 1$ and so $|a_n| < 1 + |A|$ for all $n > N$. Choose $M = \max(|a_1|, |a_2|, \ldots, |a_N|, 1 + |A|)$; clearly $|a_n| \leq M$ for every natural number n, and so $\{a_n\}$ is bounded.

The above lemma is also useful in proving that certain sequences diverge. For if $\{a_n\}$ is an unbounded sequence then necessarily $\lim_{n \to \infty} a_n$ does not exist. The sequence $\{\sqrt{n}\}$ is an example. Of course, there is no guarantee that a bounded sequence will converge. In fact, the sequence defined by

$$a_n = \begin{cases} 0 & \text{if } n \text{ is odd} \\ 1 & \text{if } n \text{ is even} \end{cases}$$

diverges. To verify this note that if $\{a_n\}$ converges then there would have to be some real number A with the property that for any given $\varepsilon > 0$ there is a natural number N such that $|a_n - A| < \varepsilon$ whenever $n > N$. Taking $\varepsilon = \tfrac{1}{2}$, we would necessarily then have

$$|0 - A| < \tfrac{1}{2} \quad \text{and} \quad |1 - A| < \tfrac{1}{2}$$

The first inequality implies that $|A| < \tfrac{1}{2}$, and the second inequality implies that $1 - |A| < \tfrac{1}{2}$ or $|A| > \tfrac{1}{2}$. This contradiction shows that there can be no such real number A.

Theorem 2.3 If $\lim_{n \to \infty} a_n = A$ and $\lim_{n \to \infty} b_n = B$ then
(a) $\lim_{n \to \infty} (a_n + b_n) = A + B$.

(b) $\lim_{n \to \infty} ka_n = kA$ for any real number k.
(c) $\lim_{n \to \infty} a_n b_n = AB$.
(d) $\lim_{n \to \infty} (1/b_n) = 1/B$ provided $b_n \neq 0$ and $B \neq 0$.
(e) $\lim_{n \to \infty} (1/b_n) = 1/B$ provided $b_n \neq 0$ and $B \neq 0$.

PROOF (a) Given any $\varepsilon > 0$ there is a natural number N_1 such that $|a_n - A| < \varepsilon/2$ whenever $n > N_1$, and there is a natural number N_2 such that $|b_n - B| < \varepsilon/2$ whenever $n > N_2$. Now, if $n > \max(N_1, N_2)$ then we have

$$|(a_n + b_n) - (A + B)| = |a_n - A + b_n - B| \leq |a_n - A| + |b_n - B|$$

$$< \frac{\varepsilon}{2} + \frac{\varepsilon}{2} = \varepsilon$$

(b) If $k = 0$, the result is obvious since $|ka_n - kA| = 0$ for every natural number n. Suppose $k \neq 0$ and let $\varepsilon > 0$ be given. Then there is a natural number N such that

$$|a_n - A| < \frac{\varepsilon}{|k|}$$

whenever $n > N$. Hence, if $n > N$, we have

$$|ka_n - kA| = |k(a_n - A)| = |k| \cdot |a_n - A| < |k| \frac{\varepsilon}{|k|} = \varepsilon$$

(c) By the preceding lemma the sequence $\{a_n\}$ is bounded. Say $|a_n| \leq M$ for every natural number n, where M is some positive real number. Now let $\varepsilon > 0$ be given. There is a natural number N_1 such that

$$|a_n - A| < \frac{\varepsilon}{2(1 + |B|)} \quad \text{whenever } n > N_1$$

and there is a natural number N_2 such that

$$|b_n - B| < \frac{\varepsilon}{2M} \quad \text{whenever } n > N_2$$

Now if $n > \max(N_1, N_2)$ then we have

$$|a_n b_n - AB| = |a_n b_n - a_n B + a_n B - AB|$$
$$\leq |a_n b_n - a_n B| + |a_n B - AB|$$
$$= |a_n| \cdot |b_n - B| + |B| \cdot |a_n - A|$$
$$< M \frac{\varepsilon}{2M} + |B| \frac{\varepsilon}{2(1 + |B|)} < \varepsilon$$

Note that the bound for $|a_n - A|$ was chosen to be $\varepsilon/[2(1+|B|)]$ instead of $\varepsilon/2|B|$ because B may be zero.

(d) We are assuming here that B and each b_n is nonzero. Let $\varepsilon > 0$ be given. There is a natural number N_1 such that $|b_n - B| < |B|/2$ whenever $n > N_1$. Hence if $n > N_1$ then $|B|/2 > |b_n - B| = |B - b_n| \geq |B| - |b_n|$, and so $|b_n| > |B|/2$. Also, there is a natural number N_2 such that

$$|b_n - B| < \frac{\varepsilon \cdot |B|^2}{2} \quad \text{whenever } n > N_2$$

Therefore, if $n > \max(N_1, N_2)$, we have

$$\left| \frac{1}{b_n} - \frac{1}{B} \right| = \left| \frac{B - b_n}{b_n B} \right| = \frac{|b_n - B|}{|b_n| \cdot |B|}$$

$$< \frac{(\varepsilon \cdot |B|^2)/2}{(|B|/2)|B|} = \varepsilon$$

(e) By part (d) $\lim_{n \to \infty}(1/b_n) = 1/B$. Consequently by part (c)

$$\lim_{n \to \infty} \frac{a_n}{b_n} = \lim_{n \to \infty} a_n \frac{1}{b_n} = A \frac{1}{B} = \frac{A}{B}$$

The next two examples illustrate how Theorems 2.2 and 2.3 can be used to find (and prove) limits of sequences.

Example 2.4 Find

$$\lim_{n \to \infty} \frac{4n^3 + 3n^2 + 2n + 1}{7n^3 - 2n^2 + 3n}$$

SOLUTION

$$\lim_{n \to \infty} \frac{4n^3 + 3n^2 + 2n + 1}{7n^3 - 2n^2 + 3n}$$

$$= \lim_{n \to \infty} \frac{4 + (3/n) + (2/n^2) + 1/n^3}{7 - (2/n) + 3/n^2}$$

$$= \frac{\lim_{n \to \infty}[4 + (3/n) + (2/n^2) + 1/n^3]}{\lim_{n \to \infty}[7 - (2/n) + 3/n^2]}$$

$$= \frac{\lim_{n \to \infty} 4 + 3\lim_{n \to \infty}(1/n) + 2\lim_{n \to \infty}(1/n^2) + \lim_{n \to \infty}(1/n^3)}{\lim_{n \to \infty} 7 - 2\lim_{n \to \infty}(1/n) + 3\lim_{n \to \infty}(1/n^2)} = \frac{4}{7}$$

We have used Example 2.1 and Theorem 2.3 repeatedly.

Example 2.5 Find $\lim_{n \to \infty} a^{1/n}$, where a is a fixed positive real number.

SOLUTION **Case 1**: $a \geq 1$ Then $a^{1/n} \geq 1$. Define a new sequence $b_n = a^{1/n} - 1$. Now $a^{1/n} = 1 + b_n$, where $b_n \geq 0$, and so

$$a = (1 + b_n)^n \geq 1 + nb_n \qquad \text{by Bernoulli's inequality}$$

Thus $0 \leq b_n \leq (a-1)/n$. Clearly

$$\lim_{n \to \infty} [(a-1)/n] = (a-1) \lim_{n \to \infty} (1/n) = 0$$

By the old squeeze play, Theorem 2.2, $\lim_{n \to \infty} b_n = 0$. Therefore $\lim_{n \to \infty} a^{1/n} = \lim_{n \to \infty} (1 + b_n) = 1$.

Case 2: $0 < a < 1$ Then $1/a > 1$ and

$$\lim_{n \to \infty} a^{1/n} = \lim_{n \to \infty} \frac{1}{(1/a)^{1/n}} = \frac{1}{\lim_{n \to \infty} (1/a)^{1/n}}$$

But $\lim_{n \to \infty} (1/a)^{1/n} = 1$ by case 1 above. Hence $\lim_{n \to \infty} a^{1/n} = 1$.

If a sequence converges, the definition implies that the terms a_n of the sequence $\{a_n\}$ get arbitrarily close to the limit of the sequence A as n gets large. Intuitively, then, the terms of the sequence must be clustering up, or getting closer to each other. Such sequences are called Cauchy sequences.

Definition 2.2 A sequence $\{a_n\}$ is called a *Cauchy sequence* if given any $\varepsilon > 0$ there is a natural number N such that $|a_m - a_n| < \varepsilon$ whenever $m, n > N$.

Every convergent sequence is a Cauchy sequence. Suppose $\lim_{n \to \infty} a_n = A$. Then given any $\varepsilon > 0$ there is a natural number N such that $|a_n - A| < \varepsilon/2$ whenever $n > N$. Therefore, if $m, n > N$, we have

$$|a_m - a_n| = |a_m - A + A - a_n| < |a_m - A| + |A - a_n|$$

$$= |a_m - A| + |a_n - A| < \frac{\varepsilon}{2} + \frac{\varepsilon}{2} = \varepsilon$$

and so $\{a_n\}$ is a Cauchy sequence.

In the next section we shall show that the converse of this result is also true; that is, every Cauchy sequence of real numbers converges. The proof requires the completeness of the real number system, and we shall use the completeness in the form of the least-upper-bound property. It is also true (though we shall not show it) that from the assumption that every Cauchy sequence of real numbers converges one can prove the completeness of the real number system. For this reason it is not uncommon to see the completeness of **R** stated in the following form.

Completeness of R Every Cauchy sequence of real numbers converges.

For the present we shall content ourselves to prove the following intermediate result.

Theorem 2.4 Every Cauchy sequence is bounded.

PROOF Suppose $\{a_n\}$ is a Cauchy sequence. Then for $\varepsilon = 1$ there is a natural number N such that $|a_m - a_n| < 1$ whenever $m, n > N$. Choose $n_0 > N$ and observe that $|a_n| = |a_n - a_{n_0} + a_{n_0}| \leq |a_n - a_{n_0}| + |a_{n_0}| < 1 + |a_{n_0}|$ whenever $n > N$. Let

$$M = \max(|a_1|, |a_2|, \ldots, |a_N|, 1 + |a_{n_0}|)$$

Clearly $|a_n| \leq M$ for every natural number n, and hence $\{a_n\}$ is a bounded sequence.

EXERCISES

2.1 Evaluate each of the following limits (if it exists) and prove your answer:

(a) $\lim_{n \to \infty} \dfrac{1}{n + n^3}$ (b) $\lim_{n \to \infty} (2n^3 - 3n^2 + 1)$ (c) $\lim_{n \to \infty} \dfrac{2n^3 - 3n^2 + 1}{n + n^3}$

(d) $\lim_{n \to \infty} \left(\dfrac{1}{1 + n}\right)^n$ (e) $\lim_{n \to \infty} \left(2 + \dfrac{1}{n}\right)^n$ (f) $\lim_{n \to \infty} \left(\dfrac{n}{n + 1}\right)^{1/n}$

(g) $\lim_{n \to \infty} \left(\dfrac{\sin n + \cos n}{3}\right)^n$ (h) $\lim_{n \to \infty} \dfrac{n}{3^n}$

2.2 Use Definition 2.1 to prove parts a and c of Exercise 2.1.

2.3 Prove that if $\lim_{n \to \infty} a_n = 0$ and $\{b_n\}$ is bounded (but not necessarily convergent) then $\lim_{n \to \infty} a_n b_n = 0$.

2.4 Prove that if $\{a_n\}$ converges and $\{b_n\}$ diverges then $\{a_n + b_n\}$ diverges.

2.5 Give examples of:
(a) A convergent sequence $\{a_n\}$ and a divergent sequence $\{b_n\}$ such that $\{a_n b_n\}$ converges
(b) A pair of divergent sequences $\{a_n\}$ and $\{b_n\}$ with $\{a_n + b_n\}$ convergent
(c) A pair of divergent sequences $\{a_n\}$ and $\{b_n\}$ with $\{a_n b_n\}$ convergent

2.6 A special type of divergent sequence is one which approaches infinity; that is, given any positive real number M there is a natural number N such that $a_n > M$ whenever $n > N$. We denote this by $\lim_{n \to \infty} a_n = \infty$. Prove:

(a) If $\lim_{n \to \infty} a_n = \infty$ and $\{b_n\}$ is a bounded sequence then

$$\lim_{n \to \infty} \frac{b_n}{a_n} = 0 \qquad a_n \neq 0$$

(b) If $\lim_{n \to \infty} a_n = \infty$ and if there exists a positive real number $\eta > 0$ and a natural number N such that $b_n > \eta$ whenever $n > N$ then $\lim_{n \to \infty} a_n b_n = \infty$.

(c) If $\lim_{n \to \infty} |a_n| = \infty$ then

$$\lim_{n \to \infty} \frac{1}{a_n} = 0 \qquad a_n \neq 0$$

2.7 Find examples of sequences $\{a_n\}$ and $\{b_n\}$ such that $\lim_{n\to\infty} a_n = \infty$ and $\lim_{n\to\infty} b_n = 0$ and:

(a) $\lim_{n\to\infty} a_n b_n = 0.$ (b) $\lim_{n\to\infty} a_n b_n = \infty.$

(c) $\lim_{n\to\infty} a_n b_n$ does not exist and is not ∞.

(d) $\lim_{n\to\infty} a_n b_n = c$, where c is an arbitrary real number.

2.8 Suppose $\{A_n\}$ is a sequence of countable sets; that is, A_n is countable for each natural number n. Prove that the union

$$A_1 \cup A_2 \cup A_3 \cup \cdots = \bigcup_{n=1}^{\infty} A_n$$

is a countable set.

2.9 Show that if $\{A_n\}$ is a sequence of countable sets, it may well be the case that the cartesian product

$$A_1 \times A_2 \times A_3 \times A_4 \times \cdots = \bigtimes_{n=1}^{\infty} A_n$$

is uncountable. *Note:* $\bigtimes_{n=1}^{\infty} A_n$ is the set of all sequences $\{a_n\}$, where $a_n \in A_n$ for every $n \in N$.

2.10 Let $\{a_n\}$ be any sequence and form the auxiliary sequence

$$\hat{a}_n = \frac{a_1 + a_2 + \cdots + a_n}{n} \quad \text{for every } n \in N$$

Prove that if $\lim_{n\to\infty} a_n = A$ then $\lim_{n\to\infty} \hat{a}_n = A$ but that the converse is not true.

2.2 BOUNDED SEQUENCES

In the previous section we saw that if a sequence converges then the sequence is necessarily bounded but the converse does not hold; that is, bounded sequences may diverge. In this section there are two main results on bounded sequences. Theorem 2.5 involves the idea of a monotone sequence, and Theorem 2.7 involves the notion of subsequences of a given sequence.

The sequence $\{a_n\}$ is called *monotone increasing* if $a_n \leq a_{n+1}$ for every natural number n (*strictly increasing* if $a_n < a_{n+1}$ for all $n \in N$), and $\{a_n\}$ is called *monotone decreasing* if $a_n \geq a_{n+1}$ for every natural number n (*strictly decreasing* if $a_n > a_{n+1}$ for every $n \in N$).

Theorem 2.5 If $\{a_n\}$ is monotone increasing and bounded above then $\{a_n\}$ converges and $\lim_{n\to\infty} a_n = \sup_n a_n$.

PROOF By the least-upper-bound property of real numbers the set $\{a_n \mid n \in N\}$ has a supremum, say L. Given any $\varepsilon > 0$ there is a natural number N such that $L - \varepsilon < a_N \leq L$, for otherwise $L - \varepsilon$ would be an

46 INTRODUCTION TO MATHEMATICAL ANALYSIS

upper bound of $\{a_n \mid n \in \mathbf{N}\}$ less than the least upper bound L. Now, if $n > N$ then $L - \varepsilon < a_N \leq a_n \leq L$, and so $|a_n - L| < \varepsilon$. Therefore $\lim_{n \to \infty} a_n = L = \sup_n a_n$.

A similar result holds for monotone decreasing sequences (see Exercise 2.11).

Theorem 2.5 is a consequence of the completeness of the real number system (we used the least-upper-bound property of $\mathbf{R}$ in its proof), and in our first example we use this result to establish that each "decimal" of the form $0.d_1 d_2 d_3 \cdots$ represents a unique real number in the interval $[0, 1]$.

Example 2.6 Let d_i be any one of the decimal digits 0, 1, 2, 3, 4, 5, 6, 7, 8, 9 for $i = 1, 2, 3, \cdots$. Then the sequence

$$a_1 = 0.d_1$$
$$a_2 = 0.d_1 d_2$$
$$a_3 = 0.d_1 d_2 d_3$$
$$\cdots$$
$$a_n = 0.d_1 d_2 d_3 \cdots d_n$$
$$\cdots$$

is monotone increasing (since $a_{n+1} - a_n = 0.000 \cdots 0 d_{n+1} \geq 0$) and is bounded above by 1. Thus $\{a_n\}$ has a unique limit. This limit is the real number represented by the decimal $0.d_1 d_2 d_3 \cdots$.

Sometimes it may be quite difficult to determine exactly what the limit of a given sequence is, but one can still use the previous theorem to verify that a limit does exist in certain situations.

Example 2.7 Let $b_n = (1 + 1/n)^n$ for each natural number n. By the binomial theorem (see Exercise 1.39)

$$b_n = 1 + n\frac{1}{n} + \cdots + \frac{n(n-1) \cdots (n-k+1)}{k!}\left(\frac{1}{n}\right)^k$$
$$+ \cdots + n\left(\frac{1}{n}\right)^{n-1} + \left(\frac{1}{n}\right)^n$$
$$= 1 + \sum_{k=1}^{n} \frac{n(n-1) \cdots (n-k+1)}{k!}\left(\frac{1}{n}\right)^k$$

Similarly,

$$b_{n+1} = 1 + \sum_{k=1}^{n+1} \frac{(n+1)n \cdots (n+1-k+1)}{k!}\left(\frac{1}{n+1}\right)^k$$

The kth term of $b_{n+1} = \dfrac{(n+1)n\cdots(n+1-k+1)}{k!}\left(\dfrac{1}{n+1}\right)^k$

$= \dfrac{n+1}{n+1}\dfrac{n}{n+1}\cdots\dfrac{n+1-k+1}{n+1}\dfrac{1}{k!}$

$= 1\left(1-\dfrac{1}{n+1}\right)\cdots\left(1-\dfrac{k-1}{n+1}\right)\dfrac{1}{k!}$

$\geq 1\left(1-\dfrac{1}{n}\right)\cdots\left(1-\dfrac{k-1}{n}\right)\dfrac{1}{k!}$

$= \dfrac{n}{n}\dfrac{n-1}{n}\cdots\dfrac{n-k+1}{n}\dfrac{1}{k!}$

$= \dfrac{n(n-1)\cdots(n-k+1)}{k!}\left(\dfrac{1}{n}\right)^k$

$= k$th term of $b_n \quad$ for $k = 1, 2, 3, \cdots, n$

Also, b_{n+1} has an extra positive term, namely

$$\dfrac{(n+1)!}{(n+1)!}\left(\dfrac{1}{n+1}\right)^{n+1} = \left(\dfrac{1}{n+1}\right)^{n+1}$$

It follows that $b_{n+1} \geq b_n$, and so $\{b_n\}$ is monotone increasing. To show that $\{b_n\}$ is bounded above observe that

$b_n = 1 + n\dfrac{1}{n} + \cdots + \dfrac{n(n-1)\cdots(n-k+1)}{k!}\left(\dfrac{1}{n}\right)^k + \cdots$

$\quad + n\left(\dfrac{1}{n}\right)^{n-1} + \left(\dfrac{1}{n}\right)^n$

$\leq 1 + \dfrac{1}{1!} + \cdots + \dfrac{1}{k!} + \cdots + \dfrac{1}{n!}$

$\leq 1 + \dfrac{1}{2^0} + \cdots + \dfrac{1}{2^{k-1}} + \cdots + \dfrac{1}{2^{n-1}} \quad$ see Exercise 1.16

$= 1 + \dfrac{1-(\frac{1}{2})^n}{1-\frac{1}{2}} \quad$ see Example 2.2

$< 1 + \dfrac{1}{1-\frac{1}{2}} = 3$

By Theorem 2.5 the sequence $b_n = (1 + 1/n)^n$ converges and its limit does not exceed 3. We call the limit of this sequence the real number e.

The reader is no doubt familiar with e from elementary calculus; e rounded to three decimal places is 2.718.

A sequence $\{b_k\}$ is called a *subsequence* of the sequence $\{a_n\}$ if there are natural numbers
$$n_1 < n_2 < n_3 < \cdots$$
such that $b_k = a_{n_k}$ for $k = 1, 2, 3, \cdots$. It is often important to look at various subsequences of a given sequence in order to learn more about the properties of the sequence.

Example 2.8 Let
$$c_n = \left(1 - \frac{1}{n}\right) \sin \frac{n\pi}{2}$$

The sequence $\{c_n\}$ diverges, but the following subsequences of $\{c_n\}$ converge:

$$c_{2k} = \left(1 - \frac{1}{2k}\right) \sin \frac{2k\pi}{2} = \left(1 - \frac{1}{2k}\right) \sin k\pi = \left(1 - \frac{1}{2k}\right) \cdot 0$$
$$= 0 \quad k = 1, 2, 3, \cdots$$

$$c_{4k+1} = \left(1 - \frac{1}{4k+1}\right) \sin \frac{(4k+1)\pi}{2}$$
$$= \left(1 - \frac{1}{4k+1}\right)(1) \quad k = 0, 1, 2, 3, \cdots$$

$$c_{4k-1} = \left(1 - \frac{1}{4k-1}\right) \sin \frac{(4k-1)\pi}{2}$$
$$= \left(1 - \frac{1}{4k-1}\right)(-1) \quad k = 1, 2, 3, \cdots$$

The subsequence $\{c_{2k}\}$ converges to 0 (each term in the sequence is 0; the subsequence $\{c_{4k+1}\}$ converges to 1; and the subsequence $\{c_{4k-1}\}$ converges to -1.

It should be clear that any sequence which has two subsequences converging to two different real numbers necessarily diverges (see Exercise 2.12); in fact $\lim_{n \to \infty} a_n = A$ if and only if each and every subsequence of $\{a_n\}$ converges to A.

Before proving the second main result of this section, namely that every bounded sequence has some convergent subsequence, we need the nested-interval property of the set of real numbers. The proof below is again based on the completeness of the real number system in the form of the least-upper-bound property.

Theorem 2.6: Nested-interval property Suppose that $I_1 = [a_1, b_1]$, $I_2 = [a_2, b_2]$, $I_3 = [a_3, b_3]$, $\cdots$, where $I_1 \supseteq I_2 \supseteq I_3 \supseteq \cdots$ and $\lim_{n \to \infty} (b_n - a_n) = 0$. Then there is exactly one real number common to all the intervals I_n.

PROOF $\{a_n\}$ is a sequence which is monotone increasing and bounded above (b_1 is an upper bound). Therefore $\lim_{n \to \infty} a_n = A$ exists and $A = \sup_n a_n \leq b_k$ for each natural number k. Hence $a_k \leq A \leq b_k$ for every $k \in \mathbf{N}$; that is, A is contained in each interval I_k. Now suppose $B \in I_n$ for every natural number n. Then $a_n \leq B \leq b_n$ for all n, and so $0 \leq B - a_n \leq b_n - a_n$ for each n. Thus by the old squeeze play, $\lim_{n \to \infty} (B - a_n) = 0$. It follows that $B = \lim_{n \to \infty} a_n = A$, and so A is the only real number common to all the intervals I_n.

Theorem 2.7: Bolzano-Weierstrass theorem Every bounded sequence has a convergent subsequence.

PROOF If $\{a_n\}$ is bounded, then there is a positive real number M such that $|a_n| \leq M$ for every natural number n. Hence $a_n \in [-M, M]$ for all $n \in \mathbf{N}$. Consider the intervals $[-M, 0]$ and $[0, M]$. At least one of these two intervals must contain a_n for infinitely many natural numbers n (note that the a_n are not necessarily distinct). Call such an interval I_0. Next divide I_0 into two subintervals of equal length; at least one of these subintervals must contain a_n for infinitely many natural numbers n. Call such an interval I_1. Continue in this manner to obtain a sequence of intervals $I_0, I_1, I_2, \cdots$ with $I_0 \supset I_1 \supset I_2 \supset \cdots$. The length of I_n is $M/2^n$, which goes to 0 as $n \to \infty$. By the nested-interval property there is exactly one point A common to all these intervals. Choose $a_{n_1} \in I_1$, $a_{n_2} \in I_2$ with $n_2 > n_1$, $a_{n_3} \in I_3$ with $n_3 > n_2$, $a_{n_4} \in I_4$ with $n_4 > n_3$, $\cdots$. (Note that these selections can be made since each I_k contains a_n for infinitely many natural numbers n.) Then $\{a_{n_k}\}$ is a subsequence of $\{a_n\}$, and a_{n_k} and A are both contained in I_k. Thus $|a_{n_k} - A| < M/2^k$, and so $\lim_{k \to \infty} a_{n_k} = A$.

We are now ready to establish that the Cauchy sequences of real numbers are precisely the convergent sequences. Recall that it was shown in the preceding section that every convergent sequence is a Cauchy sequence.

Theorem 2.8 Every Cauchy sequence of real numbers converges.

PROOF Suppose $\{a_n\}$ is a Cauchy sequence of real numbers. By Theorem 2.4, $\{a_n\}$ is bounded, and so by the Bolzano-Weierstrass theorem $\{a_n\}$ has a subsequence, say $\{a_{n_k}\}$, which converges. Suppose that $\lim_{k \to \infty} a_{n_k} = A$. We show that $\lim_{n \to \infty} a_n = A$. Let $\varepsilon > 0$ be given. Since $\{a_n\}$ is Cauchy, there exists a natural number N such that $|a_m - a_n| < \varepsilon/2$ whenever $m, n > N$. Choose the natural number n_k so large that $|a_{n_k} - A| < \varepsilon/2$ and $n_k > N$. Then if $n > N$, we have

$$|a_n - A| = |a_n - a_{n_k} + a_{n_k} - A| \leq |a_n - a_{n_k}| + |a_{n_k} - A| < \frac{\varepsilon}{2} + \frac{\varepsilon}{2} = \varepsilon$$

This result is an important property of **R**. Quite often it is this property that is referred to as the completeness of the real numbers.

In the following example the concepts of subsequences and monotone sequences are both used to demonstrate the convergence of a sequence.

Example 2.9 The sequence 1, 1, 2, 3, 5, 8, 13, 21, ..., defined by $a_0 = 1$, $a_1 = 1$ and the relation $a_{n+1} = a_n + a_{n-1}$ for $n = 1, 2, 3, \ldots$, is called the *Fibonacci sequence*. $\{a_n\}$ diverges, but the sequence of ratios $r_n = a_n/a_{n-1}$ does converge. We examine the sequence $\{r_n\}$:

$$1, 2, \tfrac{3}{2}, \tfrac{5}{3}, \tfrac{8}{5}, \tfrac{13}{8}, \tfrac{21}{13}, \ldots$$

or in decimal form (rounded off)

$$1, 2, 1.5, 1.67, 1.6, 1.62, 1.61, \ldots$$

To demonstrate that this sequence converges we shall show that the subsequence $\{r_{2k}\}$ is monotone decreasing and bounded below, $\{r_{2k-1}\}$ is monotone increasing and bounded above, and both subsequences converge to the same real number. (Why does this imply that $\{r_n\}$ converges?) We first use mathematical induction to prove that $1 \leq r_n \leq 2$ for each natural number n. $r_1 = 1$; suppose that $1 \leq r_k \leq 2$. Then

$$r_{k+1} = \frac{a_{k+1}}{a_k} = \frac{a_k + a_{k-1}}{a_k} = 1 + \frac{a_{k-1}}{a_k} = 1 + \frac{1}{r_k}$$

Hence $1 < 1 + 1/2 \leq 1 + 1/r_k = r_{k+1} \leq 1 + 1/1 = 2$, and so $1 \leq r_n \leq 2$ for every natural number n. Now, if $n \geq 3$, we have

$$r_n = 1 + \frac{1}{r_{n-1}} = 1 + \frac{1}{1 + 1/r_{n-2}} = 1 + \frac{r_{n-2}}{1 + r_{n-2}}$$

Thus
$$r_{n+2} - r_n = \left(1 + \frac{r_n}{1 + r_n}\right) - \left(1 + \frac{r_{n-2}}{1 + r_{n-2}}\right)$$

$$= \frac{r_n - r_{n-2}}{(1 + r_n)(1 + r_{n-2})}$$

The last equation implies that $r_{n+2} - r_n$ has the same sign as $r_n - r_{n-2}$.

Now, $r_3 - r_1 = \frac{3}{2} - 1 > 0$, and so $r_{2k+1} - r_{2k-1} > 0$ for every natural number k. It follows that $\{r_{2k-1}\}$ is a monotone increasing sequence. Similarly $r_4 - r_2 = \frac{5}{3} - 2 < 0$, and so $r_{2k+2} - r_{2k} < 0$ for every natural number k. Hence $\{r_{2k}\}$ is monotone decreasing. By Theorem 2.5 (see Exercise 2.11) $\{r_{2k-1}\}$ and $\{r_{2k}\}$ both converge. Let $l_1 = \lim_{k \to \infty} r_{2k-1}$ and $l_2 = \lim_{k \to \infty} r_{2k}$. As we saw above, $r_n = 1 + r_{n-2}/(1 + r_{n-2})$ for $n \geq 3$. Therefore,

$$l_1 = \lim_{k \to \infty} r_{2k-1} = \lim_{k \to \infty} \left(1 + \frac{r_{2k-3}}{1 + r_{2k-3}}\right) = 1 + \frac{l_1}{1 + l_1}$$

and

$$l_2 = \lim_{k \to \infty} r_{2k} = \lim_{k \to \infty} \left(1 + \frac{r_{2k-2}}{1 + r_{2k-2}}\right) = 1 + \frac{l_2}{1 + l_2}$$

Thus l_1 and l_2 both satisfy the equation $l^2 - l - 1 = 0$. The solutions to the equations are $l = (1 \pm \sqrt{5})/2$. Recall that $1 \leq r_n \leq 2$ for all n and so $l_1, l_2 > 0$. Consequently, $l_1 = (1 + \sqrt{5})/2 = l_2$. It follows that

$$\lim_{n \to \infty} r_n = \frac{1 + \sqrt{5}}{2}$$

The Bolzano-Weierstrass theorem guarantees that every bounded sequence has a convergent subsequence. Unbounded sequences may or may not have convergent subsequences.

Definition 2.3 A real number x is called a *cluster point* of the sequence $\{a_n\}$ provided some subsequence of $\{a_n\}$ converges to x.

Each bounded sequence, then, has at least one cluster point. The sequence $\{\sqrt{n}\}$ has no cluster point, and the sequence

$$1, 1, 2, 1, 2, 3, 1, 2, 3, 4, 1, 2, 3, 4, 5, \ldots$$

has each natural number as a cluster point. The set C of all cluster points of a sequence $\{a_n\}$ can be empty or it can be an infinite subset of $\mathbf{R}$. If $\lim_{n \to \infty} a_n = A$ then, of course, $C = \{A\}$. When $\{a_n\}$ is bounded, C is necessarily nonempty and bounded. In this case we write

$$\limsup a_n = \sup C \quad \text{and} \quad \liminf a_n = \inf C$$

We ask the reader to show that in this case $\limsup a_n$ is the largest cluster point of $\{a_n\}$ and $\liminf a_n$ is the smallest cluster point of $\{a_n\}$ (see Exercise 2.19). If the sequence $\{a_n\}$ is not bounded above, we write $\limsup a_n = \infty$, and if $\{a_n\}$ is not bounded below, we write $\liminf a_n = -\infty$. If $\{a_n\}$ is a convergent sequence with limit A then $\limsup a_n = \liminf a_n = A$. An important property of $\limsup a_n$ and $\liminf a_n$ is given in the following theorem.

Theorem 2.9 If $L = \limsup a_n$ and $l = \liminf a_n$ then given any $\varepsilon > 0$ there is a natural number N such that $l - \varepsilon < a_n < L + \varepsilon$ whenever $n > N$.

PROOF Since L and l are real numbers, $\{a_n\}$ is a bounded sequence. Let $\varepsilon > 0$ be given and suppose $a_n \geq L + \varepsilon$ for infinitely many natural numbers n. Then there exist natural numbers $n_1 < n_2 < n_3 < \cdots$ such that $a_{n_k} \geq L + \varepsilon$ for $k = 1, 2, 3, \ldots$. Clearly $\{a_{n_k}\}$ is a subsequence of $\{a_n\}$ and so $\{a_{n_k}\}$ is bounded. By the Bolzano-Weierstrass theorem, $\{a_{n_k}\}$ has a cluster point x, which is then a cluster point of $\{a_n\}$. But $a_{n_k} \geq L + \varepsilon$ for all k implies that $x \geq L + \varepsilon$. This contradicts the definition of $L = \limsup a_n$. Hence it must be the case that $a_n \geq L + \varepsilon$ for only finitely many natural numbers n. Consequently, there is a natural number N_1 such that $a_n < L + \varepsilon$ whenever $n > N_1$. In a similar fashion it can be shown that there is a natural number N_2 such that $a_n > l - \varepsilon$ whenever $n > N_2$. Let $N = \max(N_1, N_2)$; then clearly $l - \varepsilon < a_n < L + \varepsilon$ for all natural numbers $n > N$.

Corollary $\lim_{n \to \infty} a_n = A$ if and only if $\limsup a_n = A = \liminf a_n$.

Returning to Example 2.8, where $a_n = (1 - 1/n) \sin(n\pi/2)$, the reader should verify that $C = \{-1, 0, 1\}$ and so $\limsup a_n = 1$, $\liminf a_n = -1$, and the sequence $\{a_n\}$ necessarily diverges.

EXERCISES

2.11 Prove that if $\{a_n\}$ is monotone decreasing and bounded below then $\{a_n\}$ converges and $\lim_{n \to \infty} a_n = \inf_n a_n$.

2.12 Prove that if $\{a_n\}$ has one subsequence converging to A and a second subsequence converging to B and $A \neq B$ then $\{a_n\}$ diverges.

2.13 Find the limit of each of the following sequences:

(a) $\left(1 + \dfrac{1}{n^2}\right)^{n^2}$ (b) $\left(1 + \dfrac{1}{n}\right)^{2n}$ (c) $\left(\dfrac{1}{n} + \dfrac{5}{n^2}\right)^n$

(d) $\left(\dfrac{1 + n^2}{n^2}\right)^n$ (e) $\left(1 + \dfrac{1}{2n}\right)^n$ (f) $\left(1 - \dfrac{1}{n}\right)^n$

(g) $\left(\dfrac{n}{n+1}\right)^n$ (h) $\left(1 + \dfrac{2}{3n}\right)^{4n}$

2.14 Prove that if $\lim_{n \to \infty} a_n = A$ then every subsequence of $\{a_n\}$ is convergent with limit A.

2.15 Find the limit of each of the following sequences:

(a) $a_1 = 1$ and $a_n = \sqrt{1 + a_{n-1}}$ $(n = 2, 3, 4, \ldots)$

(b) $a_1 = \sqrt{2}$ and $a_n = \sqrt{2a_{n-1}}$ $(n = 2, 3, 4, \ldots)$

2.16 Prove that a monotone sequence can have at most one cluster point.

2.17 Prove that if $\{a_n\}$ is monotone and x is a cluster point of $\{a_n\}$ then $\lim_{n\to\infty} a_n = x$.

2.18 Find all cluster points of each of the following sequences:

(a) $\left\{\sin\dfrac{n\pi}{2} + (-1)^n\right\}$ (b) $\left\{\left(\cos\dfrac{n\pi}{4}\right)^{(-1)^n}\right\}$

(c) $\{n!\}$ (d) $\left\{\dfrac{1}{n!}\right\}$ (e) $\left\{\dfrac{n-n^2}{1+2n^2}\right\}$

2.19 Suppose $\{a_n\}$ is a bounded sequence. Prove that:
(a) $\limsup a_n$ = largest cluster point of $\{a_n\}$
(b) $\liminf a_n$ = smallest cluster point of $\{a_n\}$

2.20 Find $\limsup a_n$ and $\liminf a_n$ for each sequence $\{a_n\}$ in Exercise 2.18.

2.21 Give an example of a sequence $\{a_n\}$ with finitely many cluster points and such that $\limsup a_n = \infty$.

2.22 Give an example of a sequence $\{a_n\}$ such that its set of cluster points C is uncountable.

2.23 Show that if the intervals are not closed in the nested-interval property then the intersection of all the intervals may be empty.

2.24 If $\{a_n\}$ is bounded above and its set of cluster points C is nonempty then $\limsup a_n = \sup C$; if $C = \emptyset$ then we write $\limsup a_n = -\infty$. If $\{a_n\}$ is bounded below and $C \neq \emptyset$ then $\liminf a_n = \inf C$; if $C = \emptyset$ then we write $\liminf a_n = \infty$. Prove

(a) If $\limsup a_n = -\infty$ then $\lim_{n\to\infty} a_n = -\infty$.

(b) If $\liminf a_n = \infty$ then $\lim_{n\to\infty} a_n = \infty$.

2.3 SETS OF REAL NUMBERS

Sets of real numbers have properties analogous to those of sequences of real numbers. Both sequences and sets can be thought of as "collections" of real numbers. A set, however, has no order as part of its structure, whereas a sequence inherits an ordering from the order in the set of natural numbers. The range of a sequence is a set whose properties can be helpful in determining characteristics of the sequence. We saw in the preceding section, for example, that if the range of the sequence $\{a_n\}$, denoted $\{a_n \mid n \in \mathbb{N}\}$, is unbounded then $\{a_n\}$ must diverge. On the other hand if $\{a_n \mid n \in \mathbb{N}\}$ is bounded then the sequence $\{a_n\}$ has at least one cluster point. The set analog of a cluster point of a sequence is given in the following definition, where it is assumed that S is an arbitrary set of real numbers.

Definition 2.4 A point x_0 is called a *limit point* (or an *accumulation point*) of the set S if given any $\varepsilon > 0$ there is a point $x \in S$ with $x \neq x_0$ and $|x - x_0| < \varepsilon$.

Intuitively, x_0 is a limit point of S provided there are points in S (other than x_0) which are arbitrarily close to x_0. Points in S must cluster up around the point x_0. Note that there is no requirement given in Definition 2.4 that the point x_0 be contained in the set S. For example, if $S = (0, 1)$ then the points 0 and 1 (and all real numbers between 0 and 1) are limit points of S. However, neither 0 nor 1 is an element in S. The set $\{1, \frac{1}{2}, \frac{2}{3}, \frac{3}{4}, \frac{4}{5}, \ldots\}$ has exactly one limit point, namely $x_0 = 1$. The points in this set cluster up around the point $x_0 = 1$. The sequence $1, \frac{1}{2}, \frac{2}{3}, \frac{3}{4}, \frac{4}{5}, \ldots$ has the single cluster point $x_0 = 1$, and there is very little to distinguish between cluster point and limit point in this example, where the range of the sequence $1, \frac{1}{2}, \frac{2}{3}, \frac{3}{4}, \frac{4}{5}, \ldots$ is precisely the set $\{1, \frac{1}{2}, \frac{2}{3}, \frac{3}{4}, \frac{4}{5}, \ldots\}$. On the other hand the sequence 1, 2, 1, 2, 1, 2, ... has the two cluster points 1 and 2, but the range of the sequence, namely the set $\{1, 2\}$, has no limit points. We ask the reader to prove that if a set S is finite then S has no limit points (see Exercise 2.25). Of course, infinite sets may also fail to have limit points. The set N of all natural numbers is an example of such a set. Natural numbers do not cluster up around any real number. This very intuitive concept can be established using Definition 2.4 as follows. Let x_0 be an arbitrary real number and let ε be the distance from x_0 to the nearest natural number (other than x_0 if x_0 happens to be a natural number). It is clear that there is no natural number n with the property that $n \neq x_0$ and $|n - x_0| < \varepsilon$. Consequently N has no limit points. The following theorem, analogous to Theorem 2.7, gives a sufficient condition for an infinite set to have a limit point.

Theorem 2.10: Bolzano-Weierstrass theorem for sets Every bounded infinite set has at least one limit point.

PROOF Let S be a bounded infinite set. Since S is infinite, we can select a sequence $\{a_n\}$ of distinct elements from S; that is, $a_n \in S$ for every $n \in N$ and $a_i \neq a_j$ for $i \neq j$. This sequence is bounded (since S is bounded), and so by the Bolzano-Weierstrass theorem $\{a_n\}$ has a convergent subsequent, which we shall call $\{a_{n_k}\}$. Let $x_0 = \lim_{k \to \infty} a_{n_k}$. We show that x_0 is a limit point of S. Let $\varepsilon > 0$ be given. Then there is a natural number N such that $|a_{n_k} - x_0| < \varepsilon$ whenever $k > N$. Choose $k > N$ such that $a_{n_k} \neq x_0$ (this is possible since $\{a_n\}$, and hence $\{a_{n_k}\}$, is a sequence of distinct elements). Clearly, $|a_{n_k} - x_0| < \varepsilon$ and $a_{n_k} \in S$. Since $a_{n_k} \neq x_0$ and $\varepsilon > 0$ was arbitrary, it follows that x_0 is a limit point of S.

Of course, boundedness is not necessary in order for an infinite set S to have a limit point. The set $S = \{\frac{1}{2}, 2, \frac{1}{3}, 3, \frac{1}{4}, 4, \frac{1}{5}, 5, \ldots\}$ is unbounded and infinite and has the limit point 0. The unbounded interval $(1, \infty)$ has infinitely many limit points; in fact, each element in $[1, \infty)$ is a limit point of the set $(1, \infty)$.

The open interval $(x_0 - \varepsilon, x_0 + \varepsilon)$ is called an ε *neighborhood* (or sometimes just a neighborhood) of the point x_0. We denote an ε neighborhood of x_0 by $N_\varepsilon(x_0)$. The set $(x_0 - \varepsilon, x_0) \cup (x_0, x_0 + \varepsilon)$ is called a *deleted ε neighborhood* of the point x_0 (x_0 has been deleted from the neighborhood $N_\varepsilon(x_0)$). We denote a deleted ε neighborhood of x_0 by $N_\varepsilon^*(x_0)$. In this terminology the point x_0 is a limit point of the set S provided every deleted ε neighborhood of x_0 contains a point of S.

Definition 2.5 A set $G \subseteq \mathbf{R}$ is called *open* if for each element $x_0 \in G$ there is an $\varepsilon > 0$ such that $N_\varepsilon(x_0) \subseteq G$. A set $F \subseteq \mathbf{R}$ is called *closed* provided its complement F' is an open set.

It is clear that $\mathbf{R}$ is an open set and $\emptyset$ is open since there are no points in $\emptyset$ for which the condition stated in Definition 2.5 can fail. Therefore $\mathbf{R}$ and $\emptyset$ are each closed sets as well. The open interval (a, b) is an open set, as is the set $(-\infty, a) \cup (b, \infty)$. Thus the closed interval $[a, b]$ is a closed set. The intervals $(a, b]$ and $[a, b)$ are neither open nor closed. A nonempty open set G must contain an interval and so cannot be countable. For each $x \in G$ there is an $\varepsilon_x > 0$ such that $N_{\varepsilon_x}(x) \subseteq G$. Thus G is the union of all these ε_x neighborhoods; that is,

$$G = \bigcup_{x \in G} N_{\varepsilon_x}(x)$$

A nonempty open set is therefore always representable as the union of open intervals. Actually, it can be shown (see Exercise 2.35) that every nonempty open set can be expressed as the union of countably many pairwise disjoint open intervals.

Theorem 2.11 A set F is closed if and only if F contains all its limit points.

PROOF Assume F is closed. Let $x_0 \in F'$; it suffices to show that x_0 is not a limit point of F. Since F' is open, there is an $\varepsilon > 0$ such that $N_\varepsilon(x_0) \subseteq F'$. Hence there is no $x \in F$ with $|x - x_0| < \varepsilon$, and so x_0 is not a limit point of F.

Conversely, suppose F is a set which contains all its limit points. Let $x_0 \in F'$. Since x_0 is then not a limit point of F, there must exist an $\varepsilon > 0$ for which no element $x \in F$ satisfies $|x - x_0| < \varepsilon$. Hence $N_\varepsilon(x_0) \subseteq F'$, and so F' is an open set. It follows that F is closed.

The Heine-Borel theorem, also known as the Heine-Borel covering theorem, is an extremely useful result in analysis. It plays an important role in the study of uniform continuity (Chap. 4), and we shall have occasion to use it in the study of the Riemann integral (Chap. 6). It involves the concept of an open covering. A collection of open sets G_i is called an *open covering* of the set S provided $S \subseteq \bigcup_i G_i$.

Theorem 2.12: Heine-Borel theorem If $\mathcal{G} = \{G_i\}$ is an open covering of the closed bounded set F then there is a finite subset of $\mathcal{G}$ which is also an open covering of F.

PROOF We assume that no finite subset of $\mathcal{G}$ covers F and show that this leads to a contradiction. Since F is bounded, there is a positive real number M such that $F \subseteq [-M, M]$. Consider the two intervals $[-M, 0]$ and $[0, M]$; at least one of these intervals must contain a portion of F which cannot be covered by a finite number of sets from $\mathcal{G}$. Otherwise, if $\mathcal{G}_1 \subseteq \mathcal{G}$ is finite and covers that part of F in $[-M, 0]$ and $\mathcal{G}_2 \subseteq \mathcal{G}$ is finite and covers that part of F in $[0, M]$ then $\mathcal{G}_1 \cup \mathcal{G}_2$ is finite and covers all of F, in contradiction to our assumption. Let I_0 be one of the intervals $[-M, 0]$ and $[0, M]$ which has the property that the portion of F inside it cannot be covered by finitely many sets from $\mathcal{G}$. Now subdivide I_0 into two closed intervals of equal length (as we did above with the interval $[-M, M]$); at least one of these two intervals must contain a portion of F which cannot be covered by a finite number of sets from $\mathcal{G}$. Call such an interval I_1. Next, subdivide I_1 into two closed intervals of equal length and let I_2 be one of these intervals which has the property that the portion of F in this interval cannot be covered by finitely many sets in $\mathcal{G}$. If we continue this process indefinitely, we obtain a sequence of closed intervals $I_0 \supset I_1 \supset I_2 \supset I_3 \supset \ldots$ with the property that the length of I_k is $M/2^k$ and that portion of F in I_k cannot be covered by finitely many sets from $\mathcal{G}$, for each $k = 0, 1, 2, 3, \ldots$. By the nested-interval property there is a unique point x_0 common to each of the closed intervals I_k.

We show x_0 is a limit point of F. Let $\varepsilon > 0$ be given. Choose the natural number n so large that $M/2^n < \varepsilon$. Then the length of I_n, namely $M/2^n$, is less than ε, and since $x_0 \in I_n$, it follows that $I_n \subset N_\varepsilon(x_0)$. But I_n contains infinitely many points of F (if $F \cap I_n$ were a finite set then surely it could be covered by finitely many sets from $\mathcal{G}$), and hence there is an $x \in F$ with $x \neq x_0$ and $|x - x_0| < \varepsilon$. Therefore x_0 is a limit point of F. Since F is closed, $x_0 \in F$.

Now, since $\mathcal{G}$ is an open covering of F, there is a set $G_{i_0} \in \mathcal{G}$ such that $x_0 \in G_{i_0}$. But G_{i_0} is an open set, and so there is an $\eta > 0$ such that $N_\eta(x_0) \subseteq G_{i_0}$. As above, choose the natural number m so large that $I_m \subset N_\eta(x_0)$. Then $I_m \subset G_{i_0}$; that is, I_m is covered by a finite number of sets (namely one) from $\mathcal{G}$, and so clearly that portion of F in I_m is covered by a finite number of sets from $\mathcal{G}$. This is a contradiction to the construction of the sequence of closed intervals $\{I_k\}$. Hence our assumption that no finite subset of $\mathcal{G}$ was capable of covering the closed bounded set F has led us to a contradiction, and this establishes the theorem.

The reader should observe that both properties of F, that it be bounded and then that it be closed, were used in the above proof. That these properties are necessary is evident from the following examples.

Example 2.10 Consider the bounded set $A = (0, 1]$. A is not closed since 0 is a limit point of A which is not contained in A. Let $\mathcal{G} = \{(1/n, 2) \mid n \in \mathbf{N}\}$. $\mathcal{G}$ is an open covering of A, and there is no finite subset of $\mathcal{G}$ which is a covering of A.

Example 2.11 Let $B = [0, \infty)$. B is a closed set but it is not bounded. Consider the family of open sets

$$\mathcal{G} = \{(n - 2, n) \mid n \in \mathbf{N}\}$$

Clearly $\mathcal{G}$ is an open covering of B, but no finite subset of $\mathcal{G}$ covers B.

A set $K \subseteq \mathbf{R}$ is called *compact* if every open covering $\mathcal{G}$ of K admits a finite subcovering, that is, if there is a finite subset of $\mathcal{G}$ which covers K. In this terminology the Heine-Borel theorem says that every closed, bounded subset of $\mathbf{R}$ is compact. We note in Exercise 2.36 that the converse to the Heine-Borel theorem is also true.

EXERCISES

2.25 Prove that if S is a finite set then S has no limit points.
2.26 Prove the following: x_0 is a limit point of S if and only if every neighborhood of x_0 contains infinitely many points of S.
2.27 Find the set of all limit points of each of the following sets:
 (a) The set of all integers $\mathbf{J}$
 (b) The set of all rationals $\mathbf{Q}$
 (c) $\{(-1)^k(1 - 1/k) \mid k \in \mathbf{J} \text{ and } k \neq 0\}$
 (d) $\{x \in \mathbf{R} \mid 4k < x < 4k + 1 \text{ for some integer } k\}$
 (e) $\{x \in \mathbf{R} \mid 1/(4n + 1) < x < 1/4n \text{ for some } n \in \mathbf{N}\}$
2.28 Prove that the union of any number of open sets is again an open set.
2.29 Prove that the intersection of finitely many open sets is again an open set.
2.30 Prove the following:
 (a) The intersection of any number of closed sets is a closed set.
 (b) The union of finitely many closed sets is a closed set.
2.31 Find two disjoint sets which have the same nonempty set of limit points.
2.32 Let $\hat{S}$ be the set of all limit points of S. Prove that $\hat{S}$ is a closed set.
2.33 Give an example of a sequence $\{A_n\}$ of closed sets for which $\bigcup_{n=1}^{\infty} A_n$ is not closed.
2.34 Recall that a set $D \subseteq \mathbf{R}$ is dense in $\mathbf{R}$ provided $D \cap (a, b) \neq \emptyset$ for every open interval (a, b). Prove that D is dense in $\mathbf{R}$ if and only if every real number x is a limit point of D.
2.35 Show that every open set can be represented as a countable union of pairwise disjoint open intervals.

2.36 Prove that if K is a compact subset of $\mathbf{R}$ then K is both closed and bounded.

2.37 If A and B are nonempty sets, we define the *distance between A and B* to be
$$d(A, B) = \inf_{a \in A, b \in B} |a - b|$$

(a) Prove that if $d(A, B) = 0$, where A is closed and B is compact, then $A \cap B \neq \emptyset$.

(b) Show that if A and B are each closed then we can have $d(A, B) = 0$ with $A \cap B = \emptyset$.

CHAPTER
THREE

FUNCTIONS AND LIMITS

3.1 BOUNDED FUNCTIONS

The central topic of this chapter is the notion of limit of a function. Indeed, this idea is basic to all of calculus. Before presenting this concept a related property of functions will be discussed—boundedness. Recall that a set $S \subset \mathbf{R}$ is bounded if there is a real number $M > 0$ such that $-M \leq x \leq M$ (equivalently, $|x| \leq M$) for every $x \in S$.

Definition 3.1 A function $f : A \to B$ is called *bounded* provided there is a real number $M > 0$ such that $|f(x)| \leq M$ for every $x \in A$.

Observe that f is bounded if and only if its range $f(A)$ is a bounded set. The function $f : A \to B$ is said to be *bounded on the set* $C \subseteq A$ provided there is a real number $M > 0$ such that $|f(x)| \leq M$ for every $x \in C$. f is then bounded on C if the restriction of f to the set C is a bounded function. It is clear that if $f : A \to B$ is a bounded function then f is bounded on C for every set $C \subseteq A$. If a function assumes only finitely many values, that is, has a finite range, then it is bounded. In particular, every constant function $f(x) = k$ is bounded. The identity function $f(x) = x$ is bounded on a set A if and only if A is a bounded subset of $\mathbf{R}$. Thus $f(x) = x$ is not bounded on $\mathbf{R}$, but it is bounded on every bounded interval I.

Example 3.1 Show that the function $f(x) = x^2$ is unbounded on $\mathbf{R}$ but is bounded on each bounded interval I.

SOLUTION If f is bounded on $\mathbf{R}$ then there exists a bound $M > 0$; but $f(M + 1) = (M + 1)^2 > M + 1 > M$, and this contradiction shows that there can be no bound for $f(x) = x^2$ on **R** and so f is unbounded. If we restrict f to a bounded interval, say $[a, b]$, then $|f(x)| \le M$, where $M = \max(a^2, b^2)$, for every $x \in [a, b]$. Hence $f(x) = x^2$ is bounded on $[a, b]$.

The function $f: A \to B$ is *bounded above* if there is a real number M_1 such that $f(x) \le M_1$ for every $x \in A$; f is *bounded below* if there is a real number M_2 such that $f(x) \ge M_2$ for every $x \in A$. Then f is bounded if and only if it is both bounded above and bounded below. If f is bounded above then the least upper bound of the range of f, $f(A) = \{f(x) \mid x \in A\}$, is designated $\sup_{x \in A} f(x)$; similarly if f is bounded below, $\inf_{x \in A} f(x)$ is the greatest lower bound of $f(A)$. If f is bounded then both $\sup_{x \in A} f(x)$ and $\inf_{x \in A} f(x)$ exist as real numbers. If f is not bounded above, we write $\sup_{x \in A} f(x) = \infty$; and if f is not bounded below, we write $\inf_{x \in A} f(x) = -\infty$.

Example 3.2 Show that $f(x) = 1/x$ is unbounded for $0 < x < \infty$, $\inf_{x>0} f(x) = 0$, and f is bounded on (a, ∞) for any $a > 0$ (see Fig. 3.1).

SOLUTION As in Example 3.1, we show that f is unbounded by assuming that there is a bound $M > 0$ and then arriving at a contradiction. From the graph of $f(x) = 1/x$ it is clear that $f(x)$ is large for values of x which are close to zero. Choose $x > 0$ such that $x < 1/M$; then $f(x) = 1/x > M$, and so M is not a bound for the function. Therefore f has no bound and so is an unbounded function. On the other hand f is bounded below since $f(x) = 1/x > 0$ for every $x \in (0, \infty)$. We show that $\inf_{x>0} f(x) = 0$. Since $0 < f(x)$ for every $x \in (0, \infty)$, 0 is a lower bound of f on $(0, \infty)$. Suppose δ is any positive real number. If we choose $x > 1/\delta$, then $f(x) = 1/x < \delta$ and so δ is not a lower bound of f on $(0, \infty)$. Since $\delta > 0$ was arbitrary, it follows that $0 = \inf_{x>0} f(x)$.

If we restrict the domain of $f(x) = 1/x$ to the interval (a, ∞), where a is a fixed positive number, we still have $\inf_{x>a} f(x) = 0$ but now f is bounded. To see this we merely observe that if $x > a$ then $f(x) = 1/x < 1/a$, and so $1/a$ is an upper bound for f on (a, ∞). If $0 < b < 1/a$, then for x satisfying $a < x < 1/b$ we have $f(x) = 1/x > b$. This shows that b is not an upper bound of f on (a, ∞) and hence $\sup_{x>a} f(x) = 1/a$.

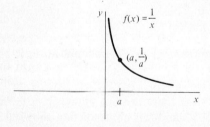

Figure 3.1

Our first theorem is useful in determining that certain combinations of bounded functions are again bounded.

Theorem 3.1 If f and g are each bounded on A and k is any real number then the functions $f + g$, kf, and $f \cdot g$ are each bounded on A.

PROOF Suppose that $|f(x)| \leq M_1$ and $|g(x)| \leq M_2$ for every $x \in A$. Then

$$|(f+g)(x)| = |f(x) + g(x)| \leq |f(x)| + |g(x)| \leq M_1 + M_2$$
$$|(kf)(x)| = |k \cdot f(x)| \leq |k| \cdot |f(x)| \leq |k|M_1$$

and
$$|(f \cdot g)(x)| = |f(x) \cdot g(x)| \leq |f(x)| \cdot |g(x)| \leq M_1 M_2$$

for every $x \in A$.

We observed above that the constant function $f(x) = k$ is bounded on **R** and that the identity function $f(x) = x$, though not bounded on **R**, is bounded on every bounded interval I. These facts, together with Theorem 3.1, imply that every polynomial function

$$p(x) = a_0 x^n + a_1 x^{n-1} + \cdots + a_{n-1} x + a_n$$

where $n \in \mathbf{N}$ and $a_i \in \mathbf{R}$ for $i = 0, 1, 2, \cdots, n$, is bounded on every bounded interval I. We leave the details to the reader (see Exercise 3.6).

If f and g are each bounded on the set A, the function f/g may fail to be bounded on A. For example, $f(x) = 1$ and $g(x) = x$ are bounded functions on $(0, 1)$, but $(f/g)(x) = f(x)/g(x) = 1/x$ is unbounded on $(0, 1)$. A different kind of condition is needed to ensure boundedness of the quotient function. If f is bounded on A and g is "bounded away from zero" on A, that is, there is an $\alpha > 0$ such that $|g(x)| \geq \alpha$ for every $x \in A$, then f/g is bounded on A. To see this suppose that $|f(x)| \leq M$ and $|g(x)| \geq \alpha$ for all $x \in A$, where M and α are positive real numbers. Then

$$\left|\frac{f}{g}(x)\right| = \left|\frac{f(x)}{g(x)}\right| = \frac{|f(x)|}{|g(x)|} \leq \frac{M}{\alpha}$$

and so f/g is bounded on A. A function g is bounded away from zero if and only if the function $1/g$ is bounded. Hence the above result is really a consequence of Theorem 3.1 by writing $f/g = f(1/g)$.

With few exceptions the functions we shall be dealing with in this and the next four chapters are functions defined on intervals or unions of intervals. Suppose f is defined on such a set A. f is said to be *bounded at* $x_0 \in A$ provided there is an open interval J containing x_0 such that f is bounded on $J \cap A$. If f is bounded (on A), then f is bounded at each $x_0 \in A$. On the other hand, a function can be bounded at each point of A without being bounded on A. In Example 3.1, $f(x) = x^2$ is bounded on every bounded interval I and hence is bounded at each real number x_0, yet this function is not bounded on **R**. In Example 3.2, $f(x) = 1/x$ is bounded at each $x_0 \in (0, 1)$, yet it is not bounded on $(0, 1)$. The domains in these two examples, namely **R** and $(0, 1)$, are not compact (closed and bounded) sets. The following theorem (our first application of the Heine-Borel theorem) gives a condition for which boundedness at each point of a set implies boundedness on the set.

Theorem 3.2 If f is bounded at each point $x_0 \in A$ and A is compact then f is bounded on A.

PROOF We assume that f is bounded at each point of the compact set A. Then for each $x \in A$ there is an open interval $I_x = (x - \delta_x, x + \delta_x)$ such that f is bounded on $A \cap I_x$. The set of open intervals $\{I_x \mid x \in A\}$ is an open covering of A, and so by the Heine-Borel theorem there are finitely many of these open intervals, say $I_{x_1}, I_{x_2}, \ldots, I_{x_n}$, such that $A \subset \bigcup_{k=1}^{n} I_{x_k}$. For each $k = 1, 2, \ldots, n$, f is bounded on $A \cap I_{x_k}$, and so there is an $M_k > 0$ such that $|f(x)| \le M_k$ for all $x \in A \cap I_{x_k}$. Let $M = \max_{1 \le k \le n} M_k$. Now, for any $x \in A$ there is a j with $1 \le j \le n$ such that $x \in I_{x_j}$. Hence $x \in A \cap I_{x_j}$, and so $|f(x)| \le M_j \le M$. Therefore f is bounded on A.

A function f defined on **R** is called an *even function* if $f(-x) = f(x)$ for all $x \in \mathbf{R}$; f is called an *odd function* if $f(-x) = -f(x)$ for all $x \in \mathbf{R}$. If f is either even or odd and if f is bounded on $(0, \infty)$ then it is immediate from the definition that f is bounded on **R**. If f is odd and bounded above then f is necessarily bounded. For suppose $f(x) \le M$ for all $x \in \mathbf{R}$. Then $-f(x) = f(-x) \le M$ for every $x \in \mathbf{R}$, and so $|f(x)| \le M$.

The function $f(x) = x^k$ is even if k is an even integer and odd if k is an odd integer. The graph of $f(x) = x^2$ is shown in Fig. 3.2. Note that its graph is a reflection of itself in the y axis. This is the case for every even function. The graph of an odd function is symmetric about the origin; that is, an odd

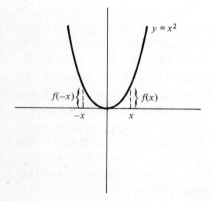

Figure 3.2

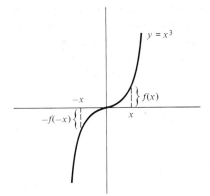

Figure 3.3

function is a double reflection of itself in the coordinate axes. $f(x) = x^3$ is graphed in Fig. 3.3. The following examples are of two even functions, the first bounded and the second unbounded.

Example 3.3 Let $f(x) = (\sin x)/x$ for $x \neq 0$ and $f(0) = 1$. Since $|\sin x| \leq |x|$ for every real number x (see Exercise 3.1), $|f(x)| \leq 1$ and so f is bounded on **R**.

Note that $\sin x$ is bounded on $(0, 1)$ (in fact, $|\sin x| \leq 1$ for every $x \in \mathbf{R}$), $1/x$ is unbounded on $(0, 1)$, and yet the product $(\sin x)/x$ is bounded on $(0, 1)$. Of course, the product of a bounded function with an unbounded function may well be unbounded. This will result, for example, if we take the bounded function to be the constant function $f(x) = 1$.

Example 3.4 Let $f(x) = x \sin x$ on **R**. Show that f is unbounded.

SOLUTION Suppose M is an arbitrary real number. Choose integers k_1 and k_2 such that $2\pi k_1 + \pi/2 > M$ and $\pi/2 - 2\pi k_2 < M$ (how do we know that such integers exist?). Now $\sin(2\pi k_1 + \pi/2) = 1$ and $\sin(2\pi k_2 - \pi/2) = -1$; hence $f(2\pi k_1 + \pi/2) = 2\pi k_1 + \pi/2 > M$ and $f(2\pi k_2 - \pi/2) = \pi/2 - 2\pi k_2 < M$. By the arbitrariness of M it follows that $f(x) = x \sin x$ is neither bounded above nor bounded below on **R**.

The following results on combining even and odd functions are left to the reader (see Exercise 3.13):

1. If f and g are even functions and if k is an arbitrary real number then $f + g$, kf, and $f \cdot g$ are even functions.
2. If f and g are odd functions and if k is an arbitrary real number then $f + g$ and kf are odd functions and $f \cdot g$ is an even function.

3. If f is an odd function and g is an even function then $f \cdot g$ is an odd function.

Further properties of even and odd functions can be found in the exercises.

EXERCISES

3.1 Give a geometrical argument to verify that $|\sin x| \leq |x|$ for every real number x.

3.2 Prove that if f is bounded on A and f is also bounded on B then f is bounded on $A \cup B$.

3.3 Prove that if f and g are each bounded above (below) on A then $f + g$ is bounded above (below) on A.

3.4 Prove: If f is bounded above (below) on A and $k > 0$ then $k \cdot f$ is bounded above (below) on A; if f is bounded above (below) on A and $k < 0$ then $k \cdot f$ is bounded below (above) on A.

3.5 Show that if f and g are both bounded above on A the product $f \cdot g$ may fail to be bounded above on A.

3.6 Prove that each polynomial function
$$p(x) = a_0 x^n + a_1 x^{n-1} + \cdots + a_{n-1} x + a_n$$
is bounded on every bounded interval I.

3.7 Prove that each of the following functions is bounded on the indicated interval:

(a) $f(x) = \dfrac{\sin x}{1 + x^2}$ on $(-\infty, \infty)$

(b) $f(x) = \dfrac{\sin 1/x}{x + 2}$ on $(0, 2)$

(c) $f(x) = \dfrac{x^4 - 3x^2 + 2}{2 - \cos x}$ on $[0, 2\pi]$

(d) $f(x) = \dfrac{\sin x}{3x - 2x \sin x}$ on $(0, \infty)$

(e) $f(x) = \dfrac{\cos x}{x^2 - 2x + 2}$ on $(-\infty, \infty)$

(f) $f(x) = \dfrac{5x^2 + 3x + 1}{x^2 - 2}$ on $[-1, 1]$

(g) $f(x) = \dfrac{\sin x}{\sqrt{x}}$ on $(0, \infty)$

(h) $f(x) = \dfrac{1 - \cos x}{x^2}$ on $(-\infty, \infty)$

(i) $f(x) = \dfrac{1 + x^2}{1 + x^3}$ on $[0, \infty)$

(j) $f(x) = \dfrac{1 - x^2}{1 + x^3}$ on $(-1, 1)$

3.8 Prove that a nonconstant polynomial cannot be bounded on an unbounded interval.

3.9 Suppose that f is bounded on A and g is unbounded on A. Prove that $f + g$ fails to be bounded on A.

3.10 Find functions f and g neither of which is bounded on A but such that the product $f \cdot g$ is bounded on A.

3.11 If f is bounded on A and g is unbounded on A, what can be said regarding the boundedness of $f \cdot g$? Explain.

3.12 Find $\sup f(x)$ and $\inf f(x)$ for each of the following functions on the indicated domain:

(a) $f(x) = 3 + 2x - x^2$ on $(0, 4)$

(b) $f(x) = 2 - |x - 1|$ on $(-2, 2)$

(c) $f(x) = -e^{-|x|}$ on $(-\infty, \infty)$

(d) $f(x) = \dfrac{x}{x - 2}$ on $(-\infty, 2) \cup (2, \infty)$

(e) $f(x) = e^{-1/x}$ on $(-\infty, 0) \cup (0, \infty)$ (f) $f(x) = x \sin \dfrac{1}{x}$ on $(0, \infty)$

(g) $f(x) = \dfrac{1 - x^2}{1 + x^2}$ on $(-\infty, \infty)$ (h) $f(x) = x \sin \dfrac{1}{\sqrt{x}}$ on $(0, \infty)$

3.13 Prove results 1 to 3 in the paragraph following Example 3.4.

3.14 Show that any function $f: \mathbf{R} \to \mathbf{R}$ can be decomposed into the sum of an even function and an odd function. *Hint*: Consider $e(x) = \tfrac{1}{2}(f(x) + f(-x))$ and $o(x) = \tfrac{1}{2}(f(x) - f(-x))$.

3.15 Prove that if f is an even function and $f(x) \neq 0$ for all $x \in \mathbf{R}$ then $1/f$ is an even function. Why isn't a similar statement for odd functions meaningful?

3.16 A rational number $r = p/q$, where $p, q \in \mathbf{J}$ and $q \neq 0$, is said to be *properly reduced* if p and q ($q > 0$) have no common integral factor other than ± 1. Define the function f as follows:

$$f(x) = \begin{cases} q & \text{if } x = p/q, \text{ properly reduced} \\ 0 & \text{if } x \text{ is irrational} \end{cases}$$

Prove that for every real number x_0, f fails to be bounded at x_0.

3.2 LIMITS OF FUNCTIONS

The notion of limit is fundamental to all calculus. Our purpose in this section is to acquire an intuitive feeling for, and understanding of, limits of functions and to develop an ability to work with the methods and techniques used in their proofs. It would be difficult to exaggerate the importance of this section to the calculus student; a good understanding here is indispensable to success with the concepts that lie ahead.

As we stated in the previous section, we shall be dealing with functions whose domain is either an interval or a union of intervals (perhaps the union of more than two intervals). It will be important to investigate the behavior of functions $f: A \to B$ at points $x = a$, where either a is contained in a neighborhood $N_\delta(a)$ which is contained in A or else a is the center of a deleted neighborhood $N_\delta^*(a)$ which is contained in A. In other words, we require that f be defined at all points which are just to the left of $x = a$ (throughout some open interval $(a - \delta, a)$) and at all points just to the right of $x = a$ (throughout some open interval $(a, a + \delta)$). In the next section we consider the behavior of functions at points $x = a$ where f is defined at all points throughout the interval $(a - \delta, a)$ or at all points throughout the interval $(a, a + \delta)$ but where f is not necessarily defined on both intervals.

Definition 3.2 The limit of f is L (some real number) as x approaches a, written $\lim_{x \to a} f(x) = L$, provided that given any $\varepsilon > 0$ there is a $\delta > 0$ such that if $0 < |x - a| < \delta$ then $|f(x) - L| < \varepsilon$.

Equivalently, $\lim_{x \to a} f(x) = L$ if given any neighborhood $N_\varepsilon(L)$ there is a deleted neighborhood $N_\delta^*(a)$ of a such that if $x \in N_\delta^*(a)$ then $f(x) \in N_\varepsilon(L)$.

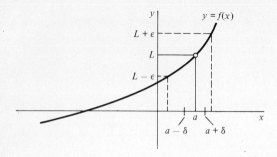

Figure 3.4

Intuitively, given any $\varepsilon > 0$ there is a $\delta > 0$ such that if x is within δ of a but not equal to a then its image $f(x)$ will be within ε of L (see Fig. 3.4).

It is important to note that when $\lim_{x \to a} f(x) = L$, for a given ε, the δ whose existence is assured by the definition of limit is not unique. In fact, if $0 < |x - a| < \delta$ implies $|f(x) - L| < \varepsilon$ and if $0 < \hat{\delta} < \delta$ then clearly $0 < |x - a| < \hat{\delta}$ implies $|f(x) - L| < \varepsilon$. Or, equivalently, if every $x \in N_\delta^*(a)$ has its image $f(x) \in N_\varepsilon(L)$ and if $0 < \hat{\delta} < \delta$ then $N_{\hat{\delta}}^*(a) \subset N_\delta^*(a)$ and so every $x \in N_{\hat{\delta}}^*(a)$ has its image $f(x) \in N_\varepsilon(L)$. It is common to say that if for a given $\varepsilon > 0$, a $\delta > 0$ "works," that is, forces the image of every x within δ of a (but not equal to a) to be within ε of L, then any smaller positive real number $\hat{\delta}$ works.

We proceed to give several examples which illustrate some of the techniques used in proving limits.

Example 3.5 Prove that $\lim_{x \to 3}(2x - 1) = 5$.

SOLUTION Let $\varepsilon > 0$ be given. We want to find a $\delta > 0$ such that if $0 < |x - 3| < \delta$ then $|f(x) - 5| < \varepsilon$, where $f(x) = 2x - 1$. Now

$$|f(x) - 5| = |(2x - 1) - 5| = |2x - 6| = 2|x - 3|$$

and this will be less than ε if $|x - 3| < \varepsilon/2$. Thus we take $\delta = \varepsilon/2$ and observe that if $0 < |x - 3| < \delta = \varepsilon/2$ then $|f(x) - 5| = 2|x - 3| < 2\delta = \varepsilon$. Hence given $\varepsilon > 0$, by choosing $\delta = \varepsilon/2$ we ensure that $f(x) \in N_\varepsilon(5)$ whenever $x \in N_\delta^*(3)$. This proves that $\lim_{x \to 3}(2x - 1) = 5$ (see Fig. 3.5).

By our remarks preceding Example 3.5 we know that any positive $\delta < \varepsilon/2$ will suffice to prove that $\lim_{x \to 3}(2x - 1) = 5$. For if $\delta < \varepsilon/2$ and $x \in N_\delta^*(3)$ then surely $f(x) \in N_\varepsilon(5)$, and the definition of limit is satisfied.

Example 3.6 Prove that $\lim_{x \to 2}[(x^2 - 4)/(x - 2)] = 4$.

SOLUTION The function $f(x) = (x^2 - 4)/(x - 2)$ is not defined at $x = 2$, but it is defined at every other real number and hence in a deleted

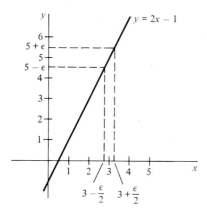

Figure 3.5

neighborhood of $x = 2$. Let $\varepsilon > 0$ be given. We must find a $\delta > 0$ such that if $0 < |x - 2| < \delta$ then $|f(x) - 4| < \varepsilon$. Now, $f(x) = x + 2$ for every real number $x \neq 2$; and so for $x \neq 2$

$$|f(x) - 4| = |(x + 2) - 4| = |x - 2|$$

Hence, if we take $\delta = \varepsilon$ (or δ any positive real number less than ε) then if $0 < |x - 2| < \delta \leq \varepsilon$, we have $|f(x) - 4| = |x - 2| < \delta \leq \varepsilon$, and this completes the proof (see Fig. 3.6).

Observe that in Example 3.6, unlike the situation in Example 3.5, the condition $x \neq a$ is necessary. The next two examples involve somewhat more work in completing the proof, but the reasoning is much the same.

Example 3.7 Prove that $\lim_{x \to 3} x^2 = 9$.

SOLUTION Let $\varepsilon > 0$ be given. We want to find a $\delta > 0$ such that if $0 < |x - 3| < \delta$ then $|f(x) - 9| < \varepsilon$. Now

$$|f(x) - 9| = |x^2 - 9| = |x + 3||x - 3|$$

and so we want to make the product $|x + 3||x - 3|$ less than ε by

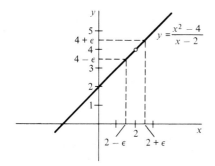

Figure 3.6

taking x sufficiently close to 3. Intuitively, if x is close to 3, $|x + 3|$ will be close to 6 and $|x - 3|$ will be close to 0. Hence the product should be close to 0. Thus we should be able to get $|f(x) - 9| < \varepsilon$ by requiring that x be sufficiently close to 3. How close? That is the problem. As a first step we bound the factor $|x + 3|$. This can be done in many different ways. If we require that $|x - 3| < 1$, that is, $x \in (2, 4)$, then $|x + 3| \leq |x| + 3 < 4 + 3 = 7$. Thus, with the initial restriction $|x - 3| < 1$ we have

$$|f(x) - 9| = |x + 3||x - 3| \leq 7|x - 3|$$

It is now apparent that we shall get $|f(x) - 9| < \varepsilon$ by further requiring that $|x - 3| < \varepsilon/7$. Therefore, if we choose $\delta = \min(1, \varepsilon/7)$ or any smaller positive real number, we have that if $0 < |x - 3| < \delta$ then $|x + 3| < 7$ and $|x - 3| < \varepsilon/7$, and so $|f(x) - 9| = |x + 3||x - 3| < 7(\varepsilon/7) = \varepsilon$, and the proof is complete.

To show the arbitrariness involved in making the initial restriction $|x - 3| < 1$ in order to bound $|x + 3|$, suppose we had decided to impose the restriction $|x - 3| < \frac{1}{2}$ instead. Then we would have $x \in (5/2, 7/2)$, and so $|x + 3| < 13/2$. Now to get $|f(x) - 9| = |x + 3||x - 3| < \varepsilon$ we would require that $|x - 3| < 2\varepsilon/13$. Therefore, we would obtain $|f(x) - 9| < \varepsilon$ by having $|x - 3| < \frac{1}{2}$ and $|x - 3| < 2\varepsilon/13$, and so we would choose $\delta > 0$ such that $\delta \leq \min(1/2, 2\varepsilon/13)$. The reader should attempt to prove this limit by making other initial restrictions; for example, require first that $|x - 3| < 2$. What δ will work in this case?

Example 3.8 Prove that $\lim_{x \to -1}[x/(2x + 1)] = 1$.

SOLUTION The function $f(x) = [x/(2x + 1)]$ is defined for all real numbers $x \neq -\frac{1}{2}$. Thus it is defined throughout the deleted neighborhood $N_\eta^*(-1)$ provided $0 < \eta \leq \frac{1}{2}$ (see Fig. 3.7). Let $\varepsilon > 0$ be given. We must find a $\delta > 0$ such that if $0 < |x - (-1)| = |x + 1| < \delta$ then $|f(x) - 1| < \varepsilon$. Now

$$|f(x) - 1| = \left|\frac{x}{2x + 1} - 1\right| = \left|\frac{-x - 1}{2x + 1}\right|$$

$$= \left|\frac{1}{2x + 1}\right||x + 1|$$

Intuitively, if x is close to -1 then $|1/(2x + 1)|$ is close to 1 and $|x + 1|$ is close to 0. Hence by taking x sufficiently close to -1 we should be able to get the product $|1/(2x + 1)||x + 1|$ less than the given ε. As in the previous example, we need to make an initial restriction which will bound the factor $|1/(2x + 1)|$. Here we must be careful, however, since

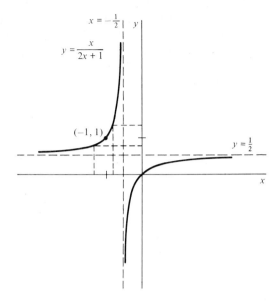

Figure 3.7

$|1/(2x + 1)|$ becomes unbounded if x is close to $-\frac{1}{2}$. The restriction $|x + 1| < \frac{1}{2}$ will not suffice to bound $|1/(2x + 1)|$. Suppose we require that $|x + 1| < \frac{1}{4}$, that is, $x \in (-\frac{5}{4}, -\frac{3}{4})$. Then $|2x + 1| > |2x| - 1 > 2(\frac{3}{4}) - 1 = \frac{1}{2}$, and so $(1/|2x + 1|) < 2$. Hence

$$|f(x) - 1| = \left|\frac{1}{2x + 1}\right||x + 1| \le 2|x + 1|$$

and it is clear that $|f(x) - 1| < \varepsilon$ if we impose the additional restriction that $|x + 1| < \varepsilon/2$. Therefore, if we choose $\delta = \min(1/4, \varepsilon/2)$, we have that if $0 < |x + 1| < \delta$ then $|x + 1| < 1/4$, which gives $|1/(2x + 1)| < 2$, and $|x + 1| < \varepsilon/2$, and so

$$|f(x) - 1| = \left|\frac{1}{2x + 1}\right||x + 1| < 2\frac{\varepsilon}{2} = \varepsilon$$

We note that our initial restriction $|x + 1| < \frac{1}{4}$, could have been $|x + 1| < \eta$ for any positive $\eta < \frac{1}{2}$. The reader should find it instructive to complete the proof by making an initial restriction $|x + 1| < \eta$, where $0 < \eta < \frac{1}{2}$; for example, $\eta = \frac{3}{8}$ or $\eta = \frac{1}{5}$. What δ do you obtain in these cases?

In the next three examples we prove that the limit exists as x approaches an arbitrary point $x = a$.

Example 3.9 Prove that $\lim_{x \to a} \sin x = \sin a$, where $a \in \mathbf{R}$.

SOLUTION The function $f(x) = \sin x$ is defined for every real number x.

We make use of the trigonometric identity

$$\sin x - \sin y = 2 \sin \frac{x-y}{2} \cos \frac{x+y}{2} \quad \text{for all } x, y \in \mathbf{R}$$

together with the two inequalities

$$|\sin x| \le |x| \quad \text{and} \quad |\cos x| \le 1 \quad \text{for all } x \in \mathbf{R}$$

Let $\varepsilon > 0$ be given. Now

$$|\sin x - \sin a| = 2 \left| \sin \frac{x-a}{2} \cos \frac{x+a}{2} \right|$$

$$= 2 \left| \sin \frac{x-a}{2} \right| \left| \cos \frac{x+a}{2} \right|$$

$$\le 2 \left| \frac{x-a}{2} \right| (1) = |x-a|$$

If we choose the positive real number δ to be less than or equal to ε then the condition $0 < |x - a| < \delta$ implies that $|\sin x - \sin a| < \varepsilon$. Hence $\lim_{x \to a} \sin x = \sin a$.

Example 3.10 Prove that $\lim_{x \to a} \sqrt{x} = \sqrt{a}$, where $a > 0$.

SOLUTION The function $f(x) = \sqrt{x}$ is defined for all nonnegative real numbers, and hence throughout a δ neighborhood of $x = a$, if $0 < \delta \le a$. Now $\sqrt{x} + \sqrt{a} > \sqrt{a}$, and so

$$|\sqrt{x} - \sqrt{a}| = \left| \frac{x-a}{\sqrt{x} + \sqrt{a}} \right| < \frac{1}{\sqrt{a}} |x - a|$$

This expression will be less than ε if $|x - a| < \varepsilon\sqrt{a}$. Therefore, we choose $\delta = \min(a, \varepsilon\sqrt{a})$ (or any smaller positive real number). Then if $0 < |x - a| < \delta$, we have

$$|\sqrt{x} - \sqrt{a}| < \frac{1}{\sqrt{a}} |x - a| < \frac{\delta}{\sqrt{a}} \le \frac{\varepsilon\sqrt{a}}{\sqrt{a}} = \varepsilon$$

Example 3.11 Prove that $\lim_{x \to a}(1/x) = 1/a$, where $a \ne 0$.

SOLUTION The function $f(x) = 1/x$ is defined for all real numbers $x \ne 0$, and so it is defined throughout a sufficiently small neighborhood of $x = a$. Let $\varepsilon > 0$ be given. Then

$$\left| \frac{1}{x} - \frac{1}{a} \right| = \left| \frac{x-a}{xa} \right| = \frac{1}{|xa|} |x - a|$$

To force this expression to be less than ε we shall need to bound the factor $1/|xa|$. We note that if x is close to 0 then $1/|xa|$ becomes large. Therefore we have to bound x away from 0. One way of doing this is to require that $|x - a| < |a|/2$, for then $|x| > |a|/2 > 0$. We then obtain the inequality $1/|xa| < 2/|a|^2$ since $1/|x| < 2/|a|$. Hence if $|x - a| < |a|/2$, we have

$$\left|\frac{1}{x} - \frac{1}{a}\right| \leq \frac{2}{|a|^2}|x - a|$$

It should now be clear that if we impose the additional restriction that $|x - a| < \varepsilon|a|^2/2$ then we shall have $|1/x - 1/a| < \varepsilon$. Hence we choose $\delta = \min(|a|/2, \varepsilon|a|^2/2)$ (or any smaller positive real number). Then when $0 < |x - a| < \delta$, we have $|x - a| < |a|/2$ (and so $1/|xa| < 2/|a|^2$) and $|x - a| < \varepsilon|a|^2/2$. Thus

$$\left|\frac{1}{x} - \frac{1}{a}\right| = \frac{1}{|xa|}|x - a| < \frac{2}{|a|^2}\left(\frac{\varepsilon|a|^2}{2}\right) = \varepsilon$$

and the proof that $\lim_{x \to a}(1/x) = 1/a$ ($a \neq 0$) is complete.

Having examined in detail the proofs of several limits, we now prove the following theorem, which says that if a limit exists then it is unique.

Theorem 3.3 If $\lim_{x \to a} f(x) = L_1$ and if $\lim_{x \to a} f(x) = L_2$, then $L_1 = L_2$.

PROOF Let $\varepsilon > 0$ be arbitrary. Since $\lim_{x \to a} f(x) = L_1$, there is a $\delta_1 > 0$ such that if $0 < |x - a| < \delta_1$ then $|f(x) - L_1| < \varepsilon/2$; similarly, since $\lim_{x \to a} f(x) = L_2$, there is a $\delta_2 > 0$ such that if $0 < |x - a| < \delta_2$ then $|f(x) - L_2| < \varepsilon/2$. Choose $x_0 \in N^*_{\delta_1}(a) \cap N^*_{\delta_2}(a)$; that is, $0 < |x_0 - a| < \delta_1$ and $0 < |x_0 - a| < \delta_2$. Then we have both $|f(x_0) - L_1| < \varepsilon/2$ and $|f(x_0) - L_2| < \varepsilon/2$, and so

$$|L_1 - L_2| = |L_1 - f(x_0) + f(x_0) - L_2|$$

$$\leq |L_1 - f(x_0)| + |f(x_0) - L_2| < \frac{\varepsilon}{2} + \frac{\varepsilon}{2} = \varepsilon$$

Since $\varepsilon > 0$ was arbitrary and $|L_1 - L_2| < \varepsilon$, we have $|L_1 - L_2| = 0$. Therefore $L_1 = L_2$.

Theorem 3.3 can be used to show that certain limits do not exist. For example, consider $\lim_{x \to 0} \sin(1/x)$. The function $f(x) = \sin(1/x)$ is defined for all real numbers $x \neq 0$. We observe that $f(x) = 0$ for the values $x = 1/n\pi$ ($n = 1, 2, 3, \ldots$) and that $f(x) = 1$ for the values $x = 2/(4n - 1)\pi$ ($n = 1, 2, 3, \ldots$). Thus in each deleted neighborhood of 0 there are points x

with $f(x) = 0$ and there are points $\hat{x}$ with $f(\hat{x}) = 1$. By Theorem 3.3 the limit of f cannot be both 0 and 1. Hence $\lim_{x \to 0} \sin(1/x)$ does not exist.

Lemma If $\lim_{x \to a} f(x) = L$ then f is bounded on some deleted neighborhood of $x = a$; that is, there is a $\delta > 0$ such that f is bounded on $N_\delta^*(a)$.

PROOF If $\lim_{x \to a} f(x) = L$ then there is a $\delta > 0$ such that $|f(x) - L| < 1$ whenever $0 < |x - a| < \delta$. Thus, for each $x \in N_\delta^*(a)$ we have

$$|f(x)| = |f(x) - L + L| \le |f(x) - L| + |L| < 1 + |L|$$

and so f is bounded on $N_\delta^*(a)$.

Note that if $f(a)$ exists, that is, a is in the domain of f, and if $\lim_{x \to a} f(x) = L$ then f is bounded at $x = a$. In addition to being used in the proof of the next theorem, this lemma can be used to show that certain limits do not exist.

Example 3.12 Prove that $\lim_{x \to 1} [1/(x-1)^2]$ does not exist.

SOLUTION The function $f(x) = 1/(x-1)^2$ is defined for all real numbers $x \ne 1$. Given any $N_\delta^*(1)$ and any $M > 0$ choose $x \in N_\delta^*(1)$ such that $|x - 1| < 1/\sqrt{M}$. Then $(x-1)^2 < 1/M$, and so $f(x) = 1/(x-1)^2 > M$. Hence f is not bounded on $N_\delta^*(1)$. Since $N_\delta^*(1)$ was arbitrary, it follows that f fails to be bounded on each deleted neighborhood of $x = 1$. It now follows from the preceding lemma that $\lim_{x \to 1} [1/(x-1)^2]$ does not exist.

Of course, boundedness at $x = a$ does not guarantee the existence of the limit there. The function $f(x) = \sin(1/x)$ for $x \ne 0$ and $f(0) = 0$ is bounded on R (and hence at every real number $x = a$), but, as we saw above, $\lim_{x \to 0} \sin(1/x)$ does not exist. Furthermore the existence of the limit, as the definition suggests, is completely independent of the value $f(a)$ if in fact $f(a)$ exists. The limit $\lim_{x \to 0} \sin(1/x)$ does not exist no matter how $f(0)$ is defined or even if we leave it undefined.

Theorem 3.4: Fundamental limit theorem If $\lim_{x \to a} f(x) = L$ and $\lim_{x \to a} g(x) = \hat{L}$ then
(a) $\lim_{x \to a} [f(x) + g(x)] = L + \hat{L}$.
(b) $\lim_{x \to a} f(x) g(x) = L\hat{L}$.
(c) $\lim_{x \to a} [1/g(x)] = 1/\hat{L}$ if $\hat{L} \ne 0$.

PROOF (a) Let $\varepsilon > 0$ be given. Since $\lim_{x \to a} f(x) = L$, there is a $\delta_1 > 0$

such that if $0 < |x - a| < \delta_1$ then $|f(x) - L| < \varepsilon/2$; similarly, since $\lim_{x \to a} g(x) = \hat{L}$, there is a $\delta_2 > 0$ such that if $0 < |x - a| < \delta_2$ then $|g(x) - \hat{L}| < \varepsilon/2$. Let $\delta = \min(\delta_1, \delta_2)$; then if $0 < |x - a| < \delta$, we have $|f(x) - L| < \varepsilon/2$ and $|g(x) - \hat{L}| < \varepsilon/2$, and so

$$|[f(x) + g(x)] - (L + \hat{L})| \leq |f(x) - L| + |g(x) - \hat{L}| < \frac{\varepsilon}{2} + \frac{\varepsilon}{2} = \varepsilon$$

Thus $\lim_{x \to a} [f(x) + g(x)] = L + \hat{L}$.

(b) Let $\varepsilon > 0$ be given. By the preceding lemma there is a $\delta_1 > 0$ such that g is bounded on $N^*_{\delta_1}(a)$. Thus there is an $M > 0$ such that $|g(x)| \leq M$ for all $x \in N^*_{\delta_1}(a)$. There is a $\delta_2 > 0$ such that if $0 < |x - a| < \delta_2$ then $|f(x) - L| < \varepsilon/2M$, and there is a $\delta_3 > 0$ such that if $0 < |x - a| < \delta_3$ then $|g(x) - \hat{L}| < \varepsilon/2(1 + |L|)$. Choose $\delta = \min(\delta_1, \delta_2, \delta_3)$; then if $0 < |x - a| < \delta$, we have $|g(x)| \leq M$, $|f(x) - L| < \varepsilon/2M$, and $|g(x) - \hat{L}| < \varepsilon/2(1 + |L|)$. Thus for $0 < |x - a| < \delta$

$$|f(x) \cdot g(x) - L\hat{L}| = |f(x)g(x) - Lg(x) + Lg(x) - L\hat{L}|$$
$$\leq |f(x)g(x) - Lg(x)| + |Lg(x) - L\hat{L}|$$
$$= |g(x)| |f(x) - L| + |L| |g(x) - \hat{L}|$$
$$< M \frac{\varepsilon}{2M} + |L| \frac{\varepsilon}{2(1 + |L|)} < \frac{\varepsilon}{2} + \frac{\varepsilon}{2} = \varepsilon$$

It follows that $\lim_{x \to a} f(x)g(x) = L\hat{L}$.

(c). Let $\varepsilon > 0$ be given. Since $\lim_{x \to a} g(x) = \hat{L} \neq 0$, there is a $\delta_1 > 0$ such that if $0 < |x - a| < \delta_1$ then $|g(x) - \hat{L}| < |\hat{L}|/2$, and so $|g(x)| > |\hat{L}|/2$. Also, there is a $\delta_2 > 0$ such that if $0 < |x - a| < \delta_2$ then $|g(x) - \hat{L}| < \varepsilon |\hat{L}|^2/2$. Choose $\delta = \min(\delta_1, \delta_2)$; then if $0 < |x - a| < \delta$, we have $|g(x)| \geq |\hat{L}|/2$ and $|g(x) - \hat{L}| < \varepsilon |L|^2/2$ and so

$$\left| \frac{1}{g(x)} - \frac{1}{\hat{L}} \right| = \frac{1}{|\hat{L}g(x)|} |g(x) - \hat{L}| \leq \frac{2}{|\hat{L}|^2} |g(x) - \hat{L}|$$
$$< \frac{2}{|\hat{L}|^2} \frac{\varepsilon |\hat{L}|^2}{2} = \varepsilon$$

Consequently, $\lim_{x \to a} [1/g(x)] = 1/\hat{L}$.

Corollary If $\lim_{x \to a} f(x) = L$, $\lim_{x \to a} g(x) = \hat{L}$, and $k \in \mathbf{R}$ then
(a) $\lim_{x \to a} kf(x) = kL$.
(b) $\lim_{x \to a} [f(x) - g(x)] = L - \hat{L}$.
(c) $\lim_{x \to a} [f(x)/g(x)] = L/\hat{L}$ provided $\hat{L} \neq 0$.

Consider the polynomial function
$$p(x) = c_0 x^n + c_1 x^{n-1} + \cdots + c_{n-1} x + c_n$$
where $c_i \in \mathbf{R}$ ($i = 0, 1, 2, \ldots, n$). If $f(x) = k$, a constant function, and $g(x) = x$, the identity function, then $\lim_{x \to a} f(x) = k$ and $\lim_{x \to a} g(x) = a$ (see Exercise 3.17). Thus by the fundamental limit theorem and its corollary
$$\lim_{x \to a} p(x) = \lim_{x \to a} [c_0 x^n + c_1 x^{n-1} + \cdots + c_{n-1} x + c_n]$$
$$= c_0 a^n + c_1 a^{n-1} + \cdots + c_{n-1} a + c_n = p(a)$$

In addition, if $q(x)$ is a polynomial with $q(a) \neq 0$ then for the rational function $r(x) = p(x)/q(x)$ we have $\lim_{x \to a} r(x) = \lim_{x \to a} [p(x)/q(x)] = p(a)/q(a) = r(a)$. If $q(a) = 0$, then of course, $r(a)$ does not exist, but it is still possible that $\lim_{x \to a} r(x)$ exists. A necessary (but not sufficient) condition for this limit to exist is that $p(a) = 0$. A function f defined in a neighborhood of $x = a$ and satisfying $\lim_{x \to a} f(x) = f(a)$ is said to be *continuous* at $x = a$. Our remarks indicate that a polynomial is continuous at each real number $x = a$ and a rational function is continuous at every point in its domain. Continuity will be discussed in detail in the next chapter.

The following theorem tells us that order is preserved while taking the limit.

Theorem 3.5 If $\lim_{x \to a} f(x) = L$ and $\lim_{x \to a} g(x) = \hat{L}$ and if $f(x) \leq g(x)$ for every x in some deleted neighborhood $N_\delta^*(a)$ of $x = a$ then $L \leq \hat{L}$.

PROOF We assume that $\lim_{x \to a} f(x) = L$, $\lim_{x \to a} g(x) = \hat{L}$, $f(x) \leq g(x)$ for all $x \in N_\delta^*(a)$, and $\hat{L} < L$ and show that this leads to a contradiction. Now
$$\lim_{x \to a} [f(x) - g(x)] = L - \hat{L} > 0$$
For $\varepsilon = \frac{1}{2}(L - \hat{L}) > 0$ there is an $N_\delta^*(a)$ such that
$$|f(x) - g(x) - (L - \hat{L})| < \tfrac{1}{2}(L - \hat{L})$$
or equivalently
$$\tfrac{1}{2}(L - \hat{L}) < f(x) - g(x) < \tfrac{3}{2}(L - \hat{L}) \qquad \text{for all } x \in N_\delta^*(a)$$
But the left side of this inequality is positive, and so
$$f(x) - g(x) > 0$$
Hence $f(x) > g(x)$ for all $x \in N_\delta^*(a)$, which contradicts the inequality $f(x) \leq g(x)$ for all $x \in N_\delta^*(a)$.

We conclude this section with the following useful result, which allows us to examine $\lim_{x \to a} f(x)$ by considering sequences of real numbers which converge to $x = a$. We are still assuming that f is defined in some deleted neighborhood of $x = a$. For the sake of brevity the arrow $\to$ denotes convergence of the sequence.

Theorem 3.6 $\lim_{x \to a} f(x) = L$ if and only if $f(x_n) \to L$ for every sequence $\{x_n\}$ in the domain of f with $x_n \to a$ and $x_n \neq a$ for $n = 1, 2, 3, \ldots$.

PROOF To prove this theorem we first assume that $\lim_{x \to a} f(x) = L$ and then we show that for an arbitrary sequence $\{x_n\}$ in the domain of $f(x)$ with $x_n \neq a$ ($n = 1, 2, 3, \ldots$) and $x_n \to a$ necessarily $f(x_n) \to L$. Then we assume that $\lim_{x \to a} f(x) \neq L$ and construct a sequence $\{x_n\}$ in the domain of $f(x)$ with $x_n \neq a$ ($n = 1, 2, 3, \ldots$) and $x_n \to a$ but $f(x_n) \not\to L$. Thus we shall show that both statements either stand or fall together and hence are equivalent. Suppose $\lim_{x \to a} f(x) = L$ and $\{x_n\}$ is a sequence in the domain of f with $x_n \to a$ and $x_n \neq a$ for $n = 1, 2, 3, \ldots$. Let $\varepsilon > 0$ be given. There is a $\delta > 0$ such that if $0 < |x - a| < \delta$ then $|f(x) - L| < \varepsilon$. Since $x_n \to a$ and $x_n \neq a$ for $n = 1, 2, 3, \ldots$, there is an $N \in \mathbf{N}$ such that $n > N$ implies $0 < |x_n - a| < \delta$. Hence $n > N$ implies $|f(x_n) - L| < \varepsilon$, and therefore $f(x_n) \to L$.

Conversely, suppose $\lim_{x \to a} f(x) \neq L$. Then there is an $\varepsilon > 0$ such that for each $\delta > 0$ there is an $x \in N_\delta^*(a)$ with $|f(x) - L| \geq \varepsilon$ (recall that f is defined throughout some deleted neighborhood of $x = a$). Now for each natural number n choose x_n in the domain of f such that $x_n \in N_{1/n}^*(a)$ and $|f(x_n) - L| \geq \varepsilon$. Then for each natural number n we have $0 < |x_n - a| < 1/n$ and $|f(x_n) - L| \geq \varepsilon$. Clearly $x_n \to a$ but $f(x_n) \not\to L$.

Example 3.13 Let

$$f(x) = \begin{cases} 1 & \text{if } x \text{ is rational} \\ 0 & \text{if } x \text{ is irrational} \end{cases}$$

We show that $\lim_{x \to a} f(x)$ does not exist for any real number a.

SOLUTION Let $\{x_n\}$ be a sequence of rational numbers such that $x_n \to a$ and $x_n \neq a$ for $n = 1, 2, 3, \ldots$; let $\{\hat{x}_n\}$ be a sequence of irrational numbers such that $\hat{x}_n \to a$ and $\hat{x}_n \neq a$ for $n = 1, 2, 3, \ldots$. Clearly $f(x_n) = 1$ and $f(\hat{x}_n) = 0$ for $n = 1, 2, 3, \ldots$; therefore $f(x_n) \to 1$ and $f(\hat{x}_n) \to 0$. $\lim_{x \to a} f(x)$ does not exist, by Theorem 3.6. Since a was arbitrary, the limit fails to exist at every real number.

Example 3.14 Let

$$f(x) = \begin{cases} x & \text{if } x \text{ is rational} \\ 0 & \text{if } x \text{ is irrational} \end{cases}$$

Show that $\lim_{x \to a} f(x)$ exists only if $a = 0$.

SOLUTION Suppose $a \neq 0$. As in the previous example, choose sequences $\{x_n\}$ and $\{\hat{x}_n\}$ each converging to $x = a$ such that x_n is rational and not equal to a and $\hat{x}_n$ is irrational and not equal to a for $n = 1, 2, 3, \ldots$. Then $f(x_n) = x_n \to a$ and $f(\hat{x}_n) = 0 \to 0$, and since $a \neq 0$, it follows from Theorem 3.6 that $\lim_{x \to a} f(x)$ does not exist. Thanks to the arbitrariness of $x = a$, the limit fails to exist at every nonzero real number. On the other hand, it is clear that for any sequence $\{x_n\}$, $f(x_n)$ is either x_n or 0, and so if $x_n \to 0$ then surely $f(x_n) \to 0$. Therefore, again by Theorem 3.6, $\lim_{x \to 0} f(x) = 0$.

Example 3.15 Prove that $\lim_{x \to a} b^x = b^a$, where $b > 0$ and a is any real number.

SOLUTION If $b = 1$ then $f(x) = b^x = 1$ for every x and so the result is immediate. Suppose $b > 1$. We consider first the case that $a = 0$. By Theorem 3.6 it suffices to show that for each *null sequence* $x_n \to 0$ we have that $f(x_n) = b^{x_n} \to b^0 = 1$. From Example 2.5 we know that $b^{1/n} \to 1$. Let $\{x_n\}$ be an arbitrary null sequence and let $\varepsilon > 0$ be given. Choose the natural number n_0 so that $b^{1/n_0} - 1 < \varepsilon$ and choose $N \in \mathbf{N}$ such that $n > N$ implies that $|x_n| < 1/n_0$. Now if $n > N$, we have

$$|f(x_n) - 1| = |b^{x_n} - 1| = \begin{cases} b^{x_n} - 1 < b^{1/n_0} - 1 < \varepsilon & x_n > 0 \\ 1 - b^{x_n} < b^{-x_n} - 1 < b^{1/n_0} - 1 < \varepsilon & x_n < 0 \\ 0 < \varepsilon & x_n = 0 \end{cases}$$

It follows that $f(x_n) = b^{x_n} \to 1$, and so $\lim_{x \to 0} b^x = 1 = b^0$. Next suppose that $a \neq 0$. If $\{x_n\}$ is any sequence with $x_n \to a$ then $x_n - a \to 0$. By the above, $f(x_n - a) = b^{x_n - a} \to 1$, and so $b^{x_n} \to b^a$; that is, $f(x_n) \to b^a$. Again by Theorem 3.6, we have that $\lim_{x \to a} b^x = b^a$.

Finally, for the case where $0 < b < 1$ we observe that $1/b > 1$. By the above case $\lim_{x \to a} (1/b)^x = (1/b)^a$. By the fundamental limit theorem we have

$$\lim_{x \to a} b^x = \lim_{x \to a} \frac{1}{(1/b)^x} = \frac{1}{(1/b)^a} = b^a$$

Therefore, for each real number a and every $b > 0$ we have $\lim_{x \to a} b^x = b^a$.

EXERCISES

3.17 Prove, using Definition 3.2, that if $f(x) = k$, a constant function, and $g(x) = x$, the identity function, then for any real number a, $\lim_{x \to a} f(x) = k$ and $\lim_{x \to a} g(x) = a$.

3.18 Use Definition 3.2 to prove the following:

(a) $\lim_{x \to 2} (3x - 1) = 5$ (b) $\lim_{x \to -2} \dfrac{x^2 + x - 2}{x + 2} = -3$ (c) $\lim_{x \to 1/2} \dfrac{1}{x^2} = 4$

(d) $\lim_{x \to 1} x^3 = 1$ (e) $\lim_{x \to 0} \dfrac{1}{8x+1} = 1$ (f) $\lim_{x \to 8} \sqrt[3]{x} = 2$

3.19 Prove that for every real number a, $\lim_{x \to a} \cos x = \cos a$.

3.20 Prove $\lim_{x \to a} 1/\sqrt{x} = 1/\sqrt{a}$ where $a > 0$.

3.21 Let
$$f(x) = \begin{cases} 2x - 1 & \text{if } x \text{ is rational} \\ 5 - x & \text{if } x \text{ is irrational} \end{cases}$$

Prove (a) $\lim_{x \to 2} f(x) = 3$; (b) $\lim_{x \to a} f(x)$ does not exist if $a \ne 2$.

3.22 Prove that the following limits do not exist:

(a) $\lim_{x \to 0} \dfrac{|x|}{x}$ (b) $\lim_{x \to 0} \left(e^x \cos \dfrac{1}{x} \right)$ (c) $\lim_{x \to 0} e^{-1/x}$

3.23 Evaluate each of the following limits:

(a) $\lim_{x \to 0} \dfrac{\sqrt{x+1}-1}{x}$ (b) $\lim_{x \to 0} \dfrac{1}{x}\left[\dfrac{1}{(x+1)^2} - 1\right]$ (c) $\lim_{x \to 1} \dfrac{x^n - 1}{x - 1},\ n \in \mathbf{N}$

(d) $\lim_{x \to 64} \dfrac{\sqrt[3]{x}-4}{x-64}$ (e) $\lim_{x \to 0} \left(e^{-1/x^2} \sin \dfrac{1}{x} \right)$

3.24 Prove: $\lim_{x \to a} f(x) = L$ implies $\lim_{x \to a} |f(x)| = |L|$.

3.25 Show that the converse to Exercise 3.24 does not hold.

3.26 Prove: $\lim_{x \to a} |f(x)| = 0$ implies $\lim_{x \to a} f(x) = 0$.

3.27 Prove: If $\lim_{x \to a} f(x) = \lim_{x \to a} h(x) = L$ and for all $x \in N_\delta^*(a)$, $f(x) \le g(x) \le h(x)$ then $\lim_{x \to a} g(x) = L$.

3.28 Use Exercises 3.26 and 3.27 to prove that $\lim_{x \to 0} \left(x \sin \dfrac{1}{x} \right) = 0$.

3.29 Prove: $\lim_{x \to 0} [(\sin x)/x] = 1$. *Hint*: First give a geometric argument to show that $0 < \sin x < x < \tan x$ for every $x \in (0, \pi/2)$ and then observe that $(\sin x)/x$ is an even function. Finally apply the result of Exercise 3.27.

3.30 Using Exercise 3.29, find the following limits:

(a) $\lim_{x \to 0} \dfrac{1 - \cos x}{x^2}$ (b) $\lim_{x \to 0} \dfrac{\sin 2x}{x}$

(c) $\lim_{x \to 0} \left(\dfrac{1}{x} \tan \dfrac{x}{2} \right)$ (d) $\lim_{x \to 0} \dfrac{\sqrt{1 - \cos x}}{|\sin x|}$

3.31 Prove: If $\lim_{x \to a} f(x) = 0$ and g is bounded on some deleted neighborhood of $x = a$ then $\lim_{x \to a} [f(x) \cdot g(x)] = 0$.

3.32 Suppose f is defined on (α, β) and f is unbounded on $[a, b] \subset (\alpha, \beta)$. Prove that for some $x_0 \in [a, b]$, $\lim_{x \to x_0} f(x)$ does not exist.

3.3 ONE-SIDED LIMITS, INFINITE LIMITS, AND LIMITS AT INFINITY

In this section we extend the notion of limit to include some additional cases. Initially, we restrict the approach of x to a by confining x to one side of a; later we consider cases where the function f becomes arbitrarily large numerically and where the variable x, rather than approaching a real number a, becomes arbitrarily large positively or negatively. First we assume that f is defined on $(a, a + \eta)$ for some $\eta > 0$.

Definition 3.3 $\lim_{x \to a+} f(x) = L$ if given any $\varepsilon > 0$ there is a $\delta > 0$ such that if $a < x < a + \delta$ then $|f(x) - L| < \varepsilon$.

Equivalently, $\lim_{x \to a+} f(x) = L$ if given any neighborhood of L, say $N_\varepsilon(L)$, there is a $\delta > 0$ such that $x \in (a, a + \delta)$ implies that $f(x) \in N_\varepsilon(L)$. If $\lim_{x \to a+} f(x) = L$, we say that the limit of f as x approaches a from the right is L or L is the right-hand limit of f at $x = a$.

Example 3.16 Prove that $\lim_{x \to 0+} \dfrac{x^2}{x + |x|} = 0$.

SOLUTION If $x \in (0, \infty)$ then $|x| = x$ and

$$f(x) = \frac{x^2}{x + |x|} = \frac{x^2}{x + x} = \frac{x^2}{2x} = \frac{x}{2}$$

Let $\varepsilon > 0$ be given. If $0 < x < 2\varepsilon$ then $0 < x/2 < \varepsilon$. Therefore we let $\delta = 2\varepsilon$ (any smaller positive δ will suffice). Now if $0 < x < \delta$ then $|f(x) - 0| = |f(x)| = |x/2| = x/2 < \delta/2 = \varepsilon$. Therefore,

$$\lim_{x \to 0+} \frac{x^2}{x + |x|} = 0$$

Notice that $f(x)$ is not defined at $x = 0$; that is, $f(0)$ does not exist.

Assume now that f is defined on $(a - \eta, a)$ for some $\eta > 0$.

Definition 3.4 $\lim_{x \to a-} f(x) = L$ if given any $\varepsilon > 0$ there is a $\delta > 0$ such that if $a - \delta < x < a$ then $|f(x) - L| < \varepsilon$.

Equivalently, $\lim_{x \to a-} f(x) = L$ if given any neighborhood of L, say $N_\varepsilon(L)$, there is a $\delta > 0$ such that $x \in (a - \delta, a)$ implies $f(x) \in N_\varepsilon(L)$. If $\lim_{x \to a-} f(x) = L$, we say that the limit of f as x approaches a from the left is L or that L is the left-hand limit of f at $x = a$. For the function in Example 3.16 we would not consider the limit $\lim_{x \to 0-} f(x)$ since f is not defined for real numbers $x < 0$.

Example 3.17 Let

$$f(x) = \begin{cases} x + 1 & \text{if } x \leq 2 \\ 2x - 3 & \text{if } x > 2 \end{cases}$$

Prove that $\lim_{x \to 2+} f(x) = 1$ and $\lim_{x \to 2-} f(x) = 3$.

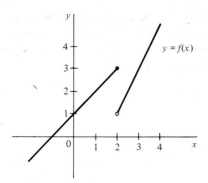

Figure 3.8

SOLUTION Let $\varepsilon > 0$ be given. Choose $\delta = \varepsilon/2$. If $2 < x < 2 + \delta$ then
$$|f(x) - 1| = |(2x - 3) - 1| = |2x - 4| = 2|x - 2| < 2\delta = \varepsilon$$
Thus $\lim_{x \to 2+} f(x) = 1$. We note that f is defined at $x = 2$ but $f(2) = 3 \neq 1$. Again, let $\varepsilon > 0$ be given. Choose $\delta = \varepsilon$; if $2 - \delta < x < 2$ then
$$|f(x) - 3| = |(x + 1) - 3| = |x - 2| < \delta = \varepsilon$$
Hence $\lim_{x \to 2-} f(x) = 3$. In this case we have
$$\lim_{x \to 2-} f(x) = f(2)$$
See Fig. 3.8.

Example 3.18 Let $f(x) = x - [\![x]\!]$, where $[\![x]\!]$ is the greatest-integer function. Prove that $\lim_{x \to 1+} f(x) = 0$ and $\lim_{x \to 1-} f(x) = 1$.

SOLUTION First observe that on the interval $[0, 1)$, $f(x) = x$ whereas $f(x) = x - 1$ on the interval $[1, 2)$. The graph of f on $[-1, 3]$ is given in Fig. 3.9. Functions like this are called *sawtooth functions* and have applications in engineering. Let $\varepsilon > 0$ be given. Choose $\delta = \min(1, \varepsilon)$. If $1 < x < 1 + \delta$ then $|f(x) - 0| = |x - 1| < \delta \leq \varepsilon$. Therefore $\lim_{x \to 1+} f(x) = 0$. Also, if $1 - \delta < x < 1$ then $|f(x) - 1| = |x - 1| < \delta \leq \varepsilon$ and consequently $\lim_{x \to 1-} f(x) = 1$.

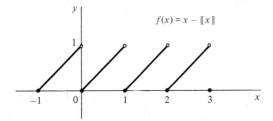

Figure 3.9

In the previous section we saw that if a limit exists it must be unique. Similarly if a right-hand limit (or a left-hand limit) exists, it is unique. The next theorem relates the one-sided limits to the limit.

Theorem 3.7 $\lim_{x \to a} f(x) = L$ if and only if

$$\lim_{x \to a^+} f(x) = \lim_{x \to a^-} f(x) = L.$$

PROOF If $\lim_{x \to a} f(x) = L$ then given any $\varepsilon > 0$ there is a $\delta > 0$ such that if $0 < |x - a| < \delta$ then $|f(x) - L| < \varepsilon$. Clearly then $a < x < a + \delta$ implies $|f(x) - L| < \varepsilon$, and $a - \delta < x < a$ implies $|f(x) - L| < \varepsilon$, and so $\lim_{x \to a^-} f(x) = L$ and $\lim_{x \to a^-} f(x) = L$. Conversely, let $\lim_{x \to a^+} f(x) = \lim_{x \to a^-} f(x) = L$ and let $\varepsilon > 0$ be given. There is a $\delta_1 > 0$ such that if $a < x < a + \delta_1$ then $|f(x) - L| < \varepsilon$; there is a $\delta_2 > 0$ such that if $a - \delta_2 < x < a$ then $|f(x) - L| < \varepsilon$. The δ's may be the same, as in Example 3.18, or they may be different, as in Example 3.17. In either case let $\delta = \min(\delta_1, \delta_2)$. Now if $0 < |x - a| < \delta$ then either $x \in (a, a + \delta) \subseteq (a, a + \delta_1)$ or $x \in (a - \delta, a) \subseteq (a - \delta_2, a)$. Both imply $|f(x) - L| < \varepsilon$ and consequently $\lim_{x \to a} f(x) = L$.

By Theorem 3.7, $\lim_{x \to 2} f(x)$ in Example 3.17 and $\lim_{x \to 1} f(x)$ in Example 3.18 do not exist. We use this theorem in the following example to show that the limit does exist.

Example 3.19 Let

$$f(x) = \begin{cases} x^2 & \text{if } x \geq 0 \\ -x & \text{if } x < 0 \end{cases}$$

SOLUTION See Fig. 3.10. Show that the limit exists.

Let $\varepsilon > 0$ be given. Choose $\delta = \sqrt{\varepsilon}$; if $0 < x < \delta$ then $|f(x) - 0| = x^2 < \delta^2 = \varepsilon$. Thus $\lim_{x \to 0^+} f(x) = 0$. Again, given $\varepsilon > 0$ choose $\delta = \varepsilon$. If $-\delta < x < 0$ then $|f(x) - 0| = |-x| = -x < \delta = \varepsilon$, and hence $\lim_{x \to 0^-} f(x) = 0$. By Theorem 3.7, $\lim_{x \to 0} f(x) = 0$.

In the previous section we found that if the $\lim_{x \to a} f(x)$ exists then f is bounded on some deleted neighborhood of $x = a$. A similar result holds for

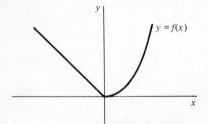

Figure 3.10

one-sided limits (see Exercise 3.35). In the next example we prove that a right-hand limit fails to exist by showing that the function is not bounded on any interval of the form $(a, a + \eta)$, $\eta > 0$.

Example 3.20 Let $f(x) = 2^{1/(x-1)}$. Show that the right-hand limit does not exist at $x = 1$.

SOLUTION We show that $\lim_{x \to 1-} 2^{1/(x-1)} = 0$ but that $\lim_{x \to 1+} 2^{1/(x-1)}$ does not exist. First we consider x to the left of 1. Given any $\varepsilon > 0$ choose the natural number n_0 so large that $1/2^{n_0} < \varepsilon$. Then pick $\delta = 1/n_0$ and let x satisfy $1 - \delta < x < 1$. Now $-\delta < x - 1 < 0$, and so $1/(x - 1) < -1/\delta < 0$. Thus $|2^{1/(x-1)} - 0| = 2^{1/(x-1)} < 2^{-1/\delta} = 2^{-n_0} < \varepsilon$. It follows that $\lim_{x \to 1-} 2^{1/(x-1)} = 0$.

Next, we consider x to the right of 1. Let $\delta > 0$ be arbitrary and choose the natural number N such that $1/N < \delta$. Then if $n > N$, $1 + 1/n \in (1, 1 + \delta)$ and $2^{1/(1+1/n)-1} = 2^n$. Since $\{2^n \mid n > N\}$ is unbounded, $\lim_{x \to 1+} 2^{1/(x-1)}$ does not exist.

Suppose $f : [a, b] \to \mathbf{R}$. If $\lim_{x \to a+} f(x)$ and $\lim_{x \to b-} f(x)$ both exist and if $\lim_{x \to x_0+} f(x)$ and $\lim_{x \to x_0-} f(x)$ both exist for each $x_0 \in (a, b)$ then f is bounded at each point of the compact set $[a, b]$, and so, by Theorem 3.2, f is bounded on $[a, b]$.

If a function is defined to the left of $x = a$ by a formula different from that defining it to the right of $x = a$, the one-sided limits, if they exist, may or may not be equal. Such functions may appear artificial or contrived, but they arise quite naturally in applications. The following is an example of such a case.

Example 3.21 Let x represent the central angle of a wedge-shaped piece of pie cut from a whole circular pie of radius $r > 0$. Let y be the radius of the smallest circular plate the piece of pie will fit on without slopping over the sides; y is a function of x, where $0 \le x \le 2\pi$. It is clear from the definition of y that $0 \le y \le r$. In Fig. 3.11 we indicate the best way of positioning the pie on a plate to give the plate minimal radius. Observe that the plates in Fig. 3.11 have minimal radius for the particular values of x indicated. From the diagrams and some elementary geometry we obtain

$$y = f(x) = \begin{cases} 0 & \text{if } x = 0 \\ \dfrac{r}{2} \sec \dfrac{x}{2} & \text{if } 0 < x \le \dfrac{\pi}{2} \\ r \sin \dfrac{x}{2} & \text{if } \dfrac{\pi}{2} < x \le \pi \\ r & \text{if } \pi < x \le 2\pi \end{cases}$$

82 INTRODUCTION TO MATHEMATICAL ANALYSIS

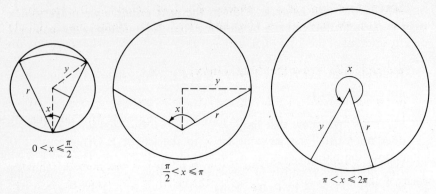

Figure 3.11

Therefore
$$\lim_{x \to 0^+} f(x) = \lim_{x \to 0^+} \frac{r}{2} \sec \frac{x}{2} = \frac{r}{2}$$
$$\lim_{x \to \pi/2^-} f(x) = \lim_{x \to \pi/2^-} \frac{r}{2} \sec \frac{x}{2} = \frac{r}{\sqrt{2}}$$
$$\lim_{x \to \pi/2^+} f(x) = \lim_{x \to \pi/2^+} r \sin \frac{x}{2} = \frac{r}{\sqrt{2}}$$

Hence
$$\lim_{x \to \pi/2} f(x) = \frac{r}{\sqrt{2}}$$
$$\lim_{x \to \pi^-} f(x) = \lim_{x \to \pi^-} r \sin \frac{x}{2} = r$$
$$\lim_{x \to \pi^+} f(x) = \lim_{x \to \pi^+} r = r$$

Hence
$$\lim_{x \to \pi} f(x) = r$$
$$\lim_{x \to 2\pi^-} f(x) = \lim_{x \to 2\pi^-} r = r$$

We shall refer to this function as the *pie function*.

There are cases where the limit of a function f does not exist, but f becomes very large positively or very large negatively as x approaches some real number a. We assume that f is defined on a deleted neighborhood of $x = a$.

Definition 3.5 $\lim_{x \to a} f(x) = \infty$ if given any $M > 0$ there is a $\delta > 0$ such that if $0 < |x - a| < \delta$ then $f(x) > M$. $\lim_{x \to a} f(x) = -\infty$ if given any $M > 0$ there is a $\delta > 0$ such that if $0 < |x - a| < \delta$ then $f(x) < -M$.

Example 3.22 Prove that $\lim_{x \to 0} (1/|x|) = \infty$.

SOLUTION Given $M > 0$ choose $\delta = 1/M$. Now if $0 < |x - 0| = |x| < \delta$, we have $|x| < 1/M$ and $1/|x| > M$. Hence $\lim_{x \to 0}(1/|x|) = \infty$.

Example 3.23 Prove that $\lim_{x \to 0} \ln|x| = -\infty$.

SOLUTION Given $M > 0$ choose $\delta = e^{-M}$. Now if $0 < |x - 0| = |x| < \delta$, we have $|x| < e^{-M}$ and so $f(x) = \ln|x| < -M$. Consequently $\lim_{x \to 0} \ln|x| = -\infty$.

We recall that the limit exists if and only if it is a real number. Thus, in each of the two preceding examples the limit of the function as x approaches 0 does not exist, but the functions do satisfy the behavior described in Definition 3.5. In our next example the function does not have a limit at $x = 0$, and in addition it fails to satisfy the behavior described in Definition 3.5. Instead, it oscillates in a neighborhood of $x = 0$.

Example 3.24 Let $f(x) = (1/x)\sin(1/x)$. The function $f(x)$ is defined for every real number $x \neq 0$. Now, for each natural number n let $x_n = 2/[\pi(1 + 4n)]$. Then

$$\frac{1}{x_n} = \frac{\pi}{2}(1 + 4n) = \frac{\pi}{2} + 2\pi n$$

and so $f(x_n) = (\pi/2)(1 + 4n) \to \infty$. Therefore $\lim_{x \to 0}[(1/x)\sin(1/x)]$ does not exist. Now let $\hat{x}_n = 1/n\pi$. Since $f(\hat{x}_n) = 0$ for every natural number n, $\lim_{x \to 0}[(1/x)\sin(1/x)] \neq \infty$. Note that for any $\delta > 0$, f is neither bounded above nor bounded below on $N_\delta^*(0)$. To verify this find a sequence $\{x_n^*\}$ such that $x_n^* \to 0$ and $f(x_n^*) \to -\infty$.

We also consider cases where the function increases or decreases without bound when x approaches the real number a from either the right or the left side. Such functions are said to have *one-sided infinite limits* at $x = a$. Consider the function $f(x) = 1/x$; f is unbounded (above and below) on each deleted neighborhood of $x = 0$, and therefore $\lim_{x \to 0}(1/x)$ does not exist. It is also easy to show that $\lim_{x \to 0}(1/x)$ is neither ∞ nor $-\infty$. In fact, as x approaches 0 from the right, $1/x$ becomes large positively; as x approaches 0 from the left, $1/x$ becomes large negatively. This behavior is denoted by $\lim_{x \to 0^+}(1/x) = \infty$ and $\lim_{x \to 0^-}(1/x) = -\infty$. Formally we have the following definition.

Definition 3.6
(a) $\lim_{x \to a^+} f(x) = \infty$ if given any $M > 0$ there is a $\delta > 0$ such that if $a < x < a + \delta$ then $f(x) > M$.
(b) $\lim_{x \to a^-} f(x) = \infty$ if given any $M > 0$ there is a $\delta > 0$ such that if $a - \delta < x < a$ then $f(x) > M$.

(c) $\lim_{x \to a^+} f(x) = -\infty$ if given any $M > 0$ there is a $\delta > 0$ such that if $a < x < a + \delta$ then $f(x) < -M$.

(d) $\lim_{x \to a^-} f(x) = -\infty$ if given any $M > 0$ there is a $\delta > 0$ such that if $a - \delta < x < a$ then $f(x) < -M$.

Analogous to Theorem 3.7 is the result that $\lim_{x \to a} f(x) = \infty$ if and only if both $\lim_{x \to a^+} f(x) = \infty$ and $\lim_{x \to a^-} f(x) = \infty$. Similarly, $\lim_{x \to a} f(x) = -\infty$ if and only if $\lim_{x \to a^+} f(x) = -\infty = \lim_{x \to a^-} f(x)$. We leave it to the reader to verify that $\lim_{x \to 0^+} (1/x) = \infty$ and $\lim_{x \to 0^-} (1/x) = -\infty$ and so $\lim_{x \to 0} (1/x)$ is neither ∞ nor $-\infty$.

Example 3.25 Let $f(x) = x/(x - 2)$. Prove that $\lim_{x \to 2^+} f(x) = \infty$.

SOLUTION Let $M > 0$ be given and choose $\delta = 2/M$. Then if $2 < x < 2 + \delta$, it follows that $x < 2 + 2/M$ and $x - 2 < 2/M$ and so $1/(x - 2) > M/2$. Hence if $2 < x < 2 + \delta$ then

$$f(x) = \frac{x}{x - 2} > \frac{2}{x - 2} > M$$

Therefore $\lim_{x \to 2^+} [x/(x - 2)] = \infty$. The reader should prove that $\lim_{x \to 2^-} f(x) = -\infty$ (see Fig. 3.12).

We conclude this section by considering the behavior of functions as the variable x becomes arbitrarily large, positively or negatively. If f approaches a real number L as x becomes "large," we say that f has the limit L at infinity. This is stated more precisely in the following definition.

Definition 3.7 $\lim_{x \to \infty} f(x) = L$ if given any $\varepsilon > 0$ there is an $M > 0$ such that if $x > M$ then $|f(x) - L| < \varepsilon$. $\lim_{x \to -\infty} f(x) = L$ if given any $\varepsilon > 0$ there is an $M > 0$ such that if $x < -M$ then $|f(x) - L| < \varepsilon$.

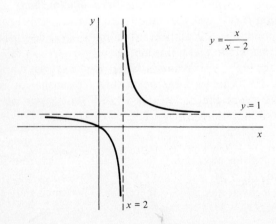

Figure 3.12

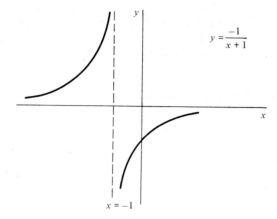

Figure 3.13

It is implicit in the definition that if $\lim_{x \to \infty} f(x) = L$ then f is defined on some set $[a, \infty)$, $a \in \mathbf{R}$; similarly, if $\lim_{x \to -\infty} f(x) = L$ then there is a $b \in \mathbf{R}$ such that f is defined on $(-\infty, b]$.

Example 3.26 Prove that $\lim_{x \to \infty} [-1/(x+1)] = 0$ and also that $\lim_{x \to -\infty} [-1/(x+1)] = 0$.

SOLUTION Let $\varepsilon > 0$ be given. Then if $x > 1/\varepsilon$, we have that

$$|f(x) - 0| = \left| -\frac{1}{x+1} \right| = \frac{1}{x+1} < \frac{1}{x} < \varepsilon$$

Therefore, $\lim_{x \to \infty} [-1/(x+1)] = 0$. Similarly, given $\varepsilon > 0$ if $x < -(1 + 1/\varepsilon)$ then $x + 1 < -1/\varepsilon$, $1/(x+1) > -\varepsilon$, and $-1/(x+1) < \varepsilon$. Thus if $x < -(1 + 1/\varepsilon)$ then $|f(x) - 0| = |-1/(x+1)| = -1/(x+1) < \varepsilon$ and so $\lim_{x \to -\infty} [-1/(x+1)] = 0$ (see Fig. 3.13).

As in previous cases, if $\lim_{x \to \infty} f(x) = L$ then f must be bounded for sufficiently large x (see Exercise 3.49).

Example 3.27 Let $f(x) = x \sin x$. We previously showed that f is not bounded on $\mathbf{R}$. Here we show that f is unbounded on $[a, \infty)$ for arbitrary $a > 0$ and hence $\lim_{x \to \infty} f(x)$ does not exist (see Fig. 3.14).

SOLUTION Consider the sequence $x_n = (\pi/2) + 2n\pi$. For some N, $(\pi/2) + 2N\pi > a$ and so for all $n > N$, $(\pi/2) + 2n\pi \in [a, \infty)$. But

$$f\left(\frac{\pi}{2} + 2n\pi\right) = \left(\frac{\pi}{2} + 2n\pi\right) \sin\left(\frac{\pi}{2} + 2n\pi\right) = \frac{\pi}{2} + 2n\pi$$

which is unbounded. Hence there is no real number L such that $\lim_{x \to \infty} f(x) = L$.

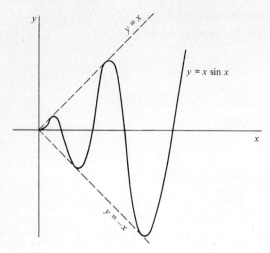

Figure 3.14

The function f can become large (positively or negatively) as x increases or decreases without bound.

✴ **Definition 3.8**
(a) $\lim_{x \to \infty} f(x) = \infty$ if given $M > 0$ there is a $K > 0$ such that if $x > K$ then $f(x) > M$.
(b) $\lim_{x \to -\infty} f(x) = \infty$ if given $M > 0$ there is a $K > 0$ such that if $x < -K$ then $f(x) > M$.
(c) $\lim_{x \to \infty} f(x) = -\infty$ if given $M > 0$ there is a $K > 0$ such that if $x > K$ then $f(x) < -M$.
(d) $\lim_{x \to -\infty} f(x) = -\infty$ if given $M > 0$ there is a $K > 0$ such that if $x < -K$ then $f(x) < -M$.

Example 3.28 Prove that $\lim_{x \to \infty} \sqrt{x} = \infty$.

SOLUTION Given any $M > 0$ let $K = M^2$. Then if $x > K$, we have $\sqrt{x} > \sqrt{K} = M$. Therefore $\lim_{x \to \infty} \sqrt{x} = \infty$.

Returning to Example 3.27, we saw that there is no real number L with $\lim_{x \to \infty} (x \sin x) = L$. For each natural number n let $x_n = n\pi$. Clearly $x_n \to \infty$, and since $f(x_n) = 0$ for all n, it follows that $\lim_{x \to \infty} (x \sin x) \neq \infty$ and $\lim_{x \to \infty} (x \sin x) \neq -\infty$.

Example 3.29 Show that $\lim_{x \to \infty} b^x = \infty$, where $b > 1$ is constant.

SOLUTION First we note that

$$b^x \geq b^{[x]} > 1 + \beta^{[x]} > \beta^{[x]} \geq \beta(x-1) \qquad \text{for } x \geq 1$$

where we have employed Bernoulli's inequality and $\beta = b - 1 > 0$. Now $\lim_{x \to \infty} \beta(x - 1) = \infty$ and $b^x > \beta(x - 1)$ for all $x \geq 1$. Hence $\lim_{x \to \infty} b^x = \infty$.

EXERCISES

3.33 Using the definitions, prove the following one-sided limits:

(a) $\lim_{x \to 1^+} [\text{sgn}(1 - x^2)] = -1$ (b) $\lim_{x \to 0^-} \dfrac{1}{[\![x]\!]} = -1$ (c) $\lim_{x \to -1^+} e^{-1/(x+1)} = 0$

3.34 Prove: If $\lim_{x \to a^+} f(x) = L$ and $\lim_{x \to a^+} g(x) = \hat{L}$ then
(a) $\lim_{x \to a^+} [f(x) + g(x)] = L + \hat{L}$.
(b) $\lim_{x \to a^+} f(x)g(x) = L\hat{L}$.
(c) $\lim_{x \to a^+} [f(x)/g(x)] = L/\hat{L}$ provided $\hat{L} \neq 0$.

3.35 Assume that f is defined on $(a, a + \eta)$ for some $\eta > 0$. Prove that if $\lim_{x \to a^+} f(x) = L$ then there is a $\delta > 0$ such that f is bounded on $(a, a + \delta)$.

3.36 Prove: $\lim_{x \to a^+} f(x) = L$ if and only if $\{f(x_n)\}$ converges to L for every sequence $\{x_n\}$ in the domain of f with $\{x_n\}$ converging to a and $x_n > a$ for $n = 1, 2, 3, \ldots$.

3.37 Evaluate the following one-sided limits:

(a) $\lim_{x \to 0^+} \dfrac{[\![x]\!]}{x}$ (b) $\lim_{x \to -\sqrt{2}^+} [x \, \text{sgn}(x^2 - 2)]$ (c) $\lim_{x \to 1^-} \dfrac{|x^2 - 1|}{x - 1}$

(d) $\lim_{x \to a^+} \dfrac{x}{[\![x]\!]}$ where $a = 4, -2, 0$ (e) $\lim_{x \to a^-} \dfrac{x}{[\![x]\!]}$, where $a = 4, -2, 0$

(f) $\lim_{x \to 0^+} \dfrac{4x[\![x]\!]}{2x + |x|}$ (g) $\lim_{x \to 0^-} \dfrac{4x[\![x]\!]}{2x + |x|}$ (h) $\lim_{x \to 2^-} \dfrac{2x - 2}{|x - 2| + [\![x - 2]\!]}$

3.38 Use Theorem 3.7 to find the following limits if they exist:

(a) $\lim_{x \to 0} \dfrac{x^2 - 6|x|}{x^2 + 2|x|}$ (b) $\lim_{x \to 4} \left([\![x]\!] - \left[\!\!\left[\dfrac{x}{4}\right]\!\!\right] \right)$ (c) $\lim_{x \to -2} \dfrac{\text{sgn}(x^2 + x - 2) - x}{1 - x^2}$

(d) $\lim_{x \to 1/3} ([\![3x]\!] + [\![-3x]\!])$ (e) $\lim_{x \to 1} \dfrac{|x^3 - 1| + 3(1 - x^2)[\![x - 1]\!]}{x - 1}$

3.39 Prove the following:
(a) If f is an even function then $\lim_{x \to 0} f(x) = L$ if and only if $\lim_{x \to 0^+} f(x) = L$.
(b) If f is an odd function then $\lim_{x \to 0} f(x) = L$ implies $L = 0$.

3.40 Using Definition 3.5, prove:

(a) $\lim_{x \to 2} \dfrac{x}{|x - 2|} = \infty$ (b) $\lim_{x \to -1} \dfrac{-1}{(x + 1)^2} = -\infty$

(c) $\lim_{x \to 0} \dfrac{1}{\sqrt{|x|}} = \infty$ (d) $\lim_{x \to -2} \dfrac{x - 2}{x^3 + 4x^2 + 4x} = \infty$

3.41 Prove or give a counterexample: *Conjecture*: If f is unbounded above on each deleted neighborhood of $x = a$ then $\lim_{x \to a} f(x) = \infty$.

3.42 Prove: If $\lim_{x \to a} f(x) = \infty$ and $g(x) \geq f(x)$ on some deleted neighborhood of $x = a$ then $\lim_{x \to a} g(x) = \infty$.

3.43 Prove:
  (a) If $\lim_{x\to a} f(x) = \infty$ then $\lim_{x\to a} [1/f(x)] = 0$.
  (b) If $\lim_{x\to a} f(x) = 0$ and $f(x) \neq 0$ for each $x \in N_\delta^*(a)$ then $\lim_{x\to a} [1/|f(x)|] = \infty$.

3.44 Prove:

  (a) $\lim_{x\to 0} |\cot x| = \infty$ (b) $\lim_{x\to 0} \dfrac{|\sin x|}{1 - \cos x} = \infty$

3.45 Prove: $\lim_{x\to a} f(x) = \infty$ if and only if $f(x_n) \to \infty$ for each sequence $\{x_n\}$ in the domain of f with $x_n \to a$ and $x_n \neq a$ for $n = 1, 2, 3, \ldots$.

3.46 Prove:

  (a) $\lim_{x\to 0^-} \left(-\dfrac{1}{x}\right) = \infty$ (b) $\lim_{x\to -1^+} \left(-\dfrac{1}{x+1}\right) = -\infty$

  (c) $\lim_{x\to -3^+} \dfrac{x-2}{x^2 + x - 6} = \infty$ (d) $\lim_{x\to 1^+} 2^{1/(x-1)} = \infty$

3.47 Prove that $\lim_{x\to a} f(x) = \infty$ if and only if $\lim_{x\to a^+} f(x) = \lim_{x\to a^-} f(x) = \infty$.

3.48 Prove:

  (a) $\lim_{x\to \infty} \dfrac{1}{x} = 0$ (b) $\lim_{x\to -\infty} \dfrac{x}{x - 2} = 1$ (c) $\lim_{x\to \infty} e^{-x} = 0$

  (d) $\lim_{x\to \infty} \dfrac{x^2 + x}{1 - 2x^2} = -\dfrac{1}{2}$ (e) $\lim_{x\to -\infty} e^{-1/(x+1)} = 1$ (f) $\lim_{x\to \infty} \dfrac{\sin x}{x} = 0$

3.49 Prove that if $\lim_{x\to \infty} f(x) = L$ then there is an $a \in \mathbf{R}$ such that f is bounded on $[a, \infty)$.

3.50 Prove that $\lim_{x\to \infty} f(x) = L$ if and only if $f(x_n) \to L$ for each sequence $\{x_n\}$ in the domain of f with $x_n \to \infty$.

3.51 Prove:

  (a) $\lim_{x\to \infty} \beta [\![x]\!] = \infty$, where $\beta > 0$ (b) $\lim_{x\to \infty} \dfrac{x}{\sin x - 2} = -\infty$

  (c) $\lim_{x\to -\infty} (x^2 - x) = \infty$ (d) $\lim_{x\to -\infty} \dfrac{x^2}{x - 3} = -\infty$

  (e) $\lim_{x\to \infty} (x + \sqrt{x} \sin x) = \infty$ but $\lim_{x\to \infty} (x + x \sin x) \neq \infty$ (f) $\lim_{x\to \infty} \dfrac{e^x}{x} = \infty$

3.52 Prove that $\lim_{x\to \infty} f(x) = \infty$ if and only if $f(x_n) \to \infty$ for each sequence $\{x_n\}$ in the domain of f with $x_n \to \infty$.

3.53 Prove that if $\lim_{x\to \infty} f(x) = \infty$ and there is an $a \in \mathbf{R}$ such that $g(x) \geq f(x)$ for each $x \geq a$ then $\lim_{x\to \infty} g(x) = \infty$.

3.54 Determine which of the following have an infinite limit:
  (a) $\lim_{x\to \infty} (2 + \sin x)^x$ (b) $\lim_{x\to -\infty} x^2 [\operatorname{sgn}(\cos x)]$
  (c) $\lim_{x\to \infty} x^{1+\sin x}$ (d) $\lim_{x\to \infty} [e^x |\operatorname{sgn}(x - [\![x]\!])|]$

3.55 Prove:
  (a) If $\lim_{x\to \infty} f(x) = \infty$ then $\lim_{x\to \infty} [1/f(x)] = 0$.
  (b) If $\lim_{x\to \infty} f(x) = 0$ and there is an $a \in \mathbf{R}$ such that $f(x) \neq 0$ for each $x \geq a$ then $\lim_{x\to \infty} [1/|f(x)|] = \infty$.

3.56 Use Example 3.29 to prove that $\lim_{x\to \infty} b^x = 0$, where $0 < b < 1$.

3.57 Prove that if $\lim_{x \to \infty} f(x) = \infty$ and $g(x) \geq \alpha > 0$ on some set $[a, \infty)$, where $a \in \mathbf{R}$, then $\lim_{x \to \infty} f(x) \cdot g(x) = \infty$.

3.58 Prove that if $\lim_{x \to \infty} f(x) = \infty$ then $\lim_{x \to \infty} \sqrt{f(x)} = \infty$.

3.59 Suppose $\lim_{x \to \infty} f(x) = \infty$. Prove that if g is positive and bounded above on $[a, \infty)$ for some real number a then $\lim_{x \to \infty} [f(x)/g(x)] = \infty$.

3.4 MONOTONE FUNCTIONS

In this section we examine some of the basic properties of monotone functions. These functions play an important role in calculus and have a special relationship with the notions of continuity, the derivative, and the Riemann integral, discussed in the next four chapters. Monotonicity is a property that is linked to the ordering of the real number system; simply stated, a function is monotone if it either preserves or reverses the order of the elements in its domain.

Definition 3.9 A function $f: A \to B$ is said to be *monotone increasing* on A if $f(x_1) \leq f(x_2)$ for all $x_1, x_2 \in A$ with $x_1 < x_2$; $f(x)$ is said to be *monotone decreasing* on A if $f(x_1) \geq f(x_2)$ for all $x_1, x_2 \in A$ with $x_1 < x_2$.

A function is called *monotone* on A if it is either monotone increasing or monotone decreasing on A. Graphically, a monotone increasing function is one whose graph does not fall as it proceeds from left to right (as the x values increase); a monotone decreasing function is one whose graph does not rise as it proceeds from left to right (see Fig. 3.15). Any constant function is one which is both monotone increasing and monotone decreasing. If $f: A \to B$ and the interval $I \subset A$, it may be that f is monotone on I but not monotone on A. For example, $f(x) = |x|$ is monotone decreasing

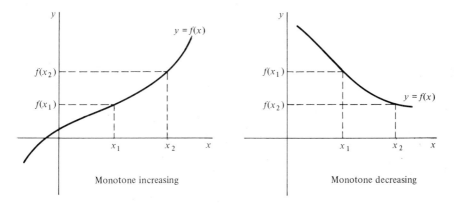

Monotone increasing   Monotone decreasing

Figure 3.15

on $(-\infty, 0]$ and is monotone increasing on $[0, \infty)$, but f is not monotone on $\mathbf{R}$. A function f is called *strictly increasing* on A if $f(x_1) < f(x_2)$ for all x_1, $x_2 \in A$ with $x_1 < x_2$; f is called *strictly decreasing* on A if $f(x_1) > f(x_2)$ for all $x_1, x_2 \in A$ with $x_1 < x_2$.

A constant function is neither strictly increasing nor strictly decreasing. $f(x) = |x|$ is strictly decreasing on $(-\infty, 0]$ and strictly increasing on $[0, \infty)$. If a function is strictly increasing or strictly decreasing on A then we say the function is *strictly monotone* on A.

If the function $f: A \to B$ is strictly monotone on A then it follows from the definition that f is a one-to-one function; that is, $f(x_1) \neq f(x_2)$ whenever $x_1 \neq x_2$. Hence the inverse function $f^{-1}: f(A) \to A$, defined by $f^{-1}(y) = x$ if and only if $f(x) = y$, exists. We shall discuss inverse functions further in the next two chapters.

Example 3.30 Consider the function $f(x) = x^3$ on $\mathbf{R}$. Show that f is strictly increasing on $\mathbf{R}$ (see Fig. 3.3).

SOLUTION We first show that for any two real numbers a and b, not both zero, $a^2 + ab + b^2 > 0$. Now $(a+b)^2 \geq 0$ and so $a^2 + b^2 \geq -2ab$; hence $a^2 + b^2 > \frac{1}{2}(a^2 + b^2) \geq -ab$, and thus $a^2 + ab + b^2 > 0$. Suppose that $x_1 < x_2$; then $x_2 - x_1 > 0$, and by the above argument $x_2^2 + x_1 x_2 + x_1^2 > 0$. Therefore

$$x_2^3 - x_1^3 = (x_2 - x_1)(x_2^2 + x_1 x_2 + x_1^2) > 0$$

and so $f(x_1) = x_1^3 < x_2^3 = f(x_2)$.

Example 3.31 Let $f(x) = 1/[\![x]\!]$ on $[1, \infty)$. Show that this nonconstant function is monotone decreasing but not strictly decreasing.

SOLUTION Suppose $x_1, x_2 \in [1, \infty)$ with $x_1 < x_2$. There exist unique natural numbers n_1 and n_2 such that $x_1 \in [n_1, n_1 + 1)$ and $x_2 \in [n_2, n_2 + 1)$. Moreover, $n_1 \leq n_2$ since $x_1 < x_2$. It follows that

$$f(x_1) = \frac{1}{[\![x_1]\!]} = \frac{1}{n_1} \geq \frac{1}{n_2} = \frac{1}{[\![x_2]\!]} = f(x_2)$$

and that f is monotone decreasing on $[1, \infty)$. f is not strictly decreasing since, for example, $\frac{9}{4} < \frac{5}{2}$ but $f(\frac{9}{4}) = \frac{1}{2} = f(\frac{5}{2})$ (see Fig. 3.16).

Theorem 3.8 If f is monotone on (a, b), then for each x_0 in the interval (a, b) $\lim_{x \to x_0^+} f(x)$ and $\lim_{x \to x_0^-} f(x)$ both exist.

PROOF We assume that f is monotone increasing on (a, b); the proof for

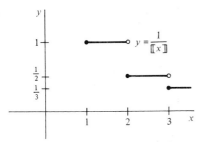

Figure 3.16

the case where f is monotone decreasing is similar and is left to the reader (see Exercise 3.62).

Let $x_0 \in (a, b)$. For every $x \in (a, x_0)$, $f(x) \leq f(x_0)$. Therefore the set $\{f(x) \mid x \in (a, x_0)\}$ is bounded above, and so $\sup_x f(x) = \alpha \in \mathbf{R}$, where $x \in (a, x_0)$. We show that $\lim_{x \to x_0^-} f(x) = \alpha$. Let $\varepsilon > 0$ be given. Since $\alpha - \varepsilon < \alpha = \sup_x f(x)$ there exists $x_1 \in (a, x_0)$ with $f(x_1) > \alpha - \varepsilon$. Let $\delta = x_0 - x_1$. Now if x satisfies $x_0 - \delta < x < x_0$ then $f(x) \geq f(x_1)$, and so

$$|f(x) - \alpha| = \alpha - f(x) \leq \alpha - f(x_1) < \varepsilon$$

Therefore $\lim_{x \to x_0^-} f(x) = \alpha$.

For every $x \in (x_0, b)$, $f(x) \geq f(x_0)$. Thus the set $\{f(x) \mid x \in (x_0, b)\}$ is bounded below and so $\inf_x f(x) = \beta \in \mathbf{R}$, where $x \in (x_0, b)$. We show that $\lim_{x \to x_0^+} f(x) = \beta$. Let $\varepsilon > 0$ be given. Since $\beta + \varepsilon > \beta = \inf_x f(x)$ there exists $x_2 \in (x_0, b)$ with $f(x_2) < \beta + \varepsilon$. Let $\delta = x_2 - x_0$. Now if x satisfies $x_0 < x < x_0 + \delta$ then $f(x) \leq f(x_2)$, and so

$$|f(x) - \beta| = f(x) - \beta \leq f(x_2) - \beta < \varepsilon$$

Therefore $\lim_{x \to x_0^+} f(x) = \beta$.

Corollary If f is monotone increasing on (a, b) then for each $x_0 \in (a, b)$

$$\lim_{x \to x_0^-} f(x) = \sup_{x \in (a, x_0)} f(x) \leq f(x_0) \leq \inf_{x \in (x_0, b)} f(x) = \lim_{x \to x_0^+} f(x)$$

Since both one-sided limits exist at each point $x_0 \in (a, b)$ when f is monotone on (a, b), it follows that f is bounded at each point $x_0 \in (a, b)$. Of course, as we saw earlier, f may well be unbounded on the interval (a, b).

Lemma If f is monotone increasing on (a, b) and if $a < x_1 < x_2 < b$ then $\lim_{x \to x_1^+} f(x) \leq \lim_{x \to x_2^-} f(x)$.

Proof Choose $\hat{x} \in (x_1, x_2)$. Since $\hat{x} \in (x_1, b)$, $f(\hat{x}) \geq \inf_{x \in (x_1, b)} f(x)$; sim-

ilarly, since $\hat{x} \in (a, x_2), f(\hat{x}) \leq \sup_{x \in (a, x_2)} f(x)$. By the above corollary
$$\lim_{x \to x_2^-} f(x) = \sup_{x \in (a, x_2)} f(x)$$
and

Therefore
$$\lim_{x \to x_1^+} f(x) = \inf_{x \in (x_1, b)} f(x).$$

$$\lim_{x \to x_1^+} f(x) \leq f(\hat{x}) \leq \lim_{x \to x_2^-} f(x)$$

The following theorem will be important for our discussion of continuous functions in the next chapter. Briefly, it says that monotone functions must not be "too discontinuous." Later we shall see that even more can be said about the behavior of monotone functions at points where such functions fail to be continuous.

Theorem 3.9 If f is monotone on (a, b) then $\lim_{x \to x_0} f(x)$ exists and equals $f(x_0)$ for all but countably many points $x_0 \in (a, b)$.

PROOF We assume first that f is monotone increasing on (a, b). By Theorem 3.8 and its corollary
$$\lim_{x \to x_0^-} f(x) \leq f(x_0) \leq \lim_{x \to x_0^+} f(x) \quad \text{for every } x_0 \in (a, b)$$
We define the set
$$D = \{x_0 \in (a, b) \mid \lim_{x \to x_0^-} f(x) < \lim_{x \to x_0^+} f(x)\}$$
If $x_0 \notin D$ then
$$\lim_{x \to x_0^-} f(x) = f(x_0) = \lim_{x \to x_0^+} f(x)$$
and so $\lim_{x \to x_0} f(x)$ exists and $\lim_{x \to x_0} f(x) = f(x_0)$. Therefore, it suffices to prove that D is countable. Define the function $\varphi: D \to \mathbf{Q}$ (recall that $\mathbf{Q}$ is the set of all rationals) as follows. For each $x_0 \in D$ choose $r_{x_0} \in \mathbf{Q}$ such that
$$\lim_{x \to x_0^-} f(x) < r_{x_0} < \lim_{x \to x_0^+} f(x)$$
and let $\varphi(x_0) = r_{x_0}$. We show that φ is one-to-one. Let $x_1, x_2 \in D$ with $x_1 < x_2$. Then
$$\lim_{x \to x_1^-} f(x) < \lim_{x \to x_1^+} f(x) \quad \text{and} \quad \lim_{x \to x_2^-} f(x) < \lim_{x \to x_2^+} f(x)$$
Now, by the above lemma $\lim_{x \to x_1^+} f(x) \leq \lim_{x \to x_2^-} f(x)$. Hence
$$\varphi(x_1) = r_{x_1} < \lim_{x \to x_1^+} f(x) \leq \lim_{x \to x_2^-} f(x) < r_{x_2} = \varphi(x_2)$$

that is, $\varphi(x_1) \neq \varphi(x_2)$, and so φ is a one-to-one function. It follows from Theorem 1.11 (and the fact that $\mathbf{Q}$ is countable) that the set D is countable.

If f is monotone decreasing on (a, b) then $-f$ is monotone increasing on (a, b) (see Exercise 3.61). By applying the result proved above to the function $g = -f$ and then using the fundamental limit theorem we again have that $\lim_{x \to x_0} f(x) = f(x_0)$ for all but countably many $x_0 \in (a, b)$.

EXERCISES

3.60 Determine which of the following functions are monotone on the indicated domains:

(a) $f(x) = x^2$ on $\mathbf{R}$ (b) $f(x) = \dfrac{1}{1 - x}$ on $(-1, 1)$

(c) $f(x) = \dfrac{1}{1 - x}$ on $(0, 1) \cup (1, 2)$ (d) $f(x) = \dfrac{1}{1 + x^2}$ on $\mathbf{R}$

(e) $f(x) = \dfrac{1}{1 + x^2}$ on $[0, \infty)$ (f) $f(x) = [\![x]\!] - \left[\!\left[\dfrac{x}{2}\right]\!\right]$ on $\mathbf{R}$

(g) $f(x) = [\![x]\!] - [\![2x]\!]$ on $\mathbf{R}$ (h) $f(x) = 1 + 4\sin^2 x - \sin^2 2x - 4\sin^4 x$ on $\mathbf{R}$

(i) $f(x) = e^{-1/x}$ on $(0, \infty)$ (j) $f(x)$ is the pie function on $[0, 2\pi]$

3.61 Prove that if f is monotone decreasing on (a, b) then $g = -f$ is monotone increasing on (a, b).

3.62 Prove that if f is monotone decreasing on (a, b) then for each $x_0 \in (a, b)$ $\lim_{x \to x_0^+} f(x)$ and $\lim_{x \to x_0^-} f(x)$ both exist and

$$\lim_{x \to x_0^-} f(x) = \inf_{x \in (a, x_0)} f(x) \geq f(x_0) \geq \sup_{x \in (x_0, b)} f(x) = \lim_{x \to x_0^+} f(x)$$

3.63 Prove that if f is monotone decreasing on (a, b) and if $a < x_1 < x_2 < b$ then $\lim_{x \to x_1^+} f(x) \geq \lim_{x \to x_2^-} f(x)$

3.64 Give an example of a function f defined on an open interval (a, b) such that $\lim_{x \to x_0} f(x) = f(x_0)$ fails for uncountably many $x_0 \in (a, b)$.

3.65 Suppose f is one-to-one on (a, b) and satisfies the following property: Whenever $f(x_1) \neq f(x_2)$ for $x_1 < x_2$, $x_1, x_2 \in (a, b)$ and k is any number between $f(x_1)$ and $f(x_2)$, there exists a $c \in (x_1, x_2)$ with $f(c) = k$. Prove that f is strictly monotone on (a, b).

CHAPTER FOUR

CONTINUOUS FUNCTIONS

4.1 CONTINUITY

What is the first idea that comes to mind when you hear the word "continuous"? Uninterrupted, remaining together, not broken, smooth-flowing—all these seem to convey a notion related to the word continuous. When we think of a continuous function, we probably get the mental image of a smooth curve in the (x, y) plane which represents the graph of such a function. Although this is not altogether correct, it puts us on the right track and it is often helpful to visualize continuous functions in this way. One of the purposes of this chapter will be to show that continuous functions are not quite as simple as our initial intuitive feelings might lead us to believe; in fact, most continuous functions have graphs which, though unbroken and connected, are not smooth at all. The depth of this remark can be illuminated by pointing out that of the continuous functions only the simplest and most well-behaved can be graphed at all.

Nevertheless, continuity is an extremely important concept in mathematics and mathematics application and will occupy a central role in the remainder of the text. We begin with the definition of continuity of a function at a point in its domain.

Definition 4.1 f is said to be *continuous* at x_0 if $\lim_{x \to x_0} f(x) = f(x_0)$; equivalently, given any $\varepsilon > 0$ there is a $\delta > 0$ such that $|f(x) - f(x_0)| < \varepsilon$ whenever $|x - x_0| < \delta$.

We first point out that in order for a function f to be continuous at x_0,

f must be defined on some δ neighborhood of x_0. This is necessary in order for $\lim_{x \to x_0} f(x)$ to exist. We summarize Definition 4.1 in three stages: f is continuous at x_0 provided

1. x_0 is in the domain of f; that is, $f(x_0)$ exists.
2. $\lim_{x \to x_0} f(x)$ exists.
3. $f(x_0)$ and $\lim_{x \to x_0} f(x)$ are equal.

If any of these three conditions fails, f is not continuous at x_0. We say that f is discontinuous at x_0 or that f has a discontinuity at x_0. There are several different ways a function can be discontinuous at a point x_0, and it is important to see exactly what it is that goes wrong when a function is not continuous.

If condition 2 above holds but 1 (and hence 3) fails or if 2 and 1 hold but 3 fails then f is said to have a *removable discontinuity* at x_0. This name suggests that in some rather simple fashion the function can be made continuous at x_0, and, indeed, if f has a removable discontinuity at x_0 then by defining, or redefining as the case may be, $f(x_0)$ to be $\lim_{x \to x_0} f(x)$ the function is made continuous at x_0; that is, the discontinuity has been removed.

Example 4.1 Let $f(x) = (x^2 - 4)/(x - 2)$ (see Fig. 4.1). f has a discontinuity at $x = 2$ because $f(2)$ does not exist ($x = 2$ is not in the domain of f). On the other hand

$$\lim_{x \to 2} f(x) = \lim_{x \to 2} \frac{x^2 - 4}{x - 2} = \lim_{x \to 2} (x + 2) = 4$$

Hence f has a removable discontinuity at $x = 2$. If we define $f(2) = 4$ then f has been made continuous at $x = 2$.

Example 4.2 Let $f(x) = \text{sgn} |x|$ (see Fig. 4.2). $f(0) = 0$ and also

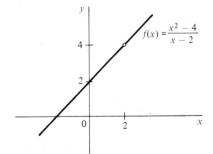

Figure 4.1

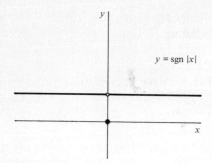

Figure 4.2

$\lim_{x \to 0} f(x) = 1$; thus f has a removable discontinuity at $x = 0$. If we redefine $f(0) = 1$, f becomes continuous at $x = 0$; that is,

$$\lim_{x \to 0} f(x) = 1 = f(0)$$

If $\lim_{x \to x_0} f(x)$ exists at a discontinuity $x = x_0$ of f, we have defined this to be a removable discontinuity of the function. On the other hand, if $\lim_{x \to x_0} f(x)$ fails to exist, the discontinuity is more substantial. There are many ways that $\lim_{x \to x_0} f(x)$ can fail to exist; f can be unbounded at x_0, f can oscillate near x_0, or perhaps there is a "jump" at x_0; that is, the one-sided limits of f at x_0 exist but they are not equal.

Example 4.3 Consider $f(x) = [\![x]\!]$ at $x = 1$ (see Fig. 4.3). $f(1) = 1$, but $\lim_{x \to 1} f(x)$ does not exist. In fact $\lim_{x \to 1^-} [\![x]\!] = 0$ and $\lim_{x \to 1^+} [\![x]\!] = 1$. Therefore f has a jump discontinuity at $x = 1$. Note that it is not possible to redefine f at $x = 1$ so that f becomes continuous there. Also, the behavior of f at each integer is similar to the behavior at $x = 1$. For any $k \in \mathbf{J}$, $f(k) = k$ while $\lim_{x \to k^-} f(x) = k - 1$ and $\lim_{x \to k^+} f(x) = k$. Thus f has a jump discontinuity at each $k \in \mathbf{J}$. On the other hand f is continuous at every noninteger real number.

There are "worse" discontinuities than those in the example above. The functions $f(x) = 1/x$ and $g(x) = \sin(1/x)$ have discontinuities at $x = 0$

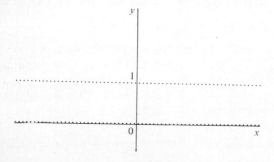

Figure 4.3

which are neither removable nor jump discontinuities. In fact, f is unbounded at $x = 0$ and g has an oscillating behavior (it oscillates between -1 and 1) at $x = 0$.

Definition 4.2 The function f has a *simple discontinuity* (or a *discontinuity of the first kind*) at $x = x_0$ if the discontinuity is either removable or a jump discontinuity. Every other discontinuity is called a *discontinuity of the second kind*.

Summarizing, if f has a discontinuity at $x = x_0$, it is called removable if $\lim_{x \to x_0} f(x)$ exists; it is called a jump discontinuity if both $\lim_{x \to x_0^+} f(x)$ and $\lim_{x \to x_0^-} f(x)$ exist but are not equal; and it is called a discontinuity of the second kind if at least one of the two one-sided limits at x_0 fails to exist. It is clear, for example, that if f is unbounded at x_0 then necessarily f has a discontinuity of the second kind at this point. We shall see later that a function can be bounded at a point where it has a discontinuity of the second kind. If a function f is monotone on an interval (a, b), we know from Theorem 3.8 that the one-sided limits exist at each $x_0 \in (a, b)$ and so f can have no discontinuities of the second kind in (a, b). Even further, Theorem 3.9 tells us that a monotone function on (a, b) can have only countably many discontinuities in (a, b) (all being jump discontinuities—why?), and so monotone functions have strong continuity properties; that is, they are continuous at "most" of the points in the interval (a, b). We shall see in Sec. 4.4 that for any function f, monotone or not, there can be only countably many points x_0 at which f has a simple discontinuity. Hence if f is a function with no discontinuities of the second kind then f is continuous for all but countably many points.

Our next example is a function whose graph is difficult to construct (but not so difficult to visualize).

Example 4.4 Let

$$f(x) = \begin{cases} 1 & \text{if } x \text{ is rational} \\ 0 & \text{if } x \text{ is irrational} \end{cases}$$

Show that f has a discontinuity of the second kind at every real number.

SOLUTION Let x_0 be an arbitrary real number. By the same argument as that used in Example 3.13 $\lim_{x \to x_0^-} f(x)$ and $\lim_{x \to x_0^+} f(x)$ both fail to exist, and so the function f has a discontinuity of the second kind at x_0. By the arbitrariness of x_0, f has a discontinuity of the second kind at every real number (see Fig. 4.3).

We used sequences of real numbers converging to x_0 for the preceding function to show that the limit does not exist. This is a common and quite useful technique in dealing with limits. The following theorem is an extension of Theorem 3.6 and follows readily from it.

Theorem 4.1 Assume that f is defined in some neighborhood of x_0. The function f is continuous at x_0 if and only if $f(x_n) \to f(x_0)$ for every sequence $\{x_n\}$ in the domain of f with $x_n \to x_0$.

From Examples 3.9 to 3.11 it follows that $f(x) = \sin x$ is continuous at each real number, $f(x) = \sqrt{x}$ is continuous at each $x > 0$, and $f(x) = 1/x$ is continuous at each $x \neq 0$. Of course, the constant functions $f(x) = k$ and the identity function $f(x) = x$ are continuous at each real number. The following theorem shows that certain combinations of continuous functions are again continuous; it is an immediate consequence of the fundamental limit theorem (see Theorem 3.4).

Theorem 4.2 If f and g are each continuous at $x = x_0$ then $f + g$, $f \cdot g$ are continuous at x_0 and f/g is continuous at x_0 provided $g(x_0) \neq 0$.

Thus every polynomial function $p(x) = a_0 x^n + a_1 x^{n-1} + \ldots + a_{n-1} x + a_n$ is continuous at each real number x_0, and every rational function $r(x) = p(x)/q(x)$ is continuous at each real number x_0 except the finitely many zeros of the polynomial $q(x)$. Of course, if x_0 is a zero of $q(x)$ for which $p(x_0) = 0$ then it can occur that $\lim_{x \to x_0} r(x)$ exists, in which case x_0 is a removable discontinuity of $r(x)$ (see Example 4.1). On the other hand, the rational function $1/x$ has a discontinuity of the second kind at $x = 0$.
Next we consider continuity of the composite of two functions.

Theorem 4.3 If f is continuous at x_0 and g is continuous at $f(x_0)$ then $(g \circ f)(x) = g(f(x))$ is continuous at x_0.

PROOF Let $\varepsilon > 0$ be given. Since g is continuous at the point $f(x_0)$, there is an $\eta > 0$ such that if $|y - f(x_0)| < \eta$ then $|g(y) - g(f(x_0))| < \varepsilon$. But since f is continuous at x_0, there is a $\delta > 0$ such that if $|x - x_0| < \delta$ then $|f(x) - f(x_0)| < \eta$. Consequently, for each x satisfying $|x - x_0| < \delta$ we have $|g(f(x)) - g(f(x_0))| < \varepsilon$. Therefore $g \circ f$ is continuous at x_0.

Example 4.5 Let $f(x) = \sin(1/x)$ and define $f(0) = 0$. The rational function $1/x$ is continuous at each nonzero real number, and $\sin x$ is continuous at each real number. Hence, by Theorem 4.3, f is continuous at each $x \neq 0$. By considering the sequence $x_n = 2/n\pi \to 0$ we see that $\lim_{x \to 0^+} f(x)$ does not exist, and so $f(x) = \sin(1/x)$ has a discontinuity of the second kind at $x = 0$ (see Fig. 4.4).

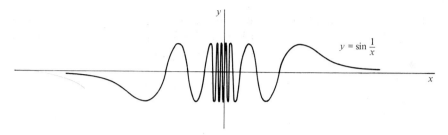

Figure 4.4

Example 4.6 Prove that $f(x) = x \sin(1/x)$ has a removable discontinuity at $x = 0$.

SOLUTION By Theorems 4.2 and 4.3 f is continuous at every nonzero real number. $f(0)$ is undefined, therefore f has a discontinuity at 0. $\lim_{x \to 0} x \sin(1/x)$ exists and equals 0 since $|\sin(1/x)| \leq 1$ in any deleted neighborhood of $x = 0$ and $\lim_{x \to 0} x = 0$ (see Exercise 3.31). Therefore f has a removable discontinuity at $x = 0$; if we extend f by defining $f(0) = 0$ then f is continuous at $x = 0$ (as well as at every other real number) (see Fig. 4.5).

A function is continuous at x_0 if the functional values are close to $f(x_0)$ when x is close to x_0. Sometimes a function may be discontinuous at a point x_0 although the functional values are all close to $f(x_0)$ when x is close to x_0 but restricted to one side of x_0. In this case we have what is known as *one-sided continuity*.

Definition 4.3 f is said to be *right-continuous* (or *continuous from the*

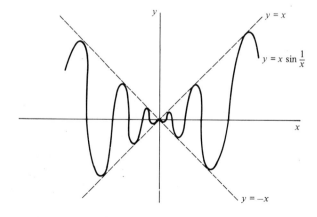

Figure 4.5

right) at x_0 if $\lim_{x \to x_0+} f(x) = f(x_0)$; similarly, f is *left-continuous* (*continuous from the left*) at x_0 if $\lim_{x \to x_0-} f(x) = f(x_0)$.

In order for f to be right-continuous at x_0 it is necessary that f be defined on an interval of the form $[x_0, x_0 + \eta)$ for some $\eta > 0$. If f is left-continuous at x_0 then f must be defined on $(x_0 - \eta, x_0]$ for some $\eta > 0$. Also, it is an immediate consequence of Theorem 3.7 that f is continuous at x_0 if and only if f is both right- and left-continuous at x_0. For example, $f(x) = |x|$ is right-continuous at $x = 0$ since $\lim_{x \to 0+} |x| = \lim_{x \to 0+} x = 0 = f(0)$; similarly, f is left-continuous at $x = 0$ since $\lim_{x \to 0-} |x| = \lim_{x \to 0-} (-x) = 0 = f(0)$. Therefore $f(x) = |x|$ is continuous at $x = 0$. Now if g is continuous at x_0 then since $f(x) = |x|$ is continuous at every real number it follows from Theorem 4.3 that $|g|$ is continuous at x_0.

Example 4.7 Let $f(x) = [\![x]\!]$. Recall that f has a jump discontinuity at every integer and is continuous at all other real numbers. Since $\lim_{x \to k+} f(x) = \lim_{x \to k+} [\![x]\!] = k = f(k)$ for every $k \in \mathbf{J}$, f is right-continuous at each integer. But f is not left-continuous at $x = k$ since $\lim_{x \to k-} f(x) = \lim_{x \to k-} [\![x]\!] = k - 1 \neq k = f(k)$.

We say that f is continuous on a set A provided f is continuous at every point in A. Of course, since this means that f must be defined in a neighborhood of each point in A, the definition is rather restrictive. It is convenient to relax the definition of continuity somewhat to allow functions to be continuous at endpoints of intervals upon which they are defined provided they have the appropriate one-sided continuity at these endpoints. For example, $f(x) = \sqrt{x}$ is not continuous at $x = 0$ (it is undefined to the left of zero). However, $\lim_{x \to 0+} \sqrt{x} = 0 = f(0)$, and so $\sqrt{x}$ is right-continuous at $x = 0$. Thus, we say that f is continuous on $[0, \infty)$.

If a function f is defined on the closed interval $[a, b]$, we say that f is *continuous on* $[a, b]$ provided that it is continuous on (a, b), it is right-continuous at the left-hand endpoint a, and it is left-continuous at the right-hand endpoint b. This modification of a definition in mathematics is not uncommon and serves the purpose of simplifying the statements of subsequent results. We shall be careful to point out when such modifications are made in what follows.

EXERCISES

4.1 Prove that if f is continuous at x_0 then f is bounded at x_0.
4.2 Find all points of discontinuity for the following functions, classify the discontinuities as

removable, jump, or second kind, and determine where the function is right- and left-continuous.

(a) $$f(x) = \begin{cases} x^2 & \text{if } x < -1 \\ 2x + 3 & \text{if } -1 \leq x \leq 0 \\ |x - 1| & \text{if } 0 < x < 2 \\ x^3 - 7 & \text{if } 2 \leq x < 3 \\ \dfrac{x - 3}{x - 4} & \text{if } 3 \leq x < 4 \\ 0 & \text{if } 4 \leq x \end{cases}$$

(b) $f(x) = x + [\![-x]\!]$ (c) $f(x) = x[\![x]\!]$ (d) $f(x) = \operatorname{sgn} [\![|x|]\!]$

(e) $f(x) = \begin{cases} [\![x + 1]\!] \sin \dfrac{1}{x} & \text{if } x \in (-1, 0) \cup (0, 1) \\ 0 & \text{otherwise} \end{cases}$

(f) $f(x) = \begin{cases} (1 + x) \operatorname{sgn} x + \operatorname{sgn} |x| - 1 & \text{if } x \text{ is rational} \\ \operatorname{sgn} x & \text{if } x \text{ is irrational} \end{cases}$

4.3 Prove that $f(x) = \cos x$ is continuous on $\mathbf{R}$.

4.4 Prove that if f is continuous at x_0 and g is discontinuous at x_0 then $f + g$ must have a discontinuity at x_0.

4.5 Show that $f + g$ can be continuous at x_0 even though both f and g have discontinuities at x_0.

4.6 Show that $f \cdot g$ can be continuous at x_0 even though both f and g have discontinuities at x_0.

4.7 If f is continuous at x_0 and g is discontinuous at x_0, what can be said about continuity of the product $f \cdot g$ at x_0?

4.8 Show that the composition function $g \circ f$ can be continuous at x_0 even though f or g or both f and g are discontinuous at x_0.

4.9 Prove that the function

$$f(x) = \begin{cases} x & \text{if } x \text{ is rational} \\ 0 & \text{if } x \text{ is irrational} \end{cases}$$

has a discontinuity of the second kind at each nonzero real number.

4.10 Prove that if f is continuous at x_0 and f is nonnegative then $h(x) = \sqrt{f(x)}$ is continuous at x_0.

4.11 Find a function f which has a discontinuity of the second kind at every real number although $f \circ f$ is continuous on $\mathbf{R}$.

4.12 If f is continuous on $(0, 1)$ and $f(x) = 1 - x$ for every rational number $x \in (0, 1)$, find $f(\pi/4)$. Explain your answer.

4.13 Prove that if f and g are each continuous on (a, b) and $f(x) = g(x)$ for every rational $x \in (a, b)$ then $f(x) = g(x)$ for every $x \in (a, b)$.

4.14 Prove: f is right-continuous at x_0 if and only if $f(x_n) \to f(x_0)$ for every sequence $\{x_n\}$ in the domain of f with $x_n \to x_0$ and $x_n \geq x_0$ for $n = 1, 2, 3, \ldots$.

4.15 Discuss one-sided continuity for the pie function.

4.16 Prove that if f is defined on $\mathbf{R}$ and continuous at $x_0 = 0$ and if $f(x_1 + x_2) = f(x_1) + f(x_2)$ for each $x_1, x_2 \in \mathbf{R}$ then f is continuous on $\mathbf{R}$.

4.17 Find all functions f which are continuous on $\mathbf{R}$ and which satisfy the equation $f(x)^2 = x^2$ for each $x \in \mathbf{R}$. *Hint*: There are four possible solutions.

4.18 Prove that if g is continuous at $x_0 = 0$, $g(0) = 0$ and for some $\delta > 0$ $|f(x)| \leq |g(x)|$ for each $x \in N_\delta(0)$ then f is continuous at $x_0 = 0$.

4.19 Prove that if f is continuous on $[a, b]$ then there exists a function g continuous on $\mathbf{R}$ such that $g(x) = f(x)$ for each $x \in [a, b]$. g is called a *continuous extension* of f to $\mathbf{R}$.

4.20 The function $f(x) = \tan x$ defined on $(-\pi/2, \pi/2)$ clearly has no continuous extension to $\mathbf{R}$. Find a bounded continuous function on (a, b) which has no continuous extension to $\mathbf{R}$.

4.21 Assume that f is continuous on (a, b). Prove that f has a continuous extension to $\mathbf{R}$ if and only if both limits $\lim_{x \to a^+} f(x)$ and $\lim_{x \to b^-} f(x)$ exist.

4.22 Prove that if f is continuous on (a, b) and both $\lim_{x \to a^+} f(x)$ and $\lim_{x \to b^-} f(x)$ exist then f is bounded on (a, b).

4.23 Suppose f is one-to-one on (a, b) and satisfies the following property: whenever $f(x_1) \neq f(x_2)$ for $x_1 < x_2$, $x_1, x_2 \in (a, b)$ and k is any number between $f(x_1)$ and $f(x_2)$, there exists a $c \in (x_1, x_2)$ with $f(c) = k$. Prove that f is continuous on (a, b) (see Exercise 3.65).

4.2 PROPERTIES OF CONTINUOUS FUNCTIONS

In this section we investigate some important properties of continuous functions. As we saw in Sec. 3.1, the functions $f(x) = x$ on $\mathbf{R}$ and $f(x) = 1/x$ on $(0, 1)$ are not bounded. Yet each of these functions is continuous on the given domain. Our first result is to show that if a function f is continuous on a closed bounded interval $[a, b]$ then f is necessarily bounded on $[a, b]$. Recall that f is continuous on $[a, b]$ if (1) f is continuous at each $x \in (a, b)$, (2) f is right-continuous at $x = a$, and (3) f is left-continuous at $x = b$.

Lemma 1 If f is continuous at x_0 then there exists a $\delta > 0$ such that f is bounded on $(x_0 - \delta, x_0 + \delta)$; that is, f is bounded at x_0.

PROOF Suppose f is continuous at x_0. Then there exists a $\delta > 0$ such that for each $x \in (x_0 - \delta, x_0 + \delta)$, $|f(x) - f(x_0)| < 1$. Thus $|f(x)| < 1 + |f(x_0)| = M$ for each $x \in (x_0 - \delta, x_0 + \delta)$, and so f is bounded on $(x_0 - \delta, x_0 + \delta)$.

The following lemma is proved in much the same way, and we leave the details to the reader (see Exercise 4.24).

Lemma 2 If f is right-continuous at x_0 then there exists a $\delta > 0$ such that f is bounded on $[x_0, x_0 + \delta)$; similarly, if f is left-continuous at x_0 then there exists a $\delta > 0$ such that f is bounded on $(x_0 - \delta, x_0]$.

Theorem 4.4 If f is continuous on the closed, bounded interval $[a, b]$ then f is bounded on $[a, b]$.

PROOF Suppose $f:[a, b] \to \mathbf{R}$ is continuous on $[a, b]$. By Lemmas 1 and 2, f is bounded at each $x \in [a, b]$. $[a, b]$ is a compact set. It follows from Theorem 3.2 that f is bounded on $[a, b]$.

Corollary If f is continuous on $\mathbf{R}$ then f is bounded on every bounded set $B \subset \mathbf{R}$. In particular, f is bounded on every bounded interval I.

Recall that each polynomial function is bounded on every bounded interval I, and hence continuous functions on $\mathbf{R}$ are like polynomials in this way. Of course, the polynomials are a small subset of the set of all continuous functions on $\mathbf{R}$.

Suppose the function f is bounded on the set A; following the notation adopted in Chap. 3, we define

$$M = \sup_{x \in A} f(x) \quad \text{and} \quad m = \inf_{x \in A} f(x)$$

In many applications of calculus we are concerned with extreme values, M and m. For example, we may be looking for the greatest force that can be applied to a structure or the least cost involved with certain expenditures or any of a number of similar problems that arise quite naturally in the physical sciences, business and economics, and engineering. In such problems it is important to find points x in the domain of the function, if such points exist, at which these extreme values are assumed.

Example 4.8 Let $f(x) = x^2$ on $(0, 2)$. f is continuous and bounded on $(0, 2)$. $M = 4$ and $m = 0$, but there are no points $x_1, x_2 \in (0, 2)$ with $f(x_1) = 4$ and $f(x_2) = 0$.

Example 4.9 Let $f(x) = x|x|/(1 + x^2)$ on $\mathbf{R}$. Note that $f(x)$ is an odd function; $f(x)$ is continuous and bounded on $\mathbf{R}$, $M = 1$ and $m = -1$, but there are no points $x_1, x_2 \in \mathbf{R}$ with $f(x_1) = 1$ and $f(x_2) = -1$ (see Fig. 4.6).

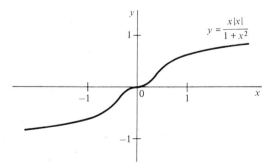

Figure 4.6

Example 4.10 Let $f(x) = x$ on $(0, 1]$ and $f(0) = 1$. f is bounded on $[0, 1]$, but it is not right-continuous at $x = 0$. Therefore, f is not continuous on $[0, 1]$. $M = 1$ and $m = 0$, but there is no $x \in [0, 1]$ with $f(x) = 0$. There is an x for which $f(x) = 1$; in fact, there are two such points, namely $x = 0$ and $x = 1$.

In these three examples, the extreme values of the function f, namely M and m, are not always assumed by the function. The following important theorem gives us conditions under which the extreme values are necessarily assumed.

Theorem 4.5: Extreme-value theorem If f is continuous on $[a, b]$ then there exists points $x_1, x_2 \in [a, b]$ such that $f(x_2) \leq f(x) \leq f(x_1)$ for all $x \in [a, b]$; that is, $f(x_1) = M$ and $f(x_2) = m$.

PROOF Suppose f is continuous on $[a, b]$; then f is bounded on $[a, b]$, and so $M = \sup_{x \in [a, b]} f(x)$ and $m = \inf_{x \in [a, b]} f(x)$ exist as real numbers. Suppose the value M is not assumed; then $f(x) < M$ for every $x \in [a, b]$. Define $g(x) = 1/[M - f(x)]$ on $[a, b]$; clearly $g(x) > 0$ for every $x \in [a, b]$ and, by Theorem 4.2, g is continuous on $[a, b]$. Then, by Theorem 4.4, g is bounded on $[a, b]$, and so there is a real number $k > 0$ such that $g(x) \leq k$ for every $x \in [a, b]$. Now for every $x \in [a, b]$

$$k \geq g(x) = \frac{1}{M - f(x)}$$

and so $M - f(x) \geq 1/k > 0$. Hence $f(x) \leq M - 1/k$ on $[a, b]$, and this contradicts the definition of M as the least upper bound of f on $[a, b]$. Therefore the value M must be assumed; i.e., there exists an $x_1 \in [a, b]$ with $f(x_1) = M$. To prove the existence of an $x_2 \in [a, b]$ with $f(x_2) = m$ we can apply the above argument to the function $-f$.

The two preceding theorems tell us that if a function f is continuous on a closed bounded interval $[a, b]$ then f is bounded on $[a, b]$ and f assumes its extreme values. The next theorem says that f also assumes all intermediate values. First we need two lemmas.

Lemma 3 If f is continuous at x_0 and $f(x_0) > 0$ then there exists a $\delta > 0$ such that for every $x \in (x_0 - \delta, x_0 + \delta)$ we have $f(x) > \frac{1}{2}f(x_0) > 0$; similarly, if f is continuous at x_0 and $f(x_0) < 0$ then there exists a $\delta > 0$ such that for every $x \in (x_0 - \delta, x_0 + \delta)$ we have $f(x) < \frac{1}{2}f(x_0) < 0$.

PROOF Suppose f is continuous at x_0 and $f(x_0) > 0$. Then there exists a $\delta > 0$ such that for every $x \in (x_0 - \delta, x_0 + \delta)$, $|f(x) - f(x_0)| < \frac{1}{2}f(x_0)$.

Hence, for every $x \in (x_0 - \delta, x_0 + \delta)$, $-\frac{1}{2}f(x_0) < f(x) - f(x_0) < \frac{1}{2}f(x_0)$, and so $f(x) > \frac{1}{2}f(x_0) > 0$. To prove the second part of the lemma, we apply the above argument to the function $-f$.

The next lemma is proved in a similar way, and we leave the details to the reader (see Exercise 4.25).

Lemma 4 If f is right-continuous at x_0 and $f(x_0) > 0$ then there exists a $\delta > 0$ such that for every $x \in [x_0, x_0 + \delta)$, $f(x) > \frac{1}{2}f(x_0) > 0$; similarly, if f is right-continuous at x_0 and $f(x_0) < 0$ then there exists a $\delta > 0$ such that for every $x \in [x_0, x_0 + \delta)$, $f(x) < \frac{1}{2}f(x_0) < 0$. If f is left-continuous at x_0 and $f(x_0) > 0$ then there exists a $\delta > 0$ such that for every $x \in (x_0 - \delta, x_0]$, $f(x) > \frac{1}{2}f(x_0) > 0$; similarly, if f is left-continuous at x_0 and $f(x_0) < 0$ then there exists a $\delta > 0$ such that for every $x \in (x_0 - \delta, x_0], f(x) < \frac{1}{2}f(x_0) < 0$.

Theorem 4.6 Intermediate-value theorem If f is continuous on $[a, b]$ and k is between $f(a)$ and $f(b)$ then there exists a $c \in (a, b)$ such that $f(c) = k$ (see Fig. 4.7).

PROOF Let f be continuous on $[a, b]$ and let k be between $f(a)$ and $f(b)$; that is, either $f(a) < k < f(b)$ or $f(b) < k < f(a)$. We assume the former, the proof for the latter being quite similar. Define the function $g(x) = f(x) - k$ for all $x \in [a, b]$. Then $g(a) = f(a) - k < 0$ and $g(b) = f(b) - k > 0$. Let $c = \sup\{x \in [a, b] \mid g(x) < 0\}$. Since g is continuous on $[a, b]$, it follows from Lemmas 3 and 4 that $g(x) < 0$ on some interval $[a, a + \delta_1)$, where $\delta_1 > 0$, and that $g(x) > 0$ on some interval $(b - \delta_2, b]$, where $\delta_2 > 0$. Thus c satisfies $a < c < b$. Now for every x satisfying $c < x \leq b$, $g(x) \geq 0$, for otherwise c would not be an upper bound of the set $\{x \in [a, b] \mid g(x) < 0\}$. Hence $\lim_{x \to c^+} g(x) \geq 0$. By continuity of g, $g(c) = \lim_{x \to c^+} g(x)$, and so $g(c) \geq 0$. Choose a sequence $\{x_n\}$ in $[a, b]$ such that $x_n < c$, $x_n \to c$, and $g(x_n) < 0$ for every $n \in \mathbb{N}$. Note that if there were no such sequence then the set $\{x \in [a, b] \mid g(x) < 0\}$ would have an upper bound less than c, in contra-

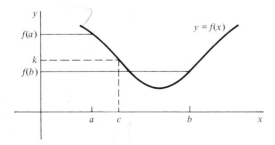

Figure 4.7

diction with the definition of c. Now, since g is continuous at c, $g(x_n) \to g(c)$. Hence $g(c) \leq 0$. It follows that $g(c) = 0$ and consequently $f(c) = k$.

Corollary If f is a nonconstant, continuous function on $[a, b]$ then the range of f is the closed, bounded interval $[m, M]$, where $M = \sup_{x \in [a, b]} f(x)$ and $m = \inf_{x \in [a, b]} f(x)$.

PROOF By Theorem 4.4 the real numbers m and M exist; by Theorem 4.5 there exists $x_1, x_2 \in [a, b]$ with $f(x_1) = M$ and $f(x_2) = m$, and $m < M$ since f is nonconstant. Now, by Theorem 4.6, every $y \in (m, M)$ is in the range of f; that is, there exists an x between x_1 and x_2 with $f(x) = y$. Therefore the range of f is $[m, M]$.

Example 4.11 Show that the function $f(x) = x + 1$ on $(0, 1]$, $f(0) = 0$ does not satisfy the conclusion of the intermediate-value theorem.

SOLUTION f is bounded but is not continuous on $[0, 1]$ since f fails to be right-continuous at $x = 0$. $M = 2$ and $m = 0$; in fact $f(0) = 0$ and $f(1) = 2$, but there is no $c \in (0, 1)$ with $f(c) = 1$, for example. It is easy to see that none of the intermediate values $x \in (0, 1]$ are assumed by f.

As a first application of the intermediate-value theorem we prove a fixed-point theorem. Fixed-point theorems are especially important in the theory of differential equations.

Theorem 4.7 If f is continuous on $[a, b]$ and $f(x) \in [a, b]$ for every $x \in [a, b]$ then f has a fixed point; that is, there exists an $x_0 \in [a, b]$ with $f(x_0) = x_0$.

PROOF Suppose f is continuous on $[a, b]$ and $f(x) \in [a, b]$ for every $x \in [a, b]$. If $f(a) = a$ or $f(b) = b$ then we are done; hence assume $f(a) > a$ and $f(b) < b$. Define $g(x) = f(x) - x$ for every $x \in [a, b]$. Clearly $g(a) > 0$, $g(b) < 0$, and g is continuous on $[a, b]$. Now 0 is an intermediate value for g on $[a, b]$. Hence by the intermediate-value theorem there exists an $x_0 \in (a, b)$ such that $g(x_0) = 0$. Then $f(x_0) = x_0$.

Theorem 4.8 If f is one-to-one and continuous on $[a, b]$ then f is strictly monotone on $[a, b]$.

PROOF Suppose that f is one-to-one and continuous on $[a, b]$. Then $f(a) \neq f(b)$. We first consider the case $f(a) < f(b)$. If f is not strictly increasing on $[a, b]$ then there exists $x_1, x_2 \in [a, b]$ with $x_1 < x_2$ and $f(x_1) \geq f(x_2)$. Equality here violates the assumption that f is one-to-one, and so we must have $f(x_1) > f(x_2)$. There are two possibilities (the reader should graph them).

(a) $f(x_1) > f(b)$. Choose $k \in (f(b), f(x_1))$. Then $k \in (f(a), f(x_1))$, and by the intermediate-value theorem there is a $c_1 \in (a, x_1)$ with $f(c_1) = k$ and there is a $c_2 \in (x_1, b)$ with $f(c_2) = k$. But $c_1 \neq c_2$, and this contradicts the fact that f is one-to-one.

(b) $f(x_1) < f(b)$. Then $f(x_2) < f(b)$. Choose $k \in (f(x_2), f(x_1))$. Clearly $k \in (f(x_2), f(b))$, and by the intermediate-value theorem there is a $c_1 \in (x_1, x_2)$ with $f(c_1) = k$ and there is a $c_2 \in (x_2, b)$ with $f(c_2) = k$. Again $c_1 \neq c_2$, and this contradicts the fact that f is one-to-one on $[a, b]$.

It follows that f is strictly increasing on $[a, b]$. Finally for the case $f(a) > f(b)$ the above argument applied to $-f$ shows that $-f$ is strictly increasing on $[a, b]$ and hence f is strictly decreasing on $[a, b]$.

Recall that if f is one-to-one on the set A then f^{-1} exists as a function defined on $f(A)$; f^{-1} is one-to-one on $f(A)$ and the range of f^{-1} is the set A.

Theorem 4.9 If f is one-to-one and continuous on $[a, b]$ then f^{-1} is continuous on $[m, M]$, where $M = \sup_{x \in [a, b]} f(x)$ and $m = \inf_{x \in [a, b]} f(x)$.

PROOF We know that the range of f is the closed, bounded interval $[m, M]$, and so f^{-1} is a well-defined one-to-one function defined on $[m, M]$ with range $[a, b]$. Let y_0 be an arbitrary element in $[m, M]$; it suffices to prove that f^{-1} is continuous at y_0. Let $\{y_n\}$ be any sequence in $[m, M]$ with $y_n \to y_0$. According to Theorem 4.1, we need to show that $f^{-1}(y_n) \to f^{-1}(y_0)$. Let $x_0 = f^{-1}(y_0)$ and $x_n = f^{-1}(y_n)$ for $n = 1, 2, 3, \ldots$ and suppose that $x_n \not\to x_0$. Then for some $\varepsilon > 0$, $|x_n - x_0| \geq \varepsilon$ for infinitely many natural numbers n. We extract a subsequence $\{x_n^*\}$ of $\{x_n\}$ such that $|x_n^* - x_0| \geq \varepsilon$ for all n. Now, $\{x_n^*\}$ is a bounded sequence and so by the Bolzano-Weierstrass theorem (for sequences) $\{x_n^*\}$ has a convergent subsequence, say $\{\hat{x}_n\}$. Then $|\hat{x}_n - x_0| \geq \varepsilon$ for all n and $\hat{x}_n \to c \in [a, b]$. Clearly $c \neq x_0$. By continuity of f at c, $f(\hat{x}_n) \to f(c)$. But $\{f(\hat{x}_n)\}$ is a subsequence of $\{f(x_n)\} = \{y_n\}$ and $y_n \to y_0 = f(x_0)$. Therefore $f(c) = f(x_0)$. Since this contradicts the assumption that f is a one-to-one function, we must have $x_n \to x_0$; that is, $f^{-1}(y_n) \to f^{-1}(y_0)$. It follows that f^{-1} is continuous on $[m, M]$.

Suppose f is one-to-one and continuous on $[a, b]$. By Theorem 4.8, f is strictly monotone on $[a, b]$. If f is strictly increasing then $m = f(a)$ and $M = f(b)$ and, by Theorem 4.9, f^{-1} is one-to-one and continuous on $[f(a), f(b)]$. Hence f^{-1} is strictly increasing on $[f(a), f(b)]$. On the other hand, if f is strictly decreasing then $m = f(b)$ and $M = f(a)$ and (again using Theorem 4.9) f^{-1} is one-to-one and continuous on $[f(b), f(a)]$. Thus f^{-1} is strictly decreasing on $[f(b), f(a)]$.

Definition 4.4 A function f defined on $[a, b]$ is said to satisfy the intermediate-value property on $[a, b]$ if for every $x_1, x_2 \in [a, b]$ with $x_1 < x_2$ and for every k between $f(x_1)$ and $f(x_2)$ there is a $c \in (x_1, x_2)$ with $f(c) = k$.

A function which satisfies the intermediate-value property on $[a, b]$ need not be continuous on $[a, b]$. For example, consider $f(x) = \sin(1/x)$ with $f(0) = 0$ defined on $[-1, 1]$. This function satisfies the intermediate-value property on $[-1, 1]$ but is discontinuous at $x = 0$.

Theorem 4.10 If f satisfies the intermediate-value property on $[a, b]$ then f has no simple discontinuities in $[a, b]$.

PROOF Suppose f has a simple discontinuity at $x_0 \in [a, b]$. Then either $\lim_{x \to x_0^+} f(x) \neq f(x_0)$ or $\lim_{x \to x_0^-} f(x) \neq f(x_0)$ (or both). We assume that $\lim_{x \to x_0^+} f(x) = l < f(x_0)$; the other cases are handled in a similar way and we omit them. l exists as a real number since the discontinuity of f at x_0 is simple. Let $\alpha = \frac{1}{2}(f(x_0) - l) > 0$. Now there exists a $\delta > 0$ such that if $x \in (x_0, x_0 + \delta)$ then $|f(x) - l| < \alpha$. Hence for every $x \in (x_0, x_0 + \delta)$, $f(x) < l + \alpha$. On the other hand $f(x_0) = l + 2\alpha > l + \alpha$. Fix $x_1 \in (x_0, x_0 + \delta)$. Then $f(x_1) < l + \alpha$, and if we choose $y \in (l + \alpha, f(x_0)) \subset (f(x_1), f(x_0))$, since each $x \in (x_0, x_1)$ satisfies $f(x) < l + \alpha$ it is clear that there is no $x \in (x_0, x_1)$ with $f(x) = y$. Therefore f fails to satisfy the intermediate-value property on $[a, b]$.

Theorem 4.11 If f is one-to-one and satisfies the intermediate-value property on $[a, b]$ then f is continuous on $[a, b]$.

PROOF In our proof to Theorem 4.8 we used only the assumption that f is one-to-one and satisfies the intermediate-value property on $[a, b]$. Therefore, it follows from the proof of Theorem 4.8 that f is (strictly) monotone on $[a, b]$. But monotone functions have no discontinuities of the second kind (see Theorem 3.8). Also, by Theorem 4.10, f has no simple discontinuities in $[a, b]$. Therefore f is continuous on $[a, b]$.

We conclude this section with a final application of the intermediate-value theorem.

Assume that at any instant in time the temperature along the equator is a continuous function of distance (a not altogether unreasonable assumption). We prove that at any instant there is a pair of antipodal points with the same temperature.

Pick any point on the equator and start measuring distance x from that point. Let L be the length of the equator and $T(x)$ the temperature at x, $0 \le x \le L$. Since $x = 0$ and $x = L$ represent the same point on the equator, $T(0) = T(L)$. Define $g(x) = T(x) - T(x + L/2)$; g is a continuous function on the interval $[0, L/2]$. Now $g(0) = T(0) - T(L/2) = T(L) - T(L/2) = -[T(L/2) - T(L)] = -g(L/2)$. If $g(0) = 0$ then $T(L/2) = T(0)$, and so $x = 0$ and $x = L/2$ are the pair of antipodal points with equal temperature. If $g(0) \ne 0$ then $g(0)$ and $g(L/2)$ have opposite signs; by the intermediate-value theorem, there exists a $c \in (0, L/2)$ such that $g(c) = 0$. Therefore $T(c) = T(c + L/2)$, and so $x = c$ and $x = c + L/2$ are the pair of antipodal points with equal temperature.

Generally, if f is continuous on $[a, b]$ and if $f(a) = f(b)$ then there are points $x \in [a, (a+b)/2]$ and $y = x + (b-a)/2 \in [(a+b)/2, b]$ such that $f(x) = f(y)$.

EXERCISES

4.24 Prove Lemma 2.

4.25 Prove Lemma 4.

4.26 Find $M = \sup_{x \in A} f(x)$ and $m = \inf_{x \in A} f(x)$ for the following bounded functions f defined on the indicated domain A and then find points $x_1, x_2 \in A$ (if they exist) such that $f(x_1) = M$ and $f(x_2) = m$.

(a) $f(x) = 3 + 2x - x^2$ on $[0, 4]$ (b) $f(x) = 2 - |x - 1|$ on $[-2, 2)$
(c) $f(x) = e^{-1/x}$ on $(0, \infty)$ (d) $f(x) = 1 - x^2$ on $(-2, 1)$

4.27 Find a function $f: A \to B$ such that f is bounded on $C \subset A$ but there exists an $x_0 \in C$ such that f is not bounded at x_0.

4.28 Suppose f is continuous on (a, b). Prove that if $\lim_{x \to a^+} f(x)$ and $\lim_{x \to b^-} f(x)$ both exist then f is bounded on (a, b) but the converse does not hold.

4.29 Suppose f is continuous on $\mathbf{R}$. Prove that if $\lim_{x \to \infty} f(x)$ and $\lim_{x \to -\infty} f(x)$ both exist then f is bounded on $\mathbf{R}$ but the converse does not hold.

4.30 Verify that the function in Example 4.5 satisfies the intermediate-value property on $[-1, 1]$.

4.31 Prove that if f is continuous on any interval then f satisfies the intermediate-value property on that interval.

4.32 Prove that if f is continuous on any interval then the range of f is again an interval.

4.33 Prove that the polynomial $p(x) = a_0 x^n + a_1 x^{n-1} + \cdots + a_{n-1} x + a_n$, where $a_0 \ne 0$ and n is an odd natural number, has at least one real root; that is, there exists $x_0 \in \mathbf{R}$ such that $p(x_0) = 0$.

4.34 Prove that there exists $x_0 \in (0, 1)$ such that $f(x_0) = 0$, where $f(x) = e^x - 3x - \sin x$.

4.35 Suppose that f is continuous on $[0, 1]$, $f(x)$ is rational for every $x \in [0, 1]$, and $f(0) = 0$. Find $f(\sqrt{2}/2)$.

4.36 Suppose that f is continuous on $(0, \infty)$, $\lim_{x \to 0^+} f(x) = 0$, and $\lim_{x \to \infty} f(x) = 1$. Prove that there exists $x_0 > 0$ such that $f(x_0) = \sqrt{3}/2$.

4.37 Prove that the function $f(x) = x^3 + x^2 - 3x - 3$ has a root between 1 and 2, between 1.5 and 2, between 1.5 and 1.75, between 1.625 and 1.75, etc. Note that if we continue this procedure we shall be able to approximate the root as closely as we wish.

4.38 Suppose that f is continuous on $[a, b]$ except at $x_0 \in (a, b)$, where f has a discontinuity. Suppose further that $f(x)$ is rational for every $x \in [a, b]$. Prove that f has a simple discontinuity at x_0.

4.39 Suppose that f is continuous on $[-1, 1]$ and $|f(x)| \le 1$ for every $x \in [-1, 1]$. Suppose that g is continuous on $[-1, 1]$ with $g(-1) = -1$ and $g(1) = 1$. Prove that there exists an $x_0 \in [-1, 1]$ with $f(x_0) = g(x_0)$.

4.40 Prove that if f is monotone on $[a, b]$ and f satisfies the intermediate-value property on $[a, b]$ then f is continuous on $[a, b]$.

4.41 Prove that there is no continuous function f on $\mathbf{R}$ such that for every real number c, $f(x) = c$ has exactly two solutions.

4.42 Find a continuous function f on $\mathbf{R}$ such that for every real number c, $f(x) = c$ has exactly three solutions.

4.43 Use the fact that $f(x) = \tan x$ is continuous and one-to-one on $(-\pi/2, \pi/2)$ to establish that $\tan^{-1} x$ is continuous on $\mathbf{R}$.

4.3 UNIFORM CONTINUITY

In the following we assume that f is defined on an interval I, which may be closed, open, or neither, bounded or unbounded. f is continuous at $x_0 \in I$ if given any $\varepsilon > 0$ there exists a $\delta > 0$ such that whenever $x \in I$ and $|x - x_0| < \delta$ then $|f(x) - f(x_0)| < \varepsilon$. In general, the δ to be found depends not only on the given ε but also on the particular point x_0 under consideration. The following example illustrates this fact.

Example 4.12 Let $f(x) = x^2$ on $\mathbf{R}$. We know from the previous section that f is continuous on $\mathbf{R}$. Suppose we try to prove this directly.

SOLUTION To begin, let x_0 be a fixed but arbitrarily chosen real number and let $\varepsilon > 0$ be given. Now

$$|f(x) - f(x_0)| = |x^2 - x_0^2|$$
$$= |x + x_0||x - x_0| \le (|x| + |x_0|)|x - x_0|$$

If we require that $\delta \le 1$ then $|x - x_0| < \delta$ implies $|x - x_0| < 1$, which implies $|x| < 1 + |x_0|$. Then $|f(x) - f(x_0)| \le (1 + 2|x_0|)|x - x_0|$. If we further require that $\delta \le \varepsilon/(1 + 2|x_0|)$ then $|x - x_0| < \delta$ implies $|x - x_0| < \varepsilon/(1 + 2|x_0|)$, and so $|f(x) - f(x_0)| < \varepsilon$. Therefore we choose $\delta > 0$ such that $\delta \le \min(1, \varepsilon/(1 + 2|x_0|))$. Here δ indeed depends not only on the given ε but also on the point x_0. In fact for a fixed given $\varepsilon > 0$ if x_0 is taken to be very large in absolute value then from our formula the δ will have to be chosen appropriately small.

By contrast, in the next example the δ depends only on ε.

Example 4.13 Let $f(x) = 3x + 2$ on **R**. Again we know that f is continuous on **R**. We prove this directly.

SOLUTION Let x_0 be a fixed but arbitrarily chosen real number and suppose that $\varepsilon > 0$ is given. Then $|f(x) - f(x_0)| = |(3x + 2) - (3x_0 + 2)| = 3|x - x_0|$. If we choose $\delta > 0$ such that $\delta \leq \varepsilon/3$ then $|x - x_0| < \delta$ implies $|x - x_0| < \varepsilon/3$, which in turn implies that $|f(x) - f(x_0)| < \varepsilon$. Therefore the continuity of f at x_0 is demonstrated by choosing the $\delta > 0$ so that $\delta \leq \varepsilon/3$. Here δ depends on the given $\varepsilon > 0$ but does not depend on the point x_0. That is, for the given $\varepsilon > 0$ the above choice of δ works regardless of the point x_0 in question.

In Example 4.13 given $\varepsilon > 0$ the same δ works for all points; in Example 4.12 apparently we are going to have to change the δ as x_0 varies. The property of having a δ which works at once for all points in I is called uniform continuity; we formalize it as follows.

Definition 4.5 f is said to be *uniformly continuous* on I if given any $\varepsilon > 0$ there exists a $\delta > 0$ such that whenever $x_1, x_2 \in I$ with $|x_1 - x_2| < \delta$ then $|f(x_1) - f(x_2)| < \varepsilon$.

We return to the preceding two examples to verify that our suspicions were correct.

First consider again $f(x) = x^2$ on **R**. The problem occurs as x_1 and x_2 become larger and larger because the $\delta > 0$ we must choose then becomes smaller and smaller (see Fig. 4.8). Since we are thinking of δ as a small positive real number, let us see what happens if we choose $x_1 = 1/\delta$ and $x_2 = 1/\delta + \delta/2$. These points are closer than δ apart, $|x_1 - x_2| = \delta/2 < \delta$ to be exact, but the smaller δ is, the larger x_1 and x_2 are. Now

$$|f(x_1) - f(x_2)| = |x_1^2 - x_2^2| = \left|\left(\frac{1}{\delta}\right)^2 - \left(\frac{1}{\delta} + \frac{\delta}{2}\right)^2\right| = 1 + \frac{\delta^2}{4} > 1$$

This shows that for $\varepsilon = 1$ there is no $\delta > 0$ such that $|x_1 - x_2| < \delta$ always guarantees $|f(x_1) - f(x_2)| < 1$. Thus $f(x) = x^2$ is not uniformly continuous on **R**. If you have a function which you suspect is not uniformly continuous, first decide where the problem is (in the above example the problem occurs when $|x|$ is large). Then try to pick pairs of points in this region which are close but whose function values stay apart.

For $f(x) = 3x + 2$ on **R**, and following the procedure of Example 4.13,

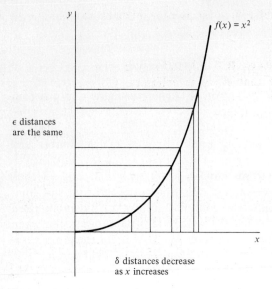

Figure 4.8

given any $\varepsilon > 0$ we choose $\delta > 0$ such that $\delta \leq \varepsilon/3$. Then $|x_1 - x_2| < \delta$ implies that

$$|f(x_1) - f(x_2)| = |(3x_1 + 2) - (3x_2 + 2)| = 3|x_1 - x_2| < 3\delta \leq 3(\varepsilon/3) = \varepsilon$$

Consequently, $f(x) = 3x + 2$ is uniformly continuous on **R**.

Example 4.14 Let $f(x) = x^2$ on $[0, b]$, where b is any positive real number. Let $\varepsilon > 0$ be given and choose $\delta = \varepsilon/2b$. Then if $x_1, x_2 \in [0, b]$ and $|x_1 - x_2| < \delta$, we have

$$|f(x_1) - f(x_2)| = |x_1^2 - x_2^2|$$
$$= |x_1 + x_2||x_1 - x_2| \leq 2b|x_1 - x_2| < 2b\delta = \varepsilon$$

Therefore $f(x) = x^2$ is uniformly continuous on $[0, b]$.

The above example demonstrates the dependence of uniform continuity on the domain of the function. A function f may fail to be uniformly continuous on an interval I but may well be uniformly continuous on an interval $I_1 \subset I$. On the other hand it is immediate from the definition that if $f(x)$ is uniformly continuous on I and if $I_1 \subset I$ then f is uniformly continuous on I_1. The following theorem is also immediate from the definition of uniform continuity, and we leave the proof to the reader (Exercise 4.44).

Theorem 4.12 If f is uniformly continuous on I then f is continuous on I.

Example 4.15 Prove that $f(x) = 1/x$ is not uniformly continuous on $(0, 1)$.

SOLUTION f is continuous on $\mathbf{R} - \{0\}$, and so f is continuous on $(0, 1)$. Let $\varepsilon = \frac{1}{2}$ and suppose that $\delta > 0$ is arbitrary. Choose the natural number $n > 1$ such that $1/n < \delta$ and let $x_1 = 1/n$ and $x_2 = 1/(n + 1)$. Clearly, $x_1, x_2 \in (0, 1)$, and $|x_1 - x_2| = 1/n - 1/(n + 1) < 1/n < \delta$. On the other hand,

$$|f(x_1) - f(x_2)| = \left| \frac{1}{x_1} - \frac{1}{x_2} \right| = |n - (n + 1)| = 1 > \varepsilon$$

Hence for $\varepsilon = \frac{1}{2}$ there is no $\delta > 0$ which satisfies Definition 4.5. Therefore $f(x) = 1/x$ is not uniformly continuous on $(0, 1)$.

Suppose we alter f somewhat by letting $f(x) = 1/x$ on (a, ∞), where $a > 0$. Let $\varepsilon > 0$ be given and choose $\delta = \varepsilon a^2$. Now if $x_1, x_2 \in (a, \infty)$ and $|x_1 - x_2| < \delta$ then we have

$$|f(x_1) - f(x_2)| = \left| \frac{1}{x_1} - \frac{1}{x_2} \right| = \frac{|x_1 - x_2|}{x_1 x_2} \leq \frac{1}{a^2}|x_1 - x_2| < \frac{\delta}{a^2} = \varepsilon$$

Hence $f(x) = 1/x$ is uniformly continuous on (a, ∞), where $a > 0$.

In Sec. 4.2 we had reason to suspect that continuous functions defined on closed, bounded intervals are special: they are necessarily bounded, they assume extreme values and all intermediate values, and under certain conditions they have fixed points. The following theorem extends this notion.

Theorem 4.13 If f is continuous on the closed, bounded interval $[a, b]$ then f is uniformly continuous on $[a, b]$.

PROOF Suppose f is continuous on $[a, b]$. Let $\varepsilon > 0$ be given. For each $x \in [a, b]$ there exists $\delta_x > 0$ such that $t \in [a, b]$ with $|t - x| < \delta_x$ implies that $|f(t) - f(x)| < \varepsilon/2$. Let $I(x) = (x - \frac{1}{2}\delta_x, x + \frac{1}{2}\delta_x)$. Clearly $\mathscr{F} = \{I(x) \mid x \in [a, b]\}$ is an open covering of $[a, b]$ and $t \in [a, b] \cap I(x)$ implies that $|f(t) - f(x)| < \varepsilon/2$. Since $[a, b]$ is compact, it follows from the Heine-Borel theorem that there exists a finite set $\hat{\mathscr{F}} \subset \mathscr{F}$ such that $\hat{\mathscr{F}}$ is also an open covering of $[a, b]$. Say $\hat{\mathscr{F}} = \{I(x_j) \mid j = 1, 2, \ldots, n\}$, where each $x_j \in [a, b]$, and let $\delta = \min_{1 \leq j \leq n} \frac{1}{2}\delta_{x_j}$. Clearly $\delta > 0$. Suppose $u, v \in [a, b]$ and $|u - v| < \delta$. Since $\hat{\mathscr{F}}$ covers $[a, b]$, there exists a k $(1 \leq k \leq n)$ with $u \in I(x_k)$. Then $|u - x_k| < \frac{1}{2}\delta_{x_k}$, and so $|f(u) - f(x_k)| < \varepsilon/2$. Also

$$|v - x_k| \leq |v - u| + |u - x_k| < \delta + \tfrac{1}{2}\delta_{x_k} \leq \delta_{x_k}$$

Therefore $|f(v) - f(x_k)| < \varepsilon/2$, and so

$$|f(u) - f(v)| \leq |f(u) - f(x_k)| + |f(x_k) - f(v)| < \frac{\varepsilon}{2} + \frac{\varepsilon}{2} = \varepsilon$$

It follows that f is uniformly continuous on $[a, b]$.

We saw in Example 4.15 that a continuous function on a bounded, open interval may fail to be uniformly continuous on that interval. The following theorem gives necessary and sufficient conditions under which a continuous function on a bounded, open interval will be uniformly continuous.

Theorem 4.14 Suppose f is continuous on (a, b). Then f is uniformly continuous on (a, b) if and only if $\lim_{x \to a^+} f(x)$ and $\lim_{x \to b^-} f(x)$ both exist.

PROOF We assume that f is a continuous function on the bounded, open interval (a, b). First, suppose that the one-sided limits $\lim_{x \to a^+} f(x)$ and $\lim_{x \to b^-} f(x)$ both exist. Define the function g on $[a, b]$ as follows. Let $g(x) = f(x)$ for every $x \in (a, b)$, $g(a) = \lim_{x \to a^+} f(x)$, and $g(b) = \lim_{x \to b^-} f(x)$. Now $g(x) = f(x)$ on (a, b), and so g is continuous on (a, b). Also, $\lim_{x \to a^+} g(x) = \lim_{x \to a^+} f(x) = g(a)$ and $\lim_{x \to b^-} g(x) = \lim_{x \to b^-} f(x) = g(b)$. Therefore g is right-continuous at $x = a$ and left-continuous at $x = b$. It follows that g is continuous on the closed, bounded interval $[a, b]$. By Theorem 4.13, g is uniformly continuous on $[a, b]$, and so g is uniformly continuous on (a, b). But g and f are identical on (a, b). Therefore f is uniformly continuous on (a, b).

Conversely, we want to show that if f is uniformly continuous on (a, b) then both one-sided limits $\lim_{x \to a^+} f(x)$ and $\lim_{x \to b^-} f(x)$ necessarily exist. Suppose that $\lim_{x \to a^+} f(x)$ does not exist (a similar argument can be made in the case where $\lim_{x \to a^+} f(x)$ exists but $\lim_{x \to b^-} f(x)$ fails to exist). Then there exists a sequence $\{x_n\}$ in (a, b) with $x_n \to a$ such that the sequence $\{f(x_n)\}$ does not converge and hence is not a Cauchy sequence. Thus there exists some $\varepsilon_0 > 0$ with the property that there is no natural number N for which $i, j \geq N$ implies that $|f(x_i) - f(x_j)| < \varepsilon_0$. Consequently, we can find arbitrarily large $i, j \in \mathbf{N}$ for which $|f(x_i) - f(x_j)| \geq \varepsilon_0$. Now, since the sequence $\{x_n\}$ is a Cauchy sequence, we have $\lim_{i, j \to \infty} |x_i - x_j| = 0$. Clearly then for this ε_0 we can find a pair of points $x_i, x_j \in (a, b)$ which are arbitrarily close and for which $|f(x_i) - f(x_j)| \geq \varepsilon_0$. Thus Definition 4.5 does not hold, and so f fails to be uniformly continuous on (a, b).

We conclude this section with the following two theorems, which give a sufficient condition for a continuous function on an unbounded interval to be uniformly continuous.

Theorem 4.15 Suppose that f is continuous on $[a, \infty)$. If $\lim_{x \to \infty} f(x)$ exists then f is uniformly continuous on $[a, \infty)$.

PROOF Let f be continuous on $[a, \infty)$ and suppose that $\lim_{x \to \infty} f(x) = L$ for some $L \in \mathbf{R}$. Let $\varepsilon > 0$ be given. Then there exists an $M > a$ such that when $x > M$ we have $|f(x) - L| < \varepsilon/2$. Since f is continuous at $x = M$, there exists $\delta_1 > 0$ such that if $|x - M| < \delta_1$ then $|f(x) - f(M)| < \varepsilon/2$. By Theorem 4.13, f is uniformly continuous on $[a, M]$. Hence there exists a $\delta_2 > 0$ such that $x_1, x_2 \in [a, M]$ with $|x_1 - x_2| < \delta_2$ implies that $|f(x_1) - f(x_2)| < \varepsilon$. Let $\delta = \min(\delta_1, \delta_2)$. Suppose that $u, v \in [a, \infty)$ with $|u - v| < \delta$. We consider three cases:

(a) $u, v > M$. Then $|f(u) - L| < \varepsilon/2$ and $|f(v) - L| < \varepsilon/2$, and so

$$|f(u) - f(v)| \leq |f(u) - L| + |L - f(v)| < \frac{\varepsilon}{2} + \frac{\varepsilon}{2} = \varepsilon$$

(b) $u, v \leq M$. Then $|u - v| < \delta \leq \delta_2$, and so

$$|f(u) - f(v)| < \varepsilon$$

(c) $u \leq M$ and $v > M$. Then since $|u - v| < \delta$, necessarily $|u - M| < \delta \leq \delta_1$ and $|v - M| < \delta \leq \delta_1$. Hence $|f(u) - f(M)| < \varepsilon/2$ and $|f(v) - f(M)| < \varepsilon/2$, and so

$$|f(u) - f(v)| \leq |f(u) - f(M)| + |f(M) - f(v)| < \frac{\varepsilon}{2} + \frac{\varepsilon}{2} = \varepsilon$$

Therefore in each case we get that $|f(u) - f(v)| < \varepsilon$. Whenever $u, v \in [a, \infty)$ with $|u - v| < \delta$, we have that $|f(u) - f(v)| < \varepsilon$. Thus f is uniformly continuous on $[a, \infty)$.

The converse to Theorem 4.15 does not hold, as is easily seen from a consideration of the function $f(x) = x$ on $[0, \infty)$. f is uniformly continuous on $[0, \infty)$ but clearly $\lim_{x \to \infty} f(x)$ does not exist. We also point out that right continuity at $x = a$ is necessary in Theorem 4.15. For example, $f(x) = 1/x$ is continuous on $(0, \infty)$ and $\lim_{x \to \infty} f(x)$ exists (and equals 0), but f is not uniformly continuous on $(0, \infty)$.

Corollary If f is continuous on $(-\infty, b]$ and $\lim_{x \to -\infty} f(x)$ exists then f is uniformly continuous on $(-\infty, b]$.

Theorem 4.16 Suppose f is continuous on **R**. If $\lim_{x \to -\infty} f(x)$ and $\lim_{x \to \infty} f(x)$ both exist then f is uniformly continuous on **R**.

EXERCISES

4.44 Prove Theorem 4.12.

4.45 Use the definition of uniform continuity to prove that each of the following functions is uniformly continuous on the indicated interval.

    (a) $f(x) = x^3$ on $[0, 1]$ (b) $f(x) = \sqrt{x}$ on $(0, 1)$
    (c) $f(x) = \sin x$ on $\mathbf{R} = (-\infty, \infty)$ (d) $f(x) = x/(1 + x^2)$ on **R**

4.46 Prove that if f and g are each uniformly continuous on I then the sum $f + g$ is uniformly continuous on I.

4.47 Show by example that if f and g are each uniformly continuous on I then the product $f \cdot g$ may fail to be uniformly continuous on I.

4.48 Prove that if f is uniformly continuous on I and $k \in \mathbf{R}$ then $k \cdot f$ is uniformly continuous on I.

4.49 Determine which of the following functions are uniformly continuous on the indicated intervals.

    (a) $f(x) = e^x$ on $(-1, 2)$ (b) $f(x) = \dfrac{1}{x-1}$ on $(0, 1)$

    (c) $f(x) = x^3 - 3x^2 + 2x - 1$ on $(0, 3)$ (d) $f(x) = \sin \dfrac{1}{x}$ on $(0, 1)$

    (e) $f(x) = x^3$ on $[0, \infty)$ (f) $f(x) = \tan^{-1} x$ on **R**

    (g) $f(x) = \dfrac{\sin x}{x}$ on $(0, \infty)$ (h) $f(x) = x \sin x$ on **R**

    (i) $f(x) = x \sin \dfrac{1}{x}$ on $(0, \infty)$ (j) $f(x) = \begin{cases} \sin \pi x & \text{for } x \in (0, 1] \\ x^2 - 1 & \text{for } x \in (1, 2) \end{cases}$ on $(0, 2)$

4.50 Prove that if f is uniformly continuous on a bounded interval I then f is bounded on I.

4.51 Prove that if f and g are each uniformly continuous on the bounded, open interval (a, b) then the product $f \cdot g$ is uniformly continuous on (a, b).

4.52 Prove that if f and g are each uniformly continuous on the interval I and in addition each function is bounded on I then the product $f \cdot g$ is uniformly continuous on I. Is the boundedness of each function on I necessary to guarantee the uniform continuity of the product? Explain and verify.

4.53 Prove Theorem 4.16.

4.54 Show by example that a continuous, bounded function on the bounded, open interval (a, b) need not be uniformly continuous on (a, b).

4.55 Prove or give a counterexample: *Conjecture*: If $f(x)$ is continuous and bounded on **R** then f is uniformly continuous on **R**.

4.56 Prove that if f is continuous on $\mathbf{R}$ then f is uniformly continuous on every bounded interval I.

4.57 Let f be defined on the interval I. Define the set

$$D = \left\{ \frac{f(x) - f(y)}{x - y} \,\Big|\, x, y \in I, x \neq y \right\}$$

(The difference quotient $[f(x) - f(y)]/(x - y)$ will be discussed in Chap. 5.) Prove that if D is bounded then f is uniformly continuous on I.

4.58 Show by example that a function can be uniformly continuous on every bounded interval and yet not be uniformly continuous on $\mathbf{R}$.

4.59 Prove that if f is continuous on (a, ∞) and $\lim_{x \to a^+} f(x)$ and $\lim_{x \to \infty} f(x)$ both exist then f is uniformly continuous on (a, ∞).

4.60 Prove the corollary to Theorem 4.15.

4.61 Let f be continuous on (a, b). Prove that f has a continuous extension to $\mathbf{R}$ if and only if f is uniformly continuous on (a, b).

4.4 FURTHER TOPICS ON CONTINUITY

We conclude this chapter with some deeper topics on continuity. As we saw in Chap. 3, monotone functions have no discontinuities of the second kind, and since they have only countably many simple discontinuities, monotone functions have only countably many discontinuities. We now show that an arbitrary function f can have only countably many simple discontinuities. Recall that if f has a simple discontinuity at x_0 then $\lim_{x \to x_0^+} f(x)$ and $\lim_{x \to x_0^-} f(x)$ both exist but either (1) $\lim_{x \to x_0} f(x)$ does not exist or (2) $\lim_{x \to x_0} f(x)$ exists but $\lim_{x \to x_0} f(x) \neq f(x_0)$.

Let f be defined on $\mathbf{R}$.

Lemma 1 There exist only countably many points x_0 for which $\lim_{x \to x_0^+} f(x)$ and $\lim_{x \to x_0^-} f(x)$ both exist but $\lim_{x \to x_0} f(x)$ does not exist; that is, $\lim_{x \to x_0^+} f(x) \neq \lim_{x \to x_0^-} f(x)$.

PROOF Let E be the set of all real numbers x_0 such that $\lim_{x \to x_0^+} f(x)$ and $\lim_{x \to x_0^-} f(x)$ both exist but are not equal. For brevity we write

$$f(x_0 +) = \lim_{x \to x_0^+} f(x) \quad \text{and} \quad f(x_0 -) = \lim_{x \to x_0^-} f(x)$$

Now for each $x_0 \in E$ we define

$$\eta(x_0) = |f(x_0 +) - f(x_0 -)| > 0$$

(see Fig. 4.9). We choose rationals r_1, r_2 such that $r_1 < x_0 < r_2$ and such that for each $x \in [r_1, x_0)$, $|f(x) - f(x_0 -)| < \tfrac{1}{6}\eta(x_0)$ and for each $x \in (x_0, r_2]$, $|f(x) - f(x_0 +)| < \tfrac{1}{6}\eta(x_0)$. That such numbers r_1 and r_2

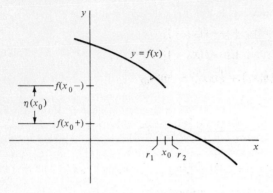

Figure 4.9

exist follows from the existence of $f(x_0 +)$ and $f(x_0 -)$. Then

$$|f(r_1) - f(r_2)| \leq |f(r_1) - f(x_0 -)|$$
$$+ |f(x_0 -) - f(x_0 +)| + |f(x_0 +) - f(r_2)|$$
$$< \tfrac{1}{6}\eta(x_0) + \eta(x_0) + \tfrac{1}{6}\eta(x_0) = \tfrac{4}{3}\eta(x_0)$$

Also

$$|f(x_0 +) - f(x_0 -)| \leq |f(x_0 +) - f(r_2)|$$
$$+ |f(r_2) - f(r_1)| + |f(r_1) - f(x_0 -)|$$

and so

$$|f(r_2) - f(r_1)| \geq |f(x_0 +) - f(x_0 -)|$$
$$- |f(x_0 +) - f(r_2)| - |f(r_1) - f(x_0 -)|$$
$$> \eta(x_0) - \tfrac{1}{6}\eta(x_0) - \tfrac{1}{6}\eta(x_0) = \tfrac{2}{3}\eta(x_0)$$

Hence

$$\tfrac{2}{3}\eta(x_0) < |f(r_1) - f(r_2)| < \tfrac{4}{3}\eta(x_0)$$

We define the function $\varphi: E \to \mathbf{Q} \times \mathbf{Q}$, where $\mathbf{Q}$ is the set of all rationals, by $\varphi(x_0) = (r_1, r_2)$. In order to prove that E is a countable set it suffices to prove that φ is one-to-one. Suppose there exists an $x_1 \in E$, $x_1 \neq x_0$ such that $\varphi(x_1) = (r_1, r_2) = \varphi(x_0)$. Then either $x_1 \in (r_1, x_0)$ or else $x_1 \in (x_0, r_2)$; we assume that $x_1 \in (r_1, x_0)$ (the other case can be handled similarly). Now $x \in [r_1, x_0)$ implies $|f(x) - f(x_0 -)| < \tfrac{1}{6}\eta(x_0)$. Thus since $x_1 \in (r_1, x_0)$, it follows that

$$|f(x_1 -) - f(x_0 -)| \leq \tfrac{1}{6}\eta(x_0) \quad \text{and} \quad |f(x_1 +) - f(x_0 -)| \leq \tfrac{1}{6}\eta(x_0)$$

Therefore
$$0 < \eta(x_1) = |f(x_1+) - f(x_1-)|$$
$$\leq |f(x_1+) - f(x_0-)| + |f(x_0-) - f(x_1-)|$$
$$\leq \tfrac{1}{6}\eta(x_0) + \tfrac{1}{6}\eta(x_0) = \tfrac{1}{3}\eta(x_0)$$

But since $\varphi(x_1) = (r_1, r_2)$,
$$\tfrac{2}{3}\eta(x_1) < |f(r_1) - f(r_2)| < \tfrac{4}{3}\eta(x_1)$$

Thus we have
$$\tfrac{2}{3}\eta(x_0) < |f(r_1) - f(r_2)| < \tfrac{4}{3}\eta(x_1) \leq \tfrac{4}{9}\eta(x_0)$$

a contradiction. Therefore we cannot have $\varphi(x_1) = \varphi(x_0)$ unless $x_1 = x_0$ and so φ is one-to-one.

Lemma 2 There exist only countably many points x_0 for which $\lim_{x \to x_0} f(x)$ exists but $\lim_{x \to x_0} f(x) \neq f(x_0)$.

PROOF Let F be the set of all real numbers x_0 such that $\lim_{x \to x_0} f(x)$ exists but $\lim_{x \to x_0} f(x) \neq f(x_0)$. For each $x_0 \in F$ we define $l(x_0) = \lim_{x \to x_0} f(x)$ and $\beta(x_0) = |f(x_0) - l(x_0)| > 0$ (see Fig. 4.10). We choose rationals r_1, r_2, r_3 such that $r_1 < x_0 < r_2$ and, for each $x \in [r_1, x_0) \cup (x_0, r_2]$, $|f(x) - l(x_0)| < \tfrac{1}{3}\beta(x_0)$ and such that $|f(x_0) - r_3| < \tfrac{1}{3}\beta(x_0)$. We define the function $\psi : F \to \mathbb{Q} \times \mathbb{Q} \times \mathbb{Q}$ by $\psi(x_0) = (r_1, r_2, r_3)$. In order to prove that F is a countable set it suffices to prove that ψ is one-to-one. Suppose there exists $x_1 \in F$ with $\psi(x_1) = (r_1, r_2, \hat{r}_3)$ and suppose further that $x_1 \neq x_0$. By the definition of ψ we have $r_1 < x_1 < r_2$ and $|f(x_1) - \hat{r}_3| < \tfrac{1}{3}\beta(x_1)$. Now $x \in [r_1, x_0) \cup (x_0, r_2]$ implies $|f(x) - l(x_0)| < \tfrac{1}{3}\beta(x_0)$, and since $x_1 \in (r_1, x_0) \cup (x_0, r_2)$, it follows that $|f(x_1) - l(x_0)| < \tfrac{1}{3}\beta(x_0)$ and $|l(x_1) - l(x_0)| \leq \tfrac{1}{3}\beta(x_0)$. Hence

$$|f(x_1) - \hat{r}_3| < \tfrac{1}{3}\beta(x_1) = \tfrac{1}{3}|f(x_1) - l(x_1)|$$
$$\leq \tfrac{1}{3}(|f(x_1) - l(x_0)| + |l(x_0) - l(x_1)|)$$
$$< \tfrac{1}{3}[\tfrac{1}{3}\beta(x_0) + \tfrac{1}{3}\beta(x_0)] = \tfrac{2}{9}\beta(x_0)$$

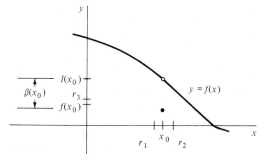

Figure 4.10

Therefore

$$|l(x_0) - \hat{r}_3| \leq |l(x_0) - f(x_1)| + |f(x_1) - \hat{r}_3|$$
$$< \tfrac{1}{3}\beta(x_0) + \tfrac{2}{9}\beta(x_0) = \tfrac{5}{9}\beta(x_0)$$

Now

$$|l(x_0) - f(x_0)| \leq |l(x_0) - \hat{r}_3| + |\hat{r}_3 - r_3| + |r_3 - f(x_0)|$$

and so

$$|\hat{r}_3 - r_3| \geq |l(x_0) - f(x_0)| - |l(x_0) - \hat{r}_3| - |r_3 - f(x_0)|$$
$$> \beta(x_0) - \tfrac{5}{9}\beta(x_0) - \tfrac{1}{3}\beta(x_0) = \tfrac{1}{9}\beta(x_0) > 0$$

Thus $\hat{r}_3 \neq r_3$; that is,

$$\psi(x_1) = (r_1, r_2, \hat{r}_3) \neq (r_1, r_2, r_3) = \psi(x_0)$$

and consequently ψ is one-to-one.

By Lemmas 1 and 2 and the remarks preceding Lemma 1 we have the following theorem.

Theorem 4.17 A function f defined on **R** can have at most countably many simple discontinuities.

Corollary If f has no discontinuities of the second kind then the set of points of discontinuity of f is countable.

Recall that a rational number p/q is called properly reduced if $q > 0$ and if p and q have no common integral factors other than ± 1. The following is an example of a function which has **Q** (the set of all rationals) as its set of points of discontinuity. We write $\mathcal{D}_f$ for the set of points of discontinuity of an arbitrary function f.

Example 4.16 Let

$$f(x) = \begin{cases} \dfrac{1}{q} & \text{if } x = \dfrac{p}{q} \neq 0 \text{ properly reduced} \\ 1 & \text{if } x = 0 \\ 0 & \text{if } x \text{ is irrational} \end{cases}$$

If x_0 is any rational number then $f(x_0) > 0$; since $f(x) = 0$ for every irrational number, it is clear that if $\lim_{x \to x_0} f(x)$ exists then it must be zero. Hence it follows that $\lim_{x \to x_0} f(x) = f(x_0)$ is impossible and so f has a discontinuity at each rational. We leave it to the reader to show that f is continuous at each irrational number (see Exercise 4.62).

In the next example we show that for an arbitrary countable set D there exists a function f which has the set D as its set of points of discontinuity; that is, $\mathcal{D}_f = D$.

Example 4.17 Let $D = \{x_1, x_2, x_3, \ldots\} \subset \mathbf{R}$ be an arbitrary countable set. We define f by

$$f(x) = \begin{cases} \dfrac{1}{n} & \text{if } x = x_n \text{ where } n = 1, 2, 3, \ldots \\ 0 & \text{otherwise} \end{cases}$$

Let x_0 be any real number and let $\varepsilon > 0$ be given. Choose $n_0 \in \mathbf{N}$ so large that $1/n_0 \leq \varepsilon$. Now choose $\delta > 0$ so small that $\{x_1, x_2, \ldots, x_{n_0}\} \cap N_\delta^*(x_0) = \varnothing$ (recall that $N_\delta^*(x_0) = (x_0 - \delta, x_0) \cup (x_0, x_0 + \delta)$). Clearly then if $x \in N_\delta^*(x_0)$, we have $|f(x)| \leq 1/(n_0 + 1) < \varepsilon$ and therefore $\lim_{x \to x_0} f(x) = 0$. Since x_0 was an arbitrary real number, it follows that f is continuous at each real number outside the set D and f has a simple discontinuity at each point of D.

The following example shows that we can even construct a monotone function which has an arbitrary countable set D as its set of points of discontinuity.

Example 4.18 Again let $D = \{x_1, x_2, x_3, \ldots\}$ be an arbitrary countable subset of $\mathbf{R}$; let $S = \{\alpha_1, \alpha_2, \alpha_3, \ldots\}$ be a set of positive real numbers which is summable; that is, $\sum_{i=1}^{\infty} \alpha_i < \infty$ (see Sec. 7.1). For each $x \in \mathbf{R}$ we define

$$f(x) = \sum_{x_j < x} \alpha_j$$

That is, $f(x)$ is the sum of all those α_j for which $x_j < x$. Now if $x < y$ then $\{x_j \in D \mid x_j < x\} \subset \{x_j \in D \mid x_j < y\}$ and so $f(x) = \sum_{x_j < x} \alpha_j \leq \sum_{x_j < y} \alpha_j = f(y)$. Hence f is monotone increasing on $\mathbf{R}$. Suppose $x_0 \notin D$ and let $\varepsilon > 0$ be given. Choose $k \in \mathbf{N}$ so large that $\sum_{j=k+1}^{\infty} \alpha_j < \varepsilon$. Now choose $\delta > 0$ so small that

$$\{x_1, x_2, \ldots, x_k\} \cap N_\delta(x_0) = \varnothing$$

Then if $x \in N_\delta(x_0)$, we have $|f(x) - f(x_0)| \leq \sum_{j=k+1}^{\infty} \alpha_j < \varepsilon$ and hence $\lim_{x \to x_0} f(x) = f(x_0)$. Therefore f is continuous at each real number outside D. We leave it to the reader (see Exercise 4.63) to show that for every $n \in \mathbf{N}$

$$f(x_n+) = \sum_{x_j \leq x_n} \alpha_j \quad \text{and} \quad f(x_n-) = \sum_{x_j < x_n} \alpha_j$$

Therefore $f(x_n +) - f(x_n -) = \alpha_n > 0$, and so f has a simple discontinuity at each element of D.

The reader may ask whether for an arbitrary set $E \subset \mathbf{R}$ there is a function which has the set E as its set of points of discontinuity. In the preceding examples we have seen that if the set E is countable then such functions do exist; in general, however, the answer to the above question is no. In the following we show that in order for a set E to be the set of points of discontinuity for some function f it is necessary that E be an F_σ set; that is, E must be equal to a countable union of closed sets.

Let $f: \mathbf{R} \to \mathbf{R}$ and let $\mathfrak{D}_f$ be the set of points of discontinuity for f. Suppose $x_0 \in \mathbf{R}$ and $\delta > 0$. We adopt the notation

$$\varphi_f[x_0 - \delta, x_0 + \delta] = \sup_x f(x) - \inf_x f(x) \quad \text{where } x \in [x_0 - \delta, x_0 + \delta]$$

It is clear that if $0 < \delta_1 \leq \delta_2$ then $[x_0 - \delta_1, x_0 + \delta_1] \subseteq [x_0 - \delta_2, x_0 + \delta_2]$ and so

$$\varphi_f[x_0 - \delta_1, x_0 + \delta_1] \leq \varphi_f[x_0 - \delta_2, x_0 + \delta_2]$$

Definition 4.6 $\omega_f(x_0) = \inf_{\delta > 0} \varphi_f[x_0 - \delta, x_0 + \delta]$ is called the *oscillation* of f at x_0.

Since $\varphi_f[x_0 - \delta, x_0 + \delta] \geq 0$ for every $\delta > 0$, it is clear that $\omega_f(x_0)$ exists and is nonnegative; also $\omega_f(x_0) < \infty$ if and only if f is bounded at x_0.

Theorem 4.18 f is continuous at x_0 if and only if $\omega_f(x_0) = 0$.

PROOF Suppose f is continuous at x_0 and let $\varepsilon > 0$ be given. There exists a $\delta > 0$ such that $|f(x) - f(x_0)| < \varepsilon/2$ for every $x \in (x_0 - \delta, x_0 + \delta)$. Consequently for all $x_1, x_2 \in [x_0 - \delta/2, x_0 + \delta/2] \subset (x_0 - \delta, x_0 + \delta)$ we have

$$|f(x_1) - f(x_2)| \leq |f(x_1) - f(x_0)| + |f(x_0) - f(x_2)| < \frac{\varepsilon}{2} + \frac{\varepsilon}{2} = \varepsilon$$

Therefore $\varphi_f[x_0 - \delta/2, x_0 + \delta/2] \leq \varepsilon$, and so $\omega_f(x_0) \leq \varepsilon$. Since $\varepsilon > 0$ was given arbitrarily, it follows that $\omega_f(x_0) = 0$.

Conversely, suppose $\omega_f(x_0) = 0$ and let $\varepsilon > 0$ be given. Since $\omega_f(x_0) < \varepsilon$, there exists $\delta > 0$ such that $\varphi_f[x_0 - \delta, x_0 + \delta] < \varepsilon$. Hence for each $x \in [x_0 - \delta, x_0 + \delta]$ we have $|f(x) - f(x_0)| < \varepsilon$. Thus $|f(x) - f(x_0)| < \varepsilon$ whenever $|x - x_0| < \delta$, and so f is continuous at x_0.

Let $\eta > 0$ be any positive real number and let $A_\eta = \{x \in \mathbf{R} \mid \omega_f(x) \geq \eta\}$.

Theorem 4.19 For each $\eta > 0$, A_η is a closed set.

PROOF Suppose x_1 is a limit point of A_η; that is, every neighborhood of x_1 contains infinitely many points of A_η. Consequently for any $\delta > 0$, $(x_1 - \delta, x_1 + \delta)$ contains a point $x_0 \in A_\eta$. Choose $\delta_1 > 0$ so small that

$$[x_0 - \delta_1, x_0 + \delta_1] \subset [x_1 - \delta, x_1 + \delta]$$

Now $\omega_f(x_0) \geq \eta$, and so $\phi_f[x_0 - \delta_1, x_0 + \delta_1] \geq \eta$. But since

$$[x_0 - \delta_1, x_0 + \delta_1] \subset [x_1 - \delta, x_1 + \delta]$$

it follows that $\varphi_f[x_1 - \delta, x_1 + \delta] \geq \eta$. Since $\delta > 0$ was arbitrary, we must have $\omega_f(x_1) \geq \eta$. Thus $x_1 \in A_\eta$, and so A_η contains all its limits points; that is, A_η is a closed set.

Theorem 4.20 For any function f, $\mathfrak{D}_f$, the set of all points of discontinuity of f, is an F_σ set.

PROOF By Theorem 4.18, $\mathfrak{D}_f = \bigcup_{n=1}^\infty A_{1/n}$, where

$$A_{1/n} = \{x \in \mathbf{R} \mid \omega_f(x) \geq 1/n\} \quad (n = 1, 2, 3, \ldots),$$

and, by Theorem 4.19, each $A_{1/n}$ is a closed set.

Theorem 4.20 shows that not every subset of $\mathbf{R}$ can be the set of points of discontinuity for some function. In the following we show that in particular the set of all irrational real numbers $\mathbf{R} - \mathbf{Q}$ is not equal to a countable union of closed sets and so is the set of points of discontinuity for no function $f: \mathbf{R} \to \mathbf{R}$.

Definition 4.7 A set $A \subseteq \mathbf{R}$ is called *nowhere dense* if for each interval I there is an interval $I_1 \subseteq I$ such that $I_1 \cap A = \emptyset$.

It is clear that if A is an interval or contains an interval then A is not nowhere dense. But there are sets which contain no interval which also fail to be nowhere dense; for example, the rationals $\mathbf{Q}$ and the irrationals $\mathbf{R} - \mathbf{Q}$.

Lemma 3 A closed set either contains an interval or else is nowhere dense.

PROOF Suppose E is a closed set and E is not nowhere dense. Then there is some interval I_0 such that for each interval $I \subseteq I_0$ we have $I \cap E \neq \emptyset$. We show that $I_0 \subseteq E$. Let $x_0 \in I_0$. Then every neighborhood of x_0 contains within it at least one point of E. But this implies that either $x_0 \in E$ or else x_0 is a limit point of E. Since E is closed, it contains all its limit points and so necessarily $x_0 \in E$.

Definition 4.8 A set $B \subseteq \mathbf{R}$ is said to be of the *first category* if B is expressible as a countable union of nowhere dense sets. If a set is not of the first category, it is said to be of the *second category*.

It is clear that if B_1 and B_2 are each of the first category then $B_1 \cup B_2$ is of the first category; in fact, if we have a countable number of sets $B_1, B_2, B_3, \ldots$ each of the first category then $\bigcup_{n=1}^{\infty} B_n$ is also of the first category.

Theorem 4.21 $\mathbf{R}$ is of the second category.

PROOF Let $B \subseteq \mathbf{R}$ be an arbitrary set of the first category. Then $B = \bigcup_{n=1}^{\infty} A_n$ where each set A_n ($n = 1, 2, 3, \ldots$) is nowhere dense. There exists a closed interval I_1 with length less than 1 such that $I_1 \cap A_1 = \emptyset$. This is immediate from the fact that A_1 is nowhere dense. Since A_2 is nowhere dense, there is a closed interval $I_2 \subset I_1$ with length less than $\frac{1}{2}$ such that $I_2 \cap A_2 = \emptyset$. Similarly since A_3 is nowhere dense, there is a closed interval $I_3 \subset I_2$ with length less than $\frac{1}{3}$ such that $I_3 \cap A_3 = \emptyset$. Continuing this process, we get a sequence of closed intervals $\{I_n\}$ such that for each natural number n we have $I_{n+1} \subset I_n$, the length of I_n is less than $1/n$, and $I_n \cap A_n = \emptyset$. By the nested-interval property $\bigcap_{n=1}^{\infty} I_n = \{x_0\}$, $x_0 \in \mathbf{R}$. Since $x_0 \in I_n$ for each $n = 1, 2, 3 \ldots$ and $I_n \cap A_n = \emptyset$, it follows that $x_0 \notin A_n$ for each n, and so $x_0 \notin B$. Hence $B \neq \mathbf{R}$, and since B was an arbitrary set of the first category, it follows that $\mathbf{R}$ is of the second category.

Corollary Every interval $I \subseteq \mathbf{R}$ is of the second category.

Lemma 4 An F_σ set either contains an interval or else is of the first category.

PROOF Suppose F is an F_σ set. Then we write $F = \bigcup_{n=1}^{\infty} F_n$, where each F_n ($n = 1, 2, 3, \ldots$) is a closed set. Now if F is not of the first category then at least one of the sets F_n is not nowhere dense. By Lemma 3 this set will necessarily contain an interval and hence so will F.

Theorem 4.22 $\mathbf{R} - \mathbf{Q}$ is of the second category.

PROOF Clearly $\mathbf{Q}$ is of the first category; in fact each countable set is of the first category since a singleton set is clearly nowhere dense. Now if $\mathbf{R} - \mathbf{Q}$ were of the first category then $\mathbf{R} = \mathbf{Q} \cup (\mathbf{R} - \mathbf{Q})$ would also be of the first category, in contradiction to Theorem 4.21. Hence $\mathbf{R} - \mathbf{Q}$ is of the second category.

Corollary $\mathbf{R} - \mathbf{Q}$ is not an F_σ set.

PROOF $\mathbf{R} - \mathbf{Q}$ is not of the first category by the preceding theorem, and since $\mathbf{R} - \mathbf{Q}$ clearly contains no interval, it follows from Lemma 4 that $\mathbf{R} - \mathbf{Q}$ cannot be an F_σ set.

Theorem 4.23 There exists no function $f : \mathbf{R} \to \mathbf{R}$ for which $\mathcal{D}_f = \mathbf{R} - \mathbf{Q}$.

PROOF This follows immediately from Theorem 4.20 and the above corollary.

EXERCISES

4.62 Prove that the function in Example 4.16 is continuous at each irrational number and has a simple discontinuity at each rational number.

4.63 For the function in Example 4.18 prove that for each natural number n

$$\lim_{x \to x_n^+} f(x) = \sum_{x_j \le x_n} \alpha_j \qquad \lim_{x \to x_n^-} f(x) = \sum_{x_j < x_n} \alpha_j$$

and hence that f is left-continuous at each x_n but is not right-continuous there.

4.64 Let E be any closed subset of $\mathbf{R}$ and define

$$f(x) = \begin{cases} 1 & \text{if } x \in E \cap \mathbf{Q} \\ -1 & \text{if } x \in E - \mathbf{Q} \\ 0 & \text{if } x \in \mathbf{R} - E \end{cases}$$

where $\mathbf{Q}$ is the set of rationals. Prove that $\mathcal{D}_f = E$.

4.65 Let F be any F_σ set. Find a function f such that $\mathcal{D}_f = F$.

4.66 Let D be an arbitrary countably infinite subset of $[0, 1]$ such that D is not a closed set. Define

$$f(x) = \begin{cases} 1 & \text{if } x \in D \\ 0 & \text{otherwise} \end{cases}$$

Show that there is a point $x_0 \notin D$ such that f is discontinuous at x_0.

4.67 Prove the corollary to Theorem 4.21.

CHAPTER
FIVE

DIFFERENTIABLE FUNCTIONS

5.1 THE DERIVATIVE

In this chapter we study the derivative of a function and the properties of differentiable functions. The reader may recall from elementary calculus that the derivative of a function f is a new function f' which represents the rate of change of f as x changes. It therefore has application in any discipline where change is measured.

We assume f is defined in a nieghborhood of x_0.

Definition 5.1 The *derivative* of f at x_0 is

$$f'(x_0) = \lim_{h \to 0} \frac{f(x_0 + h) - f(x_0)}{h}$$

provided the limit exists. When the limit exists, we say that f is *differentiable* at x_0.

The derivative of a function f is again a function f'; its domain, which is a subset of the domain of f, is the set of all points x_0 for which f is differentiable. Other notations for the derivative are $D_x f$, df/dx, and dy/dx, where $y = f(x)$.

Example 5.1 Let $f(x) = x|x|$ for all $x \in \mathbf{R}$. Then $f(x) = x^2$ if $x \geq 0$ and

$f(x) = -x^2$ for $x < 0$. If $x_0 > 0$ then

$$f'(x_0) = \lim_{h \to 0} \frac{f(x_0 + h) - f(x_0)}{h} = \lim_{h \to 0} \frac{(x_0 + h)^2 - x_0^2}{h}$$

since $x_0 + h > 0$ when $|h|$ is sufficiently small. Hence

$$f'(x_0) = \lim_{h \to 0} \frac{2x_0 h + h^2}{h} = \lim_{h \to 0} (2x_0 + h) = 2x_0$$

If $x_0 < 0$ then

$$f'(x_0) = \lim_{h \to 0} \frac{f(x_0 + h) - f(x_0)}{h} = \lim_{h \to 0} \frac{-(x_0 + h)^2 + x_0^2}{h}$$

since $x_0 + h < 0$ when $|h|$ is sufficiently small. Thus

$$f'(x_0) = \lim_{h \to 0} \frac{-2x_0 h - h^2}{h} = \lim_{h \to 0} (-2x_0 - h) = -2x_0$$

Finally,

$$f'(0) = \lim_{h \to 0} \frac{f(0 + h) - f(0)}{h} = \lim_{h \to 0} \frac{h|h|}{h} = \lim_{h \to 0} |h| = 0$$

We summarize by writing $f'(x) = 2|x|$ for every real number x.

The function f' may in turn have a derivative, denoted by f'', which is defined at all points where f' is differentiable. f'' is called the *second derivative* of f. We leave it for the reader (see Exercise 5.2) to show that for the function $f(x) = x|x|$ of Example 5.1, $f''(x) = 2$ if $x > 0$, $f''(x) = -2$ if $x < 0$, and $f''(0)$ does not exist.

Recall from the previous chapter that $f'(x) = 2|x|$ is continuous on **R**, and in particular at $x_0 = 0$. Therefore a function can be continuous at a point even though it is not differentiable there.

In the difference quotient $[f(x_0 + h) - f(x_0)]/h$ if we replace $x_0 + h$ by x, that is, let $x = x_0 + h$, we obtain

$$\frac{f(x_0 + h) - f(x_0)}{h} = \frac{f(x) - f(x_0)}{x - x_0}$$

h is within δ of 0 if and only if x is within δ of x_0. Thus an equivalent definition for the derivative of $f(x)$ at x_0 is

$$f'(x_0) = \lim_{x \to x_0} \frac{f(x) - f(x_0)}{x - x_0}$$

Theorem 5.1 If f is differentiable at x_0 then f is continuous at x_0.

PROOF Recall that f is continuous at x_0 if $\lim_{x \to x_0} f(x) = f(x_0)$ or equivalently $\lim_{x \to x_0} [f(x) - f(x_0)] = 0$.

We assume that f is differentiable at x_0. Then

$$\lim_{x \to x_0} [f(x) - f(x_0)] = \lim_{x \to x_0} \frac{f(x) - f(x_0)}{x - x_0} (x - x_0)$$

$$= \lim_{x \to x_0} \frac{f(x) - f(x_0)}{x - x_0} \lim_{x \to x_0} (x - x_0)$$

$$= f'(x_0) \cdot 0 = 0$$

We assume that the reader is familiar with the geometrical interpretation of the derivative $f'(x_0)$ as the slope of the line tangent to the graph of $y = f(x)$ at the point $(x_0, f(x_0))$. We also assume familiarity with the differentiation of polynomials and the transcendental functions from introductory calculus, and with the basic rules of differentiation:

1. $(f + g)'(x) = f'(x) + g'(x)$.
2. $(kf)'(x) = kf'(x)$.
3. $(fg)'(x) = f'(x) g(x) + f(x) g'(x)$.
4. $(f/g)'(x) = [g(x) f'(x) - f(x) g'(x)]/[g(x)]^2$

whenever the right-hand sides of the equations exist.

Example 5.2 Let

$$f(x) = \begin{cases} x^3 & \text{for } x < 1 \\ 2x - 1 & \text{for } x \geq 1 \end{cases}$$

At points $x < 1$ the derivative will involve only the x^3 formula for $f(x)$. Thus $f'(x) = 3x^2$ for $x < 1$. Similarly for points $x > 1$, $f'(x) = 2$. The derivative at $x = 1$ is a (two-sided) limit and therefore involves both formulas

$$\lim_{x \to 1^-} \frac{f(x) - f(1)}{x - 1} = \lim_{x \to 1^-} \frac{x^3 - 1}{x - 1} = \lim_{x \to 1^-} (x^2 + x + 1) = 3$$

$$\lim_{x \to 1^+} \frac{f(x) - f(1)}{x - 1} = \lim_{x \to 1^+} \frac{(2x - 1) - 1}{x - 1} = \lim_{x \to 1^+} \frac{2(x - 1)}{x - 1} = 2$$

Hence

$$f'(1) = \lim_{x \to 1} \frac{f(x) - f(1)}{x - 1}$$

does not exist. In Sec. 5.4 we discuss this type of function in connection with the notion of one-sided derivatives.

Recall that for two functions f and g the composition function

$h = g \circ f$ is defined at each x_0 in the domain of f for which $f(x_0)$ is in the domain of g and at such a point x_0

$$h(x_0) = (g \circ f)(x_0) = g[f(x_0)]$$

In Chap. 4 we showed that if f is continuous at x_0 and g is continuous at $f(x_0)$ then $h = g \circ f$ is continuous at x_0. The analogous condition for differentiability is called the *chain rule*, the precise statement of which is given in the following theorem.

Theorem 5.2 If f is differentiable at x_0 and g is differentiable at $y_0 = f(x_0)$ then $h = g \circ f$ is differentiable at x_0 and

$$h'(x_0) = (g \circ f)'(x_0) = g'[f(x_0)] \cdot f'(x_0) = g'(y_0) \cdot f'(x_0)$$

PROOF Let f be differentiable at x_0 and g be differentiable at $y_0 = f(x_0)$. It is assumed, of course, that f is defined in some neighborhood of x_0 and that g is defined in some neighborhood of $y_0 = f(x_0)$. By Theorem 5.1, f is continuous at x_0 and g is continuous at $y_0 = f(x_0)$. Thus, by Theorem 4.3, $h = g \circ f$ is continuous at x_0.

Since f is differentiable at x_0 we can write

$$f(x) - f(x_0) = (x - x_0)[f'(x_0) + \eta(x)]$$

Here we are defining

$$\eta(x) = \frac{f(x) - f(x_0)}{x - x_0} - f'(x_0)$$

$\eta(x)$ exists in a deleted neighborhood of x_0, and $\eta(x) \to 0$ as $x \to x_0$ since

$$\lim_{x \to x_0} \frac{f(x) - f(x_0)}{x - x_0} = f'(x_0)$$

exists.

Similarly, since g is differentiable at $y_0 = f(x_0)$,

$$g(y) - g(y_0) = (y - y_0)[g'(y_0) + v(y)]$$

where $v(y) \to 0$ as $y \to y_0$. Now,

$$\begin{aligned} h(x) - h(x_0) &= (g \circ f)(x) - (g \circ f)(x_0) = g[f(x)] - g[f(x_0)] \\ &= g(y) - g(y_0) = (y - y_0)[g'(y_0) + v(y)] \\ &= [f(x) - f(x_0)][g'(f(x_0)) + v(f(x))] \\ &= (x - x_0)[f'(x_0) + \eta(x)][g'(y_0) + v(y)] \end{aligned}$$

If $x \neq x_0$, we get

$$\frac{h(x) - h(x_0)}{x - x_0} = [g'(y_0) + v(y)][f'(x_0) + \eta(x)]$$

If we take the limit as $x \to x_0$ and note that f is continuous at x_0 by Theorem 5.1, we get that $y = f(x) \to f(x_0) = y_0$. Hence $\eta(x) \to 0$ and $v(y) \to 0$. Therefore

$$h'(x_0) = \lim_{x \to x_0} \frac{h(x) - h(x_0)}{x - x_0} = g'(y_0) \cdot f'(x_0)$$

EXERCISES

5.1 If $f(x) = |x^3|$, find $f'(x)$.

5.2 Let $f(x) = x|x|$; show that

$$f''(x) = \begin{cases} 2 & \text{if } x > 0 \\ -2 & \text{if } x < 0 \end{cases}$$

and that 0 is not in the domain of $f''(x)$.

5.3 Find $f'(x)$ if

$$f(x) = \begin{cases} x^2 & \text{if } x \geq 2 \\ 4x - 4 & \text{if } x < 2 \end{cases}$$

5.4 For what values of a and b is $f(x)$ differentiable at $x = 1$ if

$$f(x) = \begin{cases} x^3 & x < 1 \\ ax + b & x \geq 1 \end{cases}$$

5.5 Show that the following function is continuous at $x = 0$ but $f'(0)$ does not exist. Find $f'(x)$ for $x \neq 0$.

$$f(x) = \begin{cases} x \sin \dfrac{1}{x} & \text{if } x \neq 0 \\ 0 & \text{if } x = 0 \end{cases}$$

5.6 Show that the following function is differentiable for every $x \in \mathbf{R}$ but f' is not continuous at $x = 0$.

$$f(x) = \begin{cases} x^2 \sin \dfrac{1}{x} & \text{if } x \neq 0 \\ 0 & \text{if } x = 0 \end{cases}$$

5.7 Show that the following function is differentiable for every $x \in \mathbf{R}$ but f' is unbounded at $x = 0$.

$$f(x) = \begin{cases} x^2 \sin \dfrac{1}{x^2} & \text{if } x \neq 0 \\ 0 & \text{if } x = 0 \end{cases}$$

5.8 Find a function $f: \mathbf{R} \to \mathbf{R}$ which is differentiable at exactly one point.

5.9 Let $f(x) = e^{-|x|}$ for every $x \in \mathbf{R}$. Is f continuous at $x = 0$? Differentiable at $x = 0$?

5.10 Let $f(x) = [\![x]\!](x - 1)$ for every $x \in \mathbf{R}$. Where is f continuous? Differentiable?

5.11 (a) Define

$$f(x) = \begin{cases} \dfrac{1}{4^n} & \text{if } x = \dfrac{1}{2^n} \quad (n = 1, 2, 3, \ldots) \\ 0 & \text{otherwise} \end{cases}$$

Is f differentiable at $x = 0$? Verify.

(b) Define

$$g(x) = \begin{cases} \dfrac{1}{2^{n+1}} & \text{if } x = \dfrac{1}{2^n} \quad (n = 1, 2, 3, \ldots) \\ 0 & \text{otherwise.} \end{cases}$$

Is g differentiable at $x = 0$? Verify.

5.12 Prove that if $x = g(t)$, $y = f(t)$ are differentiable in some neighborhood of t_0, $g'(t_0) \neq 0$ and $x_0 = g(t_0)$, $y_0 = f(t_0)$ and if $y = H(x)$ in some neighborhood of x_0 then

$$\left.\frac{dy}{dx}\right|_{x_0} = \frac{f'(t_0)}{g'(t_0)} = \frac{dy/dt\,|_{t_0}}{dx/dt\,|_{t_0}}$$

5.2 PROPERTIES OF DIFFERENTIABLE FUNCTIONS

In this section we study some important properties of differentiable functions: in particular, we shall be concerned with the relationship between f and f'. We say that f is differentiable on the set A if f is differentiable at each point in A.

We begin by investigating maxima and minima of a function f and see how they are related to f'. Such considerations are extremely important in the various applications.

Definition 5.2 $f(x_0)$ is a *local maximum* of the function f if for all x in some neighborhood of x_0 we have $f(x) \leq f(x_0)$. Similarly, $f(x_0)$ is a *local minimum* of the function f if for all x in some neighborhood of x_0 we have $f(x) \geq f(x_0)$.

Definition 5.3 Let x_0 be an element in the interval I. $f(x_0)$ is the *absolute maximum* of f on I if $f(x_0) \geq f(x)$ for all $x \in I$. Similarly, $f(x_0)$ is the *absolute minimum* of f on I if $f(x_0) \leq f(x)$ for all $x \in I$.

It is clear from the above definitions that if x_0 is an interior point of the interval I and $f(x_0)$ is the absolute maximum of $f(x)$ on I then $f(x_0)$ is a local maximum of f. A similar statement holds for minima.

Theorem 5.3 If $f(x_0)$ is a local extremum (maximum or minimum) then either $f'(x_0) = 0$ or $f'(x_0)$ does not exist.

PROOF Suppose $f(x_0)$ is a local maximum (a similar proof holds for the case where $f(x_0)$ is a local minimum). Then there is a $\delta > 0$ such that for every $x \in N_\delta(x_0)$, $f(x) \leq f(x_0)$. Hence

$$\frac{f(x) - f(x_0)}{x - x_0} \begin{cases} \leq 0 & \text{if } x \text{ satisfies } x_0 < x < x_0 + \delta \\ \geq 0 & \text{if } x \text{ satisfies } x_0 - \delta < x < x_0 \end{cases}$$

Now if $f'(x_0)$ exists then necessarily

$$\lim_{x \to x_0^+} \frac{f(x) - f(x_0)}{x - x_0} = f'(x_0) = \lim_{x \to x_0^-} \frac{f(x) - f(x_0)}{x - x_0}$$

But by the above

$$\lim_{x \to x_0^+} \frac{f(x) - f(x_0)}{x - x_0} \leq 0 \quad \text{and} \quad \lim_{x \to x_0^-} \frac{f(x) - f(x_0)}{x - x_0} \geq 0$$

It follows that $f'(x_0) = 0$.

The function $f(x) = |x|$ provides an example of a function which has a local extremum at a point ($f(x) = |x|$ has a local minimum at $x = 0$) where the function fails to be differentiable (see Fig. 5.1a).

We also note that the converse to Theorem 5.3 does not hold; that is, if $f'(x_0) = 0$, we do not necessarily have a local extremum of f at $x = x_0$. To see this consider the function $f(x) = x^3$ (see Fig. 5.1b). $f'(0) = 0$, but f is strictly increasing on **R** and so has no local extremum at $x = 0$.

The following theorem forms the foundation for the rest of this section and is an important result in differential calculus.

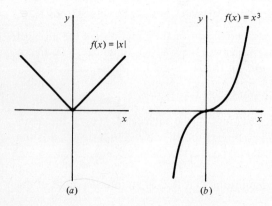

Figure 5.1

Theorem 5.4: Rolle's theorem If $f(x)$ is continuous on $[a, b]$ and differentiable on (a, b) and $f(a) = f(b)$ then there exists an $x_0 \in (a, b)$ such that $f'(x_0) = 0$.

PROOF If $f(x) = f(a)$ for every $x \in (a, b)$ then f is constant on $[a, b]$ and so $f'(x) = 0$ for every $x \in (a, b)$. Hence we can assume that there is some $x \in (a, b)$ for which $f(x) \neq f(a)$. By the extreme-value theorem f assumes its absolute-maximum and absolute-minimum values on $[a, b]$; that is, there exists $x_1, x_2 \in [a, b]$ such that $f(x_1) \geq f(x) \geq f(x_2)$ for all $x \in [a, b]$. By our assumption that f is not constant on $[a, b]$ together with the fact that $f(a) = f(b)$ it follows that there is a point $x_0 \in (a, b)$ such that f has an absolute extremum at x_0. By our remarks preceding Theorem 5.3 $f(x_0)$ is a local extremum of f. Also $f'(x_0)$ exists by hypothesis. Thus, by Theorem 5.3, $f'(x_0) = 0$.

Rolle's theorem applies to the function $f(x) = x^3 - 3x^2$ on the interval $[0, 3]$ since f is a polynomial (and hence differentiable on **R**) and $f(0) = f(3)$. Now $f'(x) = 3x^2 - 6x$ and there are two roots, $f'(0) = 0$ and $f'(2) = 0$. But $0 \notin (0, 3)$, and so $x_0 = 2$; that is, $x_0 = 2$ (but not 0) is the point whose existence is guaranteed by Rolle's theorem.

Example 5.3 Let

$$g(x) = \begin{cases} x & \text{if } x \in [0, 1) \\ 0 & \text{if } x = 1 \end{cases}$$

g satisfies all the hypotheses of Rolle's theorem except the left continuity at the right endpoint $x = 1$. Since $g'(x) = 1$ for every $x \in (0, 1)$, it is clear that there is no $x_0 \in (0, 1)$ with $g'(x_0) = 0$. In other words, the singular loss of continuity at an endpoint is enough to cause the failure of the conclusion of Rolle's theorem.

Theorem 5.5 Cauchy mean-value theorem If f and g are each continuous on $[a, b]$ and differentiable on (a, b) then there exists an $x_0 \in (a, b)$ such that

$$f'(x_0)[g(b) - g(a)] = g'(x_0)[f(b) - f(a)]$$

PROOF Let $F(x) = f(x)[g(b) - g(a)] - g(x)[f(b) - f(a)]$. F is continuous on $[a, b]$ and differentiable on (a, b). Also

$$F(b) = f(b)[g(b) - g(a)] - g(b)[f(b) - f(a)]$$
$$= f(a)g(b) - g(a)f(b)$$
$$= f(a)[g(b) - g(a)] - g(a)[f(b) - f(a)] = F(a)$$

By Rolle's theorem there exists an $x_0 \in (a, b)$ with $F'(x_0) = 0$. But

$F'(x) = f'(x)[g(b) - g(a)] - g'(x)[f(b) - f(a)]$. Hence

$$f'(x_0)[g(b) - g(a)] = g'(x_0)[f(b) - f(a)]$$

The Cauchy mean-value theorem has an interesting geometrical interpretation. If we have $x = g(t)$ and $y = f(t)$, where g and f are continuous on $[a, b]$ and differentiable on (a, b), then $(g(t), f(t))$, $a \le t \le b$, represents a "smooth curve" in the (x, y) plane. We assume that $g(a) \ne g(b)$ and that there is no $t \in (a, b)$ for which $g'(t) = f'(t) = 0$. $[f(b) - f(a)]/[g(b) - g(a)]$ is the slope of the secant line connecting the two points $(g(a), f(a))$ and $(g(b), f(b))$ on the curve. By the chain rule

$$\frac{dy}{dx} = \frac{dy/dt}{dx/dt} = \frac{f'(t)}{g'(t)}$$

which is the slope of the tangent line to the curve (see Exercise 5.12.)

The Cauchy mean-value theorem says that for some $t_0 \in (a, b)$

$$\frac{f(b) - f(a)}{g(b) - g(a)} = \frac{f'(t_0)}{g'(t_0)}$$

Consequently there is a point $(g(t_0), f(t_0))$ on the curve such that the tangent line to the curve at this point is parallel to the secant line connecting the two points $(g(a), f(a))$ and $(g(b), f(b))$ (see Fig. 5.2).

The following theorem is rich in applications, and we shall have many opportunities to apply it in the chapters ahead. The result follows immediately from the Cauchy mean value theorem by taking $g(x) = x$.

Theorem 5.6: Mean-value theorem of differential calculus If f is continuous on $[a, b]$ and differentiable on (a, b) then there exists a point $x_0 \in (a, b)$ such that

$$f'(x_0) = \frac{f(b) - f(a)}{b - a}$$

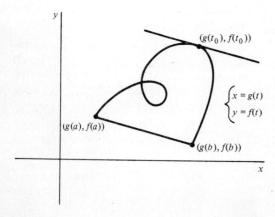

Figure 5.2

The following theorem, which is our first application of the mean-value theorem, is useful in graphing functions.

Theorem 5.7 If f is differentiable on (a, b) and $f'(x) \geq 0$ for every $x \in (a, b)$ then f is monotone increasing on (a, b).

PROOF Suppose $x_1, x_2 \in (a, b)$ with $x_1 < x_2$. Then f is continuous on $[x_1, x_2]$ and is differentiable on (x_1, x_2). By the mean value theorem there is an x_0 with $x_1 < x_0 < x_2$ such that

$$f'(x_0) = \frac{f(x_2) - f(x_1)}{x_2 - x_1}$$

Since $x_2 - x_1 > 0$ and $f'(x_0) \geq 0$ by hypothesis, it follows that $f(x_2) - f(x_1) \geq 0$; that is, $f(x_1) \leq f(x_2)$. Hence f is monotone increasing on (a, b).

A similar result holds when $f'(x) \leq 0$ on an interval (a, b) (see Exercise 5.14).

One of the important aspects of the mean-value theorem other than its wide range of applications is that it relates the derivative to the function in an intricate way. We can obtain information about the function from properties of the derivative, as evidenced by the following theorem.

Theorem 5.8 If f' exists and is bounded on some interval I then f is uniformly continuous on I.

PROOF Assume that there is an $M > 0$ such that $|f'(x)| \leq M$ for all $x \in I$. Then for any two points $x_1, x_2 \in I$ with $x_1 < x_2$ there exists $x_0 \in (x_1, x_2)$ such that $f'(x_0) = [f(x_2) - f(x_1)]/(x_2 - x_1)$. Here we have applied the mean-value theorem to f on $[x_1, x_2]$. Consequently

$$\left| \frac{f(x_2) - f(x_1)}{x_2 - x_1} \right| \leq M$$

and so $|f(x_2) - f(x_1)| \leq M|x_2 - x_1|$. Now since x_1 and x_2 were arbitrary points in I, it follows that $|f(x_2) - f(x_1)| \leq M|x_2 - x_1|$ for all $x_1, x_2 \in I$. To show that f is uniformly continuous on I let $\varepsilon > 0$ be given. Choose $\delta = \varepsilon/M$. Then if $x_1, x_2 \in I$ with $|x_2 - x_1| < \delta$, we have $|f(x_2) - f(x_1)| \leq M|x_2 - x_1| < M\delta = \varepsilon$.

Example 5.4 Let $f(x) = \tan^{-1} x$ on $\mathbf{R} = (-\infty, \infty)$. $f'(x) = 1/(x^2 + 1)$ for all $x \in I$, and clearly

$$|f'(x)| = \left| \frac{1}{x^2 + 1} \right| \leq 1$$

Thus by Theorem 5.8 f is uniformly continuous on **R**. It is also true that f has a uniformly continuous derivative on **R**. $f''(x) = -2x/(x^2 + 1)^2$ for every $x \in \mathbf{R}$. If $x \in [-1, 1]$ then

$$|f''(x)| = \left|\frac{2x}{(x^2 + 1)^2}\right| \leq |2x| \leq 2$$

and if $|x| > 1$ then

$$|f''(x)| = \left|\frac{2x}{(x^2 + 1)^2}\right| \leq \left|\frac{2x}{x^4}\right| = \left|\frac{2}{x^3}\right| < 2$$

Therefore $|f''(x)| \leq 2$ for every $x \in \mathbf{R}$, and, by Theorem 5.8, f' is uniformly continuous on $\mathbf{R} = (-\infty, \infty)$.

A useful concept in the theory of differential equations is that of a Lipschitz condition, which is also related to the mean-value theorem.

Definition 5.4 A function f is said to satisfy a *Lipschitz condition* on an interval I if there is a constant $M > 0$ such that

$$|f(x_1) - f(x_2)| \leq M|x_1 - x_2| \qquad \text{for all } x_1, x_2 \in I$$

f satifies a Lipschitz condition on I provided the difference quotient $[f(x_1) - f(x_2)]/(x_1 - x_2)$ remains bounded as x_1 and x_2 vary over I with $x_1 \neq x_2$. The following theorem tells us that for differentiable functions the Lipschitz condition is equivalent to boundedness of the derivative.

Theorem 5.9 If f' exists and is bounded on I then f satisfies a Lipschitz condition on I. Conversely, if f is differentiable on I and satisfies a Lipschitz condition on I then f' is bounded on I.

PROOF Suppose $|f'(x)| \leq M$ for every $x \in I$. Let x_1, x_2 be any two distinct points in I, say $x_1 < x_2$. By the mean-value theorem there exists $x_0 \in (x_1, x_2)$ such that

$$f'(x_0) = \frac{f(x_2) - f(x_1)}{x_2 - x_1}$$

Then

$$\left|\frac{f(x_2) - f(x_1)}{x_2 - x_1}\right| = |f'(x_0)| \leq M$$

and so

$$|f(x_2) - f(x_1)| \leq M|x_2 - x_1|$$

Hence f satisfies a Lipschitz condition on I.

Conversely suppose f is differentiable on I and $|f(x_1) - f(x_2)| \leq M|x_1 - x_2|$ for all $x_1, x_2 \in I$. Then for any $x_0 \in I$

$$f'(x_0) = \lim_{x \to x_0} \frac{f(x) - f(x_0)}{x - x_0}$$

and since

$$\left| \frac{f(x) - f(x_0)}{x - x_0} \right| \leq M \qquad \text{for all } x \neq x_0, x \in I$$

it follows that $|f'(x_0)| \leq M$. Thus f' is bounded on I.

Of course a function can satisfy a Lipschitz condition on an interval I without being differentiable on I. Consider, for example, $f(x) = |x|$ on $(-\infty, \infty)$. On the other hand, if a function satisfies a Lipschitz condition on I then it is necessarily continuous on I. In fact we have the following stronger result, the proof of which we leave to the reader (see Exercise 5.20). If f satisfies a Lipschitz condition on I then f is uniformly continuous on I.

Still another consequence of the Lipschitz condition is the following theorem.

Theorem 5.10 If f is defined on the interval $[a, b]$ and $|f(x_1) - f(x_2)| \leq k|x_1 - x_2|$ for all x_1, x_2, where $0 < k < 1$, and $f(x) \in [a, b]$ for every $x \in [a, b]$ then f has exactly one fixed point in $[a, b]$; that is, there exists uniquely an $x_0 \in [a, b]$ with $f(x_0) = x_0$.

PROOF Suppose $f: [a, b] \to [a, b]$ and $|f(x_1) - f(x_2)| \leq k|x_1 - x_2|$ for fixed $k \in (0, 1)$ and for all $x_1, x_2 \in [a, b]$. Such a function is called a *contraction map.* f satisfies a Lipschitz condition on $[a, b]$ and so is (uniformly) continuous on $[a, b]$. Now Theorem 4.7 guarantees the existence of an $x_0 \in [a, b]$ with $f(x_0) = x_0$. Suppose that $f(x^*) = x^*$. Then $|f(x_0) - f(x^*)| \leq k|x_0 - x^*|$, and so $|x_0 - x^*| \leq k|x_0 - x^*|$. Since $k < 1$, we must have $|x_0 - x^*| = 0$. Therefore $x_0 = x^*$, and this establishes the uniqueness.

EXERCISES

5.13 Give the geometrical interpretations of Rolle's theorem and the mean-value theorem.

5.14 Prove that if f is differentiable on (a, b) and $f'(x) \leq 0$ for every $x \in (a, b)$ then f is monotone decreasing on (a, b).

5.15 Suppose $f'(x) = 0$ for all x in some interval I. Prove that f is constant on I.

5.16 Suppose $f'(x) = g'(x)$ for all x in some interval I. Prove that there exists some constant k such that $f(x) = g(x) + k$ for all $x \in I$.

5.17 Suppose that $|f(x) - f(y)| \leq (x - y)^2$ for all real numbers x and y. Prove that f is a constant function.

5.18 Explain why $f(x) = 1 - |x - 1|$ does not satisfy Rolle's theorem on $[0, 2]$.

5.19 Prove that for any real number b the polynomial $f(x) = x^3 + x + b$ has exactly one real root; that is, there exists a unique $x_0 \in \mathbf{R}$ such that $f(x_0) = 0$.

5.20 Prove that if f satisfies a Lipschitz condition on I then f is uniformly continuous on I.

5.21 Prove or disprove: *Conjecture*: If f is differentiable and uniformly continuous on an interval I then f satisfies a Lipschitz condition on I.

5.22 Prove or disprove: *Conjecture*: If f is differentiable on $[a, b]$ then f satisfies a Lipschitz condition on $[a, b]$.

5.23 Prove that if $f'(x) > 0$ for all $x \in I$ then f is strictly increasing on I. Is the converse to this statement true? Justify your answer.

5.24 Prove that if the temperature $T(x)$ along the equator of the earth is a differentiable function of distance x then there is a point x_0 on the equator where $T'(x_0) = 0$.

5.25 Prove or disprove: If a person travels on Tuesday from a point P on the earth to a point Q due north of P then at some instant he must travel due north.

5.3 L'HOSPITAL'S RULE

The Cauchy mean-value theorem of the previous section is a basic tool needed to prove a collection of theorems usually called *L'Hospital rule theorems*. These results are important in the evaluation of certain kinds of limits. We begin with a particular form of the general rule.

Theorem 5.11 Assume that f' and g' exist in some deleted neighborhood of $x = a$, where $g'(x) \neq 0$, and $\lim_{x \to a} f(x) = \lim_{x \to a} g(x) = 0$ and $\lim_{x \to a} [f'(x)/g'(x)]$ exists. Then $\lim_{x \to a} [f(x)/g(x)]$ exists and equals $\lim_{x \to a} [f'(x)/g'(x)]$.

PROOF We define or redefine, as the case may be, $f(a) = g(a) = 0$. Then f and g are continuous on $N_\delta(a)$ for some $\delta > 0$. Of course $f'(a)$ and $g'(a)$ are not known to exist.

If $a < x < a + \delta$ then f and g are continuous on $[a, x]$ and differentiable on (a, x). By the Cauchy mean-value theorem there exists $t_x \in (a, x)$ such that

$$[f(x) - f(a)]g'(t_x) = [g(x) - g(a)]f'(t_x)$$

or

$$f(x)g'(t_x) = g(x)f'(t_x)$$

Also $g(x) \neq 0$ since $g(a) = 0$, and if $g(x) = 0$ then by Rolle's theorem we would have a point $c \in (a, x)$ with $g'(c) = 0$. But, as noted above, $g'(c) \neq 0$ for every $c \in N_\delta^*(a)$. Consequently

$$\frac{f(x)}{g(x)} = \frac{f'(t_x)}{g'(t_x)}$$

The same argument shows that if $a - \delta < x < a$ then there is a $t_x \in (x, a)$ such that

$$\frac{f(x)}{g(x)} = \frac{f'(t_x)}{g'(t_x)}$$

Now as $x \to a$, $t_x \to a$, and so

$$\lim_{x \to a} \frac{f(x)}{g(x)} = \lim_{x \to a} \frac{f'(t_x)}{g'(t_x)} = \lim_{x \to a} \frac{f'(x)}{g'(x)}$$

since the last limit exists by hypothesis.

Example 5.5 Evaluate $\lim_{x \to 0} [(1 - \cos x)/(\sin^2 x)]$.

SOLUTION Let $f(x) = 1 - \cos x$ and $g(x) = \sin^2 x$. Then $\lim_{x \to 0} f(x) = 0$ and $\lim_{x \to 0} g(x) = 0$. Moreover, $f'(x) = \sin x$ and $g'(x) = 2 \sin x \cos x$, and so

$$\lim_{x \to 0} \frac{f'(x)}{g'(x)} = \lim_{x \to 0} \frac{\sin x}{2 \sin x \cos x} = \lim_{x \to 0} \frac{1}{2 \cos x} = \frac{1}{2}$$

It follows from Theorem 5.11 that

$$\lim_{x \to 0} \frac{1 - \cos x}{\sin^2 x} = \frac{1}{2}$$

Example 5.6 Evaluate $\lim_{x \to 0} [(1 - \cos x^2)/(\sin^2 x)]$.

SOLUTION Let $f(x) = 1 - \cos x^2$ and $g(x) = \sin^2 x$. Then $\lim_{x \to 0} f(x) = 0$ and $\lim_{x \to 0} g(x) = 0$. Also,

$$\lim_{x \to 0} \frac{f'(x)}{g'(x)} = \lim_{x \to 0} \frac{2x \sin x^2}{2 \sin x \cos x}$$

$$= \left(\lim_{x \to 0} \frac{x}{\sin x} \right) \left(\lim_{x \to 0} \frac{\sin x^2}{\cos x} \right) = 1 \cdot 0 = 0$$

(Why does $\lim_{x \to 0} [x/(\sin x)] = 1$?) By Theorem 5.11

$$\lim_{x \to 0} \frac{1 - \cos x^2}{\sin^2 x} = 0$$

These two examples illustrate the use of L'Hospital's rule in the form given in Theorem 5.11 in evaluating certain kinds of limits. The following result is really a corollary to that theorem.

Theorem 5.12 If F is continuous at $x = a$ and if F is differentiable on some deleted neighborhood of $x = a$ and if $\lim_{x \to a} F'(x)$ exists then $F'(a)$ exists and $F'(a) = \lim_{x \to a} F'(x)$; that is, F' is continuous at $x = a$.

PROOF Let $f(x) = F(x) - F(a)$ and $g(x) = x - a$. Then $\lim_{x \to a} f(x) = 0$ and $\lim_{x \to a} g(x) = 0$. Also, $\lim_{x \to a} [f'(x)/g'(x)] = \lim_{x \to a} F'(x)$ exists. By Theorem 5.11, $\lim_{x \to a} [f(x)/g(x)]$ exists and equals

$$\lim_{x \to a} \frac{f'(x)}{g'(x)} = \lim_{x \to a} F'(x)$$

Therefore
$$\lim_{x \to a} \frac{F(x) - F(a)}{x - a} = \lim_{x \to a} F'(x)$$

and so
$$F'(a) = \lim_{x \to a} F'(x)$$

This tells us that $F(x)$ is differentiable at $x = a$ and moreover that F' is continuous at $x = a$.

A second form of L'Hospital's rule follows.

Theorem 5.13 Assume that f' and g' exist, that $g'(x) \neq 0$ for sufficiently large x, and also that $\lim_{x \to \infty} f(x) = \lim_{x \to \infty} g(x) = \infty$ and $\lim_{x \to \infty} [f'(x)/g'(x)]$ exists. Then $\lim_{x \to \infty} [f(x)/g(x)]$ exists and equals $\lim_{x \to \infty} [f'(x)/g'(x)]$.

PROOF Let $\varepsilon > 0$ be given. There exists a point a such that

$$\left| \frac{f'(x)}{g'(x)} - L \right| < \frac{\varepsilon}{2} \quad \text{for every } x > a$$

By applying the Cauchy mean-value theorem to f and g on $[a, x]$ we see that for every $x > a$

$$\left| \frac{f(x) - f(a)}{g(x) - g(a)} - L \right| < \frac{\varepsilon}{2}$$

Now $g(x) > g(a)$ for every $x > a$, for if there is $x_1 \in (a, \infty)$ with $g(x_1) \leq g(a)$ then since $\lim_{x \to \infty} g(x) = \infty$ there is also an $x_2 > x_1$ with $g(x_2) \geq g(a)$. By the intermediate-value theorem for some c satisfying $a < x_1 \leq c \leq x_2$, $g(c) = g(a)$. Finally by Rolle's theorem there is a point d satisfying $a < d < c$ such that $g'(d) = 0$. This contradicts the fact that $g'(x) \neq 0$.

Choose $x > a$ so large that $f(x) > f(a)$ and $g(x) > 0$ (recall that

DIFFERENTIABLE FUNCTIONS

$\lim_{x \to \infty} f(x) = \infty$ and $\lim_{x \to \infty} g(x) = \infty$), and

$$\frac{f(x)}{f(x)-f(a)} \frac{g(x)-g(a)}{g(x)} < 1 + \frac{\varepsilon}{2(|L|+\varepsilon)}$$

This is possible since

$$\lim_{x \to \infty} \frac{f(x)}{f(x)-f(a)} = \lim_{x \to \infty} \frac{g(x)-g(a)}{g(x)} = 1$$

Then

$$\left| \frac{f(x)}{g(x)} - \frac{f(x)-f(a)}{g(x)-g(a)} \right|$$

$$= \left| \frac{f(x)-f(a)}{g(x)-g(a)} \frac{f(x)}{f(x)-f(a)} \frac{g(x)-g(a)}{g(x)} - \frac{f(x)-f(a)}{g(x)-g(a)} \right|$$

$$= \left| \frac{f(x)-f(a)}{g(x)-g(a)} \right| \cdot \left| \frac{f(x)}{f(x)-f(a)} \frac{g(x)-g(a)}{g(x)} - 1 \right|$$

$$< \left(|L| + \frac{\varepsilon}{2} \right) \frac{\varepsilon}{2(|L|+\varepsilon)} < \frac{\varepsilon}{2}$$

Consequently

$$\left| \frac{f(x)}{g(x)} - L \right| \leq \left| \frac{f(x)}{g(x)} - \frac{f(x)-f(a)}{g(x)-g(a)} \right|$$

$$+ \left| \frac{f(x)-f(a)}{g(x)-g(a)} - L \right| < \frac{\varepsilon}{2} + \frac{\varepsilon}{2} = \varepsilon$$

Therefore

$$\lim_{x \to \infty} \frac{f(x)}{g(x)} = L$$

Example 5.7 Evaluate $\lim_{x \to \infty} (x/e^x)$.

SOLUTION Let $f(x) = x$ and $g(x) = e^x$. Then

$$\lim_{x \to \infty} f(x) = \lim_{x \to \infty} g(x) = \infty$$

and

$$\lim_{x \to \infty} \frac{f'(x)}{g'(x)} = \lim_{x \to \infty} \frac{1}{e^x} = 0$$

It follows from Theorem 5.13 that $\lim_{x \to \infty} (x/e^x) = 0$.

A third version of L'Hospital's rule is given in the following theorem.

Theorem 5.14 Assume that for some $\delta > 0$, f' and g' exist and $g'(x) \neq 0$ for all x, $a < x < a + \delta$. Suppose that

$$\lim_{x \to a^+} f(x) = \infty \qquad \lim_{x \to a^+} g(x) = \infty$$

and

$$\lim_{x \to a^+} \frac{f'(x)}{g'(x)} = L$$

Then

$$\lim_{x \to a^+} \frac{f(x)}{g(x)} = L$$

PROOF Let $x = a + 1/u$ or, equivalently, $u = 1/(x - a)$. Now

$$\lim_{x \to a^+} f(x) = \lim_{u \to \infty} f\left(a + \frac{1}{u}\right) = \infty$$

and

$$\lim_{x \to a^+} g(x) = \lim_{u \to \infty} g\left(a + \frac{1}{u}\right) = \infty$$

By Theorem 5.13 and the chain rule

$$\lim_{x \to a^+} \frac{f(x)}{g(x)} = \lim_{u \to \infty} \frac{f(a + 1/u)}{g(a + 1/u)} = \lim_{u \to \infty} \frac{f'(a + 1/u)(-1/u^2)}{g'(a + 1/u)(-1/u^2)}$$

$$= \lim_{u \to \infty} \frac{f'(a + 1/u)}{g'(a + 1/u)} = \lim_{x \to a^+} \frac{f'(x)}{g'(x)}$$

Corollary Assume that for some $\delta > 0$, f' and g' exist and $g'(x) \neq 0$ for all x, $a < x < a + \delta$. Suppose that

$$\lim_{x \to a^+} f(x) = -\infty \qquad \lim_{x \to a^+} g(x) = -\infty$$

and

$$\lim_{x \to a^+} \frac{f'(x)}{g'(x)} = L$$

Then

$$\lim_{x \to a^+} \frac{f(x)}{g(x)} = L$$

Example 5.8 Evaluate $\lim_{x \to \infty} [(\ln x)/x^\alpha]$, where $\alpha > 0$.

SOLUTION We let $u = 1/x$; then as $x \to \infty$, $u \to 0^+$. Clearly

$$\lim_{x \to \infty} \frac{\ln x}{x^\alpha} = \lim_{u \to 0^+} \frac{\ln(1/u)}{(1/u)^\alpha}$$

Let $f(u) = \ln(1/u)$ and $g(u) = (1/u)^\alpha$. Then $\lim_{u \to 0^+} f(u) = \infty$ and

$\lim_{u \to 0^+} g(u) = \infty$. Also

$$\lim_{u \to 0^+} \frac{f'(u)}{g'(u)} = \lim_{u \to 0^+} -\frac{1/u}{\alpha/u^{\alpha+1}} = \lim_{u \to 0^+} -\frac{u^\alpha}{\alpha} = 0$$

By Theorem 5.14, $\lim_{u \to 0^+} [f(u)/g(u)] = 0$. Therefore it follows that $\lim_{x \to \infty} [(\ln x)/x^\alpha)] = 0$.

Observe that the limit in Example 5.8 can also be evaluated using Theorem 5.13.

Example 5.9 Evaluate $\lim_{x \to 0^+} (\sin x)^x$.

SOLUTION We use the fact that $f(x) = e^x$ and $f^{-1}(x) = \ln x$, $x > 0$, are inverse functions which are continuous. Now

$$\ln (\sin x)^x = x \ln (\sin x) \qquad 0 < x < \pi$$

$$= \frac{\ln (\sin x)}{1/x}$$

Using L'Hospital's rule to take the limit, we obtain

$$\lim_{x \to 0^+} \frac{\ln (\sin x)}{1/x} = \lim_{x \to 0^+} \frac{(\cos x)/(\sin x)}{-1/x^2}$$

$$= \lim_{x \to 0^+} -x (\cos x) \frac{x}{\sin x} = 0$$

Finally

$$\lim_{x \to 0^+} (\sin x)^x = \exp \left[\lim_{x \to 0^+} \ln (\sin x)^x \right] = \exp (0) = 1$$

There are numerous other versions of L'Hospital's rule, many of which we leave to the exercises.

EXERCISES

Prove the following versions of L'Hospital's rule:

5.26 If $\lim_{x \to a} f(x) = \lim_{x \to a} g(x) = \infty$ and $\lim_{x \to a}[f'(x)/g'(x)] = L$ then $\lim_{x \to a} [f(x)/g(x)] = L$.

5.27 If $\lim_{x \to \infty} f(x) = \lim_{x \to \infty} g(x) = 0$ and $\lim_{x \to \infty}[f'(x)/g'(x)] = L$ then $\lim_{x \to \infty}[f(x)/g(x)] = L$.

5.28 If $\lim_{x \to a^-} f(x) = 0$, $\lim_{x \to a^+} g(x) = 0$ and $\lim_{x \to a^+}[f'(x)/g'(x)] = L$ then $\lim_{x \to a^-} [f(x)/g(x)] = L$.

5.29 If $\lim_{x \to -\infty} f(x) = \infty$, $\lim_{x \to -\infty} g(x) = \infty$ and $\lim_{x \to -\infty}[f'(x)/g'(x)] = L$ then $\lim_{x \to -\infty} [f(x)/g(x)] = L$.

5.30 Evaluate the following limits using the appropriate version of L'Hospital's rule:

(a) $\lim_{x \to 0} \dfrac{a^x - b^x}{\sin x}$ for $a > 0$ and $b > 0$ (b) $\lim_{x \to 0} \dfrac{\sin 5x}{x}$

(c) $\lim_{x \to 0} \dfrac{x}{\tan x}$ (d) $\lim_{x \to 0} \dfrac{\cos^2 x - 1}{x}$

(e) $\lim_{x \to \infty} \dfrac{x^n}{e^x}$ where $n \in \mathbb{N}$ (f) $\lim_{x \to \infty} (x - \sqrt{x^2 + x})$

(g) $\lim_{x \to 0} \dfrac{x}{\cos x + \sin x}$ (h) $\lim_{x \to 0^+} (x + \sin x)^{\tan x}$

(i) $\lim_{x \to 0} (1 - 2x)^{3/x}$ (j) $\lim_{x \to \infty} \dfrac{(\ln x)^n}{x}$ where $n \in \mathbb{N}$

5.31 Prove that if f' is continuous in a neighborhood of $x = a$ then

$$\lim_{h \to 0} \frac{f(a + h/2) - f(a - h/2)}{h} = f'(a)$$

(The term $f(a + h/2) - f(a - h/2)$, called a *central difference*, is quite useful in numerical analysis).

5.32 Prove that if f'' exists and is continuous in a neighborhood of $x = a$ then

$$\lim_{h \to 0} \frac{f(a + h) - 2f(a) + f(a - h)}{h^2} = f''(a)$$

5.33 Prove Theorem 5.12 using the mean-value theorem.

5.4 FURTHER TOPICS ON DIFFERENTIATION

We begin this section by defining the one-sided derivative of a function f. First, we suppose that f is defined in an interval $[x_0, x_0 + \delta)$ for some $\delta > 0$.

Definition 5.5 The *right-hand derivative* of f at x_0 is defined by

$$f'_+(x_0) = \lim_{h \to 0^+} \frac{f(x_0 + h) - f(x_0)}{h}$$

or equivalently $f'_+(x_0) = \lim_{x \to x_0^+} \dfrac{f(x) - f(x_0)}{x - x_0}$

provided the limit exists.

Similarly, if f is defined on an interval $(x_0 - \delta, x_0]$ for some $\delta > 0$, we have the following definition.

Definition 5.6 The *left-hand derivative* of f at x_0 is defined by

$$f'_-(x_0) = \lim_{h \to 0^-} \frac{f(x_0 + h) - f(x_0)}{h}$$

or equivalently

$$f'_-(x_0) = \lim_{x \to x_0^-} \frac{f(x) - f(x_0)}{x - x_0}$$

provided the limit exists.

Clearly $f'(x_0)$ exists if and only if $f'_+(x_0)$ and $f'_-(x_0)$ both exist and are equal.

Example 5.10 Consider

$$f(x) = \begin{cases} 1 - x & \text{if } x \leq 1 \\ (1 - x)^2 & \text{if } x > 1 \end{cases}$$

$$f'_+(1) = \lim_{x \to 1^+} \frac{f(x) - f(1)}{x - 1} = \lim_{x \to 1^+} \frac{(1 - x)^2}{x - 1} = \lim_{x \to 1^+} (x - 1) = 0$$

$$f'_-(1) = \lim_{x \to 1^-} \frac{f(x) - f(1)}{x - 1} = \lim_{x \to 1^-} \frac{1 - x}{x - 1} = -1$$

Hence $f'(1)$ does not exist; that is, f is not differentiable at $x = 1$.

Consider the pie function given by

$$f(x) = \begin{cases} \dfrac{1}{2} \sec \dfrac{x}{2} & 0 \leq x \leq \dfrac{\pi}{2} \\ \sin \dfrac{x}{2} & \dfrac{\pi}{2} < x < \pi \\ 1 & \pi \leq x \leq 2\pi \end{cases}$$

We wish to find $f'_+(\pi/2)$ and $f'_-(\pi/2)$.

$$f'_+\left(\frac{\pi}{2}\right) = \lim_{h \to 0^+} \frac{\sin(\pi/4 + h/2) - \sqrt{2}/2}{h}$$

Using L'Hospital's rule we get

$$f'_+\left(\frac{\pi}{2}\right) = \lim_{h \to 0^+} \frac{1}{2} \cos\left(\frac{\pi}{4} + \frac{h}{2}\right) = \frac{\sqrt{2}}{4}$$

Similarly,

$$f'_-\left(\frac{\pi}{2}\right) = \lim_{h \to 0^-} \frac{1/2 \sec(\pi/4 + h/2) - \sqrt{2}/2}{h}$$

Again, L'Hospital's rule yields

$$f'_-\left(\frac{\pi}{2}\right) = \lim_{h \to 0^-} \frac{1}{4} \sec\left(\frac{\pi}{4} + \frac{h}{2}\right) \tan\left(\frac{\pi}{4} + \frac{h}{2}\right) = \frac{\sqrt{2}}{4}$$

In this case f is differentiable at $x = \pi/2$ and $f'(\pi/2) = \sqrt{2}/4$. Also,

$$f'(x) = \begin{cases} \dfrac{1}{4} \sec \dfrac{x}{2} \tan \dfrac{x}{2} & 0 < x < \dfrac{\pi}{2} \\ 1/2 \cos \dfrac{x}{2} & \dfrac{\pi}{2} < x < \pi \end{cases}$$

Hence $f'_+\left(\dfrac{\pi}{2}\right) = \lim_{x \to (\pi/2)^+} f'(x)$ and $f'_-\left(\dfrac{\pi}{2}\right) = \lim_{x \to (\pi/2)^-} f'(x)$

The following example shows that these equations do not always hold.

Example 5.11 Let

$$f(x) = \begin{cases} x^2 \sin \dfrac{1}{x} & \text{if } x \neq 0 \\ 0 & \text{if } x = 0 \end{cases}$$

Then

$$f'(x) = \begin{cases} 2x \sin \dfrac{1}{x} - \cos \dfrac{1}{x} & \text{if } x \neq 0 \\ 0 & \text{if } x = 0 \end{cases}$$

$f'(0)$ exists and equals 0 (why?); therefore $f'_+(0) = f'_-(0) = 0$. On the other hand, $\lim_{x \to 0^+} f'(x)$ and $\lim_{x \to 0^-} f'(x)$ both fail to exist. Hence f' has a discontinuity of the second kind at $x = 0$.

Therefore $f'_+(x_0)$ and $f'_-(x_0)$ can exist even when $\lim_{x \to x_0^+} f'(x)$ and $\lim_{x \to x_0^-} f'(x)$ fail to exist. Later we shall show that if $f(x)$ is differentiable in a neighborhood of x_0 and if both $\lim_{x \to x_0^+} f'(x)$ and $\lim_{x \to x_0^-} f'(x)$ exist then these limits are necessarily equal. The following theorems show that if the one-sided limits of f' at x_0 exist and f is continuous there then the one-sided derivatives exist as well.

Theorem 5.15 If f is continuous on $[x_0, b)$ and differentiable on (x_0, b) and if $\lim_{x \to x_0^+} f'(x)$ exists then $f'_+(x_0)$ exists and

$$f'_+(x_0) = \lim_{x \to x_0^+} f'(x)$$

Proof Let $x_0 + h$ ($h > 0$) be between x_0 and b. Then $f(x)$ is continuous on $[x_0, x_0 + h]$ and differentiable on $(x_0, x_0 + h)$. By the mean-value theorem there exists a $c \in (x_0, x_0 + h)$ with

$$\frac{f(x_0 + h) - f(x_0)}{h} = f'(c)$$

Now

$$f'_+(x_0) = \lim_{h \to 0^+} \frac{f(x_0 + h) - f(x_0)}{h} = \lim_{h \to 0^+} f'(c)$$

where c is between x_0 and $x_0 + h$ ($h > 0$). Our assumption was that $\lim_{x \to x_0^+} f'(x)$ exists. Hence $f'_+(x_0) = \lim_{x \to x_0^+} f'(x)$.

The following theorem is proved in a similar fashion.

Theorem 5.16 If f is continuous on $(a, x_0]$ and differentiable on (a, x_0) and if $\lim_{x \to x_0^-} f'(x)$ exists then $f'_-(x_0)$ exists and

$$f'_-(x_0) = \lim_{x \to x_0^-} f'(x)$$

Theorems 5.15 and 5.16 give sufficient conditions for the existence of the one-sided derivatives; these conditions are not necessary, as shown by Example 5.11 (see also Theorem 5.12).

Just as we modified the definition of continuity in the last chapter, here we say that f is differentiable on $[a, b]$ provided it is differentiable on (a, b) and both $f'_+(a)$ and $f'_-(b)$ exist.

Derivatives need not be continuous, as Example 5.11 illustrates. The following theorem shows that although they are not necessarily continuous, derivatives (like continuous functions) satisfy the intermediate-value property.

Theorem 5.17 If f is differentiable on $[a, b]$ and if k is between $f'_+(a)$ and $f'_-(b)$ then there is a point $c \in (a, b)$ such that $f'(c) = k$.

Proof We assume that $f'_+(a) < k < f'_-(b)$; the case where $f'_-(b) < k < f'_+(a)$ is similar. Define $g(x) = f(x) - kx$ for all $x \in [a, b]$. Then g is differentiable on $[a, b]$ and $g'(x) = f'(x) - k$ on (a, b). Also

$$g'_+(a) = f'_+(a) - k < 0 \quad \text{and} \quad g'_-(b) = f'_-(b) - k > 0$$

Hence for $h > 0$ sufficiently small

$$\frac{g(a + h) - g(a)}{h} < 0$$

and so $g(a + h) < g(a)$. Similarly, for $h < 0$ sufficiently small
$$\frac{g(b + h) - g(b)}{h} > 0$$
and so $g(b + h) < g(b)$. Now, g is continuous on $[a, b]$ and by the extreme-value theorem must assume its minimum value on $[a, b]$. The above inequalities show that g does not assume its minimum at a or b. Hence there is a $c \in (a, b)$ such that $g(c) \leq g(x)$ for every $x \in [a, b]$. $g'(c)$ exists and, by Theorem 5.3, $g'(c) = 0$. Thus $f'(c) - k = 0$ and $f'(c) = k$.

A function which has a discontinuity of the first kind cannot satisfy the intermediate-value property (see Theorem 4.10). The previous theorem tells us that if f is differentiable on an interval then f' satisfies the intermediate-value property and so f' has no discontinuities of the first kind. As a consequence, if $\lim_{x \to x_0^+} f'(x)$ and $\lim_{x \to x_0^-} f'(x)$ both exist and if $f'(x_0)$ exists then $\lim_{x \to x_0} f'(x)$ exists and equals $f'(x_0)$; that is, f' is continuous at x_0.

We proved in Chap. 4 that monotone functions have no discontinuities of the second kind. It follows that monotone derivatives are necessarily continuous.

EXERCISES

5.34 Let
$$f(x) = \begin{cases} x + 2 & \text{if } x < -1 \\ x^2 & \text{if } -1 \leq x \leq 0 \\ 4x - 3 & \text{if } 0 < x \leq 2 \\ 1 + x^2 & \text{if } x > 2 \end{cases}$$
Find the one-sided derivatives $f'_+(x)$ and $f'_-(x)$. Where is f differentiable?

5.35 Prove Theorem 5.16.

5.36 Let f be the pie function. Find:

(a) $\lim_{x \to \pi^+} f'(x)$ (b) $\lim_{x \to \pi^-} f'(x)$ (c) $f'_+(\pi)$ (d) $f'_-(\pi)$

5.37 Let
$$f(x) = \begin{cases} x|x| & \text{if } x \text{ is rational} \\ 0 & \text{if } x \text{ is irrational} \end{cases}$$
Show that $f'_+(0)$ and $f'_-(0)$ exist but f is not differentiable in any neighborhood (deleted or otherwise) of $x = 0$.

5.38 Prove or give a counterexample: *Conjecture*: If $f'_+(x_0)$ and $f'_-(x_0)$ both exist then f is continuous at $x = x_0$.

5.39 Find a function f with the property that $\lim_{x \to x_0^+} f'(x)$ and $\lim_{x \to x_0^-} f'(x)$ both exist but are unequal.

CHAPTER SIX
THE RIEMANN INTEGRAL

6.1 DEFINITION OF THE INTEGRAL

The Riemann integral is an important concept for the mathematician and for those who use mathematical analysis to solve the kinds of problems for which calculus is applicable. It is of theoretical interest, as it relates to boundedness, continuity, and differentiability; and it is a useful tool in many areas of mathematics. In addition applications of the integral are found in science, engineering, and even business. In this chapter we present a detailed and rigorous development of the theory of Riemann integration.

We let $\mathcal{B}[a, b]$ be the set of all real-valued functions which are defined and bounded on the closed, bounded interval $[a, b]$. If $f \in \mathcal{B}[a, b]$, we denote

$$M = \sup_x f(x) \quad \text{and} \quad m = \inf_x f(x) \quad \text{where } x \in [a, b]$$

A *partition* of $[a, b]$ is a set of points, called the *points of subdivision*, denoted

$$P: a = x_0 < x_1 < \cdots < x_{i-1} < x_i < x_{i+1} < \cdots < x_{n-1} < x_n = b$$

For the partition P we let

$$M_i = \sup_x f(x) \quad \text{and} \quad m_i = \inf_x f(x) \quad \text{where } x \in [x_{i-1}, x_i]$$

for $i = 1, 2, \ldots, n$. Clearly $m \leq m_i \leq M_i \leq M$ for each i. The *norm* of P, written $||P||$, is defined by $||P|| = \max_{1 \leq i \leq n} \Delta x_i$, where $\Delta x_i = x_i - x_{i-1}$

for $i = 1, 2, \ldots, n$. Next we define the *upper sum* of the function f for the partition P to be

$$U(P, f) = \sum_{i=1}^{n} M_i \, \Delta x_i$$

Similarly the *lower sum* is defined by

$$L(P, f) = \sum_{i=1}^{n} m_i \, \Delta x_i$$

For a fixed $f \in \mathcal{B}[a, b]$ each partition P of $[a, b]$ has associated with it an upper sum and a lower sum. We presume that the reader has seen the geometrical interpretation of these sums as approximations to the area under the graph of f via circumscribed and inscribed rectangles, respectively. Clearly

$$L(P, f) \leq U(P, f)$$

for any partition P.

The *upper integral* of f on $[a, b]$ is defined by

$$\overline{\int} f = \inf_{P} U(P, f)$$

and the *lower integral* of f is

$$\underline{\int} f = \sup_{P} L(P, f)$$

where $\sup_P$ and $\inf_P$ mean that we take the supremum and the infimum over all possible partitions P of $[a, b]$.

These upper and lower integrals are *integrals with respect to x*, which merely means that we are using x to denote the independent variable for the function f and that $[a, b]$ is an interval on the x axis. Consequently it is not uncommon to see the upper and lower integrals written as

$$\overline{\int_a^b} f(x) \, dx \quad \text{and} \quad \underline{\int_a^b} f(x) \, dx$$

respectively. The endpoints of the interval $[a, b]$ are called the *limits of integration*.

The following lemma guarantees that under the assumptions made thus far, the upper and lower integrals of f always exist.

Lemma 1 For any partition P of $[a, b]$ we have

$$m(b - a) \leq L(P, f) \leq U(P, f) \leq M(b - a)$$

THE RIEMANN INTEGRAL

PROOF For each $i = 1, 2, \ldots, n$

$$m(x_i - x_{i-1}) \leq m_i(x_i - x_{i-1}) \leq M_i(x_i - x_{i-1}) \leq M(x_i - x_{i-1})$$

Therefore

$$\sum_{i=1}^{n} m(x_i - x_{i-1}) \leq \sum_{i=1}^{n} m_i(x_i - x_{i-1})$$

$$\leq \sum_{i=1}^{n} M_i(x_i - x_{i-1}) \leq \sum_{i=1}^{n} M(x_i - x_{i-1})$$

Thus

$$m \sum_{i=1}^{n} \Delta x_i \leq \sum_{i=1}^{n} m_i \, \Delta x_i \leq \sum_{i=1}^{n} M_i \, \Delta x_i \leq M \sum_{i=1}^{n} \Delta x_i$$

and so $\quad m(b - a) \leq L(P, f) \leq U(P, f) \leq M(b - a)$

Since the set of all lower sums $L(P, f)$ is bounded above (the real number $M(b - a)$ is an upper bound), $\underline{\int} f = \sup_P L(P, f)$ exists. Similarly, $U(P, f) \geq m(b - a)$ for every P, and so $\overline{\int} f = \inf_P U(P, f)$ exists.

Definition 6.1 f is said to be *Riemann-integrable* on $[a, b]$ provided

$$\overline{\int} f = \underline{\int} f$$

If f is Riemann-integrable on $[a, b]$, we define *the integral of f* to be the common value of the upper and lower integral; that is,

$$\int f = \overline{\int} f = \underline{\int} f$$

Again, to avoid possible ambiguity we sometimes also write $\int f$ in the form

$$\int_a^b f(x) \, dx \quad \text{or simply} \quad \int_a^b f$$

Example 6.1 We show that each constant function $f(x) = k$ is Riemann-integrable on any interval $[a, b]$.

SOLUTION Let P be an arbitrary partition of $[a, b]$

$$P: a = x_0 < x_1 < \cdots < x_{i-1} < x_i < x_{i+1} < \cdots < x_{n-1} < x_n = b$$

Then

$$U(P, f) = \sum_{i=1}^{n} M_i \, \Delta x_i = \sum_{i=1}^{n} k \, \Delta x_i = k \sum_{i=1}^{n} \Delta x_i = k(b - a)$$

and

$$L(P, f) = \sum_{i=1}^{n} m_i \, \Delta x_i = \sum_{i=1}^{n} k \, \Delta x_i = k \sum_{i=1}^{n} \Delta x_i = k(b - a)$$

Since P was arbitrary, it follows that $U(P, f) = L(P, f) = k(b - a)$ for every P, and so

$$\overline{\int} f = k(b - a) \quad \text{and} \quad \underline{\int} f = k(b - a)$$

Thus f is Riemann-integrable and

$$\int_a^b f = k(b - a)$$

The following is an example of a function $f \in \mathcal{B}[a, b]$ which is not Riemann-integrable on $[a, b]$.

Example 6.2 Let

$$f(x) = \begin{cases} 1 & \text{if } x \in \mathbf{Q} \cap [a, b] \\ 0 & \text{if } x \in (\mathbf{R} - \mathbf{Q}) \cap [a, b] \end{cases}$$

where $\mathbf{Q}$ is the set of rationals. Let P be any partition of $[a, b]$; then each $m_i = 0$ since every subinterval of P contains irrational numbers. Similarly $M_i = 1$ for $i = 1, 2, \ldots, n$, and so

$$U(P, f) = \sum_{i=1}^{n} M_i \, \Delta x_i = \sum_{i=1}^{n} 1 \cdot \Delta x_i = b - a$$

and

$$L(P, f) = \sum_{i=1}^{n} m_i \, \Delta x_i = \sum_{i=1}^{n} 0 \cdot \Delta x_i = 0$$

Since P was arbitrary, every upper sum has the value $b - a$ and every lower sum is equal to 0. It follows that

$$\overline{\int} f = b - a \quad \text{and} \quad \underline{\int} f = 0$$

and so f is not Riemann-integrable on $[a, b]$. Therefore $\int_a^b f$ does not exist.

A word of caution about these two examples: in each case the upper sum $U(P, f)$ and the lower sum $L(P, f)$ were constant as the partition P

varied. This is not usually the case, and as a result the evaluation of the upper and lower integrals can be more difficult than these examples indicate.

The partition P^* of $[a, b]$ is called a *refinement* of the partition P if each point of subdivision x_i of P is also a point of subdivision of P^*. The partition P^* is called a *common refinement* of the partitions P_1 and P_2 if P^* is a refinement of both P_1 and P_2. Every pair of partitions P_1 and P_2 has common refinements. Take for example $P^* = P_1 \cup P_2$; that is, we let P^* consist of exactly those points of P_1 together with those of P_2. Such a partition P^* is sometimes called the *first common refinement* of P_1 and P_2 in the sense that every common refinement of P_1 and P_2 is also a refinement of P^*. The following lemma says briefly that under refinement lower sums grow and upper sums shrink.

Lemma 2 If P^* is a refinement of P then $L(P^*, f) \geq L(P, f)$ and $U(P^*, f) \leq U(P, f)$.

PROOF Let P^* contain one more point than P, say x^*, $x_{i-1} < x^* < x_i$ and x_{i-1}, x_i are consecutive points of subdivision of P. We let

$$M_1^* = \sup_x f(x) \qquad m_1^* = \inf_x f(x) \qquad x \in [x_{i-1}, x^*]$$

and

$$M_2^* = \sup_x f(x) \qquad m_2^* = \inf_x f(x) \qquad x \in [x^*, x_i]$$

Then

$$m_i \Delta x_i = m_i(x_i - x^*) + m_i(x^* - x_{i-1}) \leq m_2^*(x_i - x^*) + m_1^*(x^* - x_{i-1})$$

and $M_i \Delta x_i = M_i(x_i - x^*) + M_i(x^* - x_{i-1})$

$$\geq M_2^*(x_i - x^*) + M_1^*(x^* - x_{i-1})$$

It follows that $L(P, f) \leq L(P^*, f)$ and $U(P, f) \geq U(P^*, f)$, since other terms in the lower sums and in the upper sums all agree. The general case where P^* contains k more points than P (k any natural number) is proved by mathematical induction, and we leave the details to the reader.

Lemma 3 For any bounded function f defined on $[a, b]$

$$\underline{\int} f \leq \overline{\int} f$$

PROOF Let P_1 and P_2 be arbitrary partitions of $[a, b]$ and let P^* be any common refinement of P_1 and P_2. By the preceding lemmas

$$L(P_1, f) \leq L(P^*, f) \leq U(P^*, f) \leq U(P_2, f)$$

Now since P_1 was arbitrary,

$$\underline{\int} f = \sup_P L(P,f) \leq U(P_2, f)$$

and since P_2 was arbitrary,

$$\overline{\int} f = \inf_P U(P,f) \geq \underline{\int} f$$

A function f bounded on $[a, b]$ fails to be Riemann-integrable when

$$\underline{\int} f < \overline{\int} f$$

The following theorem gives a necessary and sufficient condition for the Riemann integrability of a function f bounded on $[a, b]$; it will be useful in proving some important results that follow.

Theorem 6.1 $f \in \mathcal{B}[a, b]$ is Riemann-integrable on $[a, b]$ if and only if given any $\varepsilon > 0$ there exists a partition P of $[a, b]$ with

$$U(P, f) - L(P, f) < \varepsilon$$

PROOF First we assume that for any given $\varepsilon > 0$ there is a partition P of $[a, b]$ for which $U(P, f) - L(P, f) < \varepsilon$. Let $\varepsilon > 0$ be arbitrary. Choose P such that $U(P, f) - L(P, f) < \varepsilon$. By the preceding lemma

$$L(P, f) \leq \underline{\int} f \leq \overline{\int} f \leq U(P, f)$$

It follows that $0 \leq \overline{\int} f - \underline{\int} f \leq U(P, f) - L(P, f) < \varepsilon$, and since $\varepsilon > 0$ was arbitrary, $\overline{\int} f - \underline{\int} f = 0$. Hence $\overline{\int} f = \underline{\int} f$, and so f is Riemann-integrable on $[a, b]$.

Now suppose f is a Riemann-integrable function on $[a, b]$. Let $\varepsilon > 0$ be given. Choose P_1 such that

$$\underline{\int} f - L(P_1, f) < \frac{\varepsilon}{2}$$

and P_2 such that

$$U(P_2, f) - \overline{\int} f < \frac{\varepsilon}{2}$$

Since f is Riemann-integrable on $[a, b]$,

$$\underline{\int} f = \overline{\int} f = \int_a^b f$$

Hence

$$\int_a^b f - L(P_1, f) < \frac{\varepsilon}{2} \quad \text{and} \quad U(P_2, f) - \int_a^b f < \frac{\varepsilon}{2}$$

Let P be a common refinement of P_1 and P_2. Then

$$L(P_1, f) \le L(P, f) \le \int_a^b f \le U(P, f) \le U(P_2, f)$$

Hence

$$U(P, f) - L(P, f) \le U(P_2, f) - L(P_1, f)$$

$$< \left(\int_a^b f + \frac{\varepsilon}{2}\right) - \left(\int_a^b f - \frac{\varepsilon}{2}\right) = \varepsilon$$

We use the above theorem to establish the next two results, which are existence theorems; they give sufficient conditions for the integral to exist. Once we know that an integral exists, we can use some specific techniques to find its value.

Theorem 6.2 If f is continuous on $[a, b]$ then it is Riemann-integrable on $[a, b]$.

PROOF Suppose f is continuous on $[a, b]$ and let $\varepsilon > 0$ be given. We show there is a partition P of $[a, b]$ for which $U(P, f) - L(P, f) < \varepsilon$. By the uniform continuity of f on $[a, b]$ there is a $\delta > 0$ such that $|f(u) - f(v)| < \varepsilon/(b - a)$ whenever $u, v \in [a, b]$ with $|u - v| < \delta$. Let P be any partition of $[a, b]$ with $\|P\| < \delta$. By the extreme-value theorem there exist points $\hat{x}_i, \tilde{x}_i \in [x_{i-1}, x_i]$ for $i = 1, 2, \ldots, n$ such that $f(\hat{x}_i) = M_i$ and $f(\tilde{x}_i) = m_i$. Now $|\hat{x}_i - \tilde{x}_i| \le x_i - x_{i-1} = \Delta x_i \le \|P\| < \delta$, and so $M_i - m_i = |f(\hat{x}_i) - f(\tilde{x}_i)| < \varepsilon/(b - a)$ for $i = 1, 2, \ldots, n$. Hence

$$U(P, f) - L(P, f) = \sum_{i=1}^n M_i \Delta x_i - \sum_{i=1}^n m_i \Delta x_i$$

$$= \sum_{i=1}^n (M_i - m_i) \Delta x_i < \frac{\varepsilon}{b-a} \sum_{i=1}^n \Delta x_i$$

$$= \frac{\varepsilon}{b-a}(b-a) = \varepsilon$$

It follows from Theorem 6.1 that f is Riemann-integrable on $[a, b]$.

Theorem 6.3 If f is monotone on $[a, b]$ then it is Riemann-integrable on $[a, b]$.

PROOF If f is constant on $[a, b]$ then it is Riemann-integrable, by Example 6.1; we assume that f is monotone increasing on $[a, b]$ and $f(a) < f(b)$; the monotone-decreasing case is similar, and we leave it to the reader (see Exercise 6.1). Let $\varepsilon > 0$ be given; we show that there is a partition P of $[a, b]$ for which $U(P, f) - L(P, f) < \varepsilon$. Let P be any partition with $\|P\| < \varepsilon/[f(b) - f(a)]$. Then since f is increasing on $[a, b]$, $M_i = f(x_i)$ and $m_i = f(x_{i-1})$ for $i = 1, 2, \ldots, n$. Hence

$$U(P, f) - L(P, f) = \sum_{i=1}^{n} M_i \Delta x_i - \sum_{i=1}^{n} m_i \Delta x_i$$

$$= \sum_{i=1}^{n} (M_i - m_i) \Delta x_i$$

$$= \sum_{i=1}^{n} [f(x_i) - f(x_{i-1})] \Delta x_i$$

$$< \frac{\varepsilon}{f(b) - f(a)} \sum_{i=1}^{n} [f(x_i) - f(x_{i-1})]$$

$$= \frac{\varepsilon}{f(b) - f(a)} [f(x_n) - f(x_0)] = \varepsilon$$

It follows from Theorem 6.1 that f is Riemann-integrable on $[a, b]$.

In certain applications it is important to show that an integral exists even though the actual evaluation of the integral may be a difficult problem. If we know that an integral exists, we can often use one of the many numerical methods (trapezoidal rule, Simpson's rule, etc.) for approximating its value. In the following examples we use Theorems 6.2 and 6.3 to show existence.

Example 6.3 Prove that $\int_0^1 f(x)\, dx$ exists, where

$$f(x) = \begin{cases} \dfrac{\sin x}{x} & \text{if } x \neq 0 \\ 1 & \text{if } x = 0 \end{cases}$$

SOLUTION $(\sin x)/x$ is continuous for $x \neq 0$, and $\lim_{x \to 0} [(\sin x)/x] = 1 = f(0)$. Therefore f is continuous on $[0, 1]$, and so by Theorem 6.2 f is Riemann-integrable on $[0, 1]$. Therefore,

$$\int_0^1 f(x)\, dx$$

exists.

Example 6.4 Let $f(x) = [\![x]\!]$, the greatest-integer function, on the interval $[0, 4]$. f is not continuous on $[0, 4]$, but f is monotone, and so, by Theorem 6.3, f is Riemann-integrable on $[0, 4]$. It follows that

$$\int_0^4 [\![x]\!] \, dx$$

exists.

The next result is useful in determining the value of the integral for a number of functions which are Riemann-integrable on $[a, b]$. It is also important in the discussion of Riemann sums in the next section. We need the following: if $A \subseteq \mathbf{R}$ then we let $-A = \{x \in \mathbf{R} \mid -x \in A\}$. It is easy to see that A is bounded if and only if $-A$ is bounded, and in this case $\sup(-A) = -\inf A$ and $\inf(-A) = -\sup A$. Now if $f \in \mathcal{B}[a, b]$ then $-f \in \mathcal{B}[a, b]$. Let P be any partition of $[a, b]$. Then

$$\sup_x -f(x) = -\inf_x f(x) = -m_i \qquad x \in [x_{i-1}, x_i]$$

and

$$\inf_x -f(x) = -\sup_x f(x) = -M_i \qquad x \in [x_{i-1}, x_i]$$

hence

$$U(P, -f) = \sum_{i=1}^n (-m_i) \Delta x_i = -L(P, f)$$

and

$$L(P, -f) = \sum_{i=1}^n (-M_i) \Delta x_i = -U(P, f)$$

It follows that

$$\overline{\int}(-f) = \inf_P U(P, -f) = \inf_P -L(P, f) = -\sup_P L(P, f) = -\underline{\int} f$$

and

$$\underline{\int}(-f) = \sup_P L(P, -f) = \sup_P -U(P, f) = -\inf_P U(P, f) = -\overline{\int} f$$

Theorem 6.4 Let $f \in \mathcal{B}[a, b]$; given any $\varepsilon > 0$ there is a $\delta > 0$ such that

$$\underline{\int} f - \varepsilon < L(P, f) \leq U(P, f) < \overline{\int} f + \varepsilon$$

whenever P is a partition of $[a, b]$ with $\|P\| < \delta$.

PROOF Let $\varepsilon > 0$ be given; choose the partition P^* of $[a, b]$ such that $U(P^*, f) < \overline{\int} f + \varepsilon/2$. Suppose P^* has n subintervals; let $K = \sup_x |f(x)|$ for $x \in [a, b]$. We can assume that $K > 0$ since the theorem holds trivially for any constant function. Let $\delta_1 = \varepsilon/4Kn$. Now suppose P is a partition of $[a, b]$ with $\|P\| < \delta_1$. We call a subinterval of P an A-type subinterval if it is completely contained in one of the subintervals of P^*.

If a subinterval of P is not of A-type then it necessarily has as an interior point one (or more) of the points of subdivision of the P^* partition. Therefore, all the subintervals of P with the exception of at most $n-1$ are of A-type. We call the subintervals in the P partition which are completely contained in the P^* subinterval $[x_{i-1}, x_i]$ A_i-type subintervals $(i = 1, 2, \ldots, n)$. Clearly, each A-type subinterval is of A_i-type for exactly one i. Let s_i be the sum of the lengths of all the A_i-type subintervals of P $(i = 1, 2, \ldots, n)$. Then $(b-a) - \sum_{i=1}^{n} s_i$ is the sum of the lengths of the non-A-type subintervals of P, and this number cannot exceed $(n-1)\|P\|$. Therefore

$$U(P, f) = \sum_{j=1}^{q} \hat{M}_j \Delta y_j = \sum_{A\text{-type}} \hat{M}_j \Delta y_j + \sum_{\text{non-}A\text{-type}} \hat{M}_j \Delta y_j$$

where $P: a = y_0 < y_1 < \cdots < y_{q-1} < y_q = b$ and $\hat{M}_j = \sup_x f(x)$ for $x \in [y_{j-1}, y_j]$ and where we have separated the terms of $U(P, f)$ into those which arise from A-type subintervals and those which do not. Now

$$\sum_{A\text{-type}} \hat{M}_j \Delta y_j = \sum_{i=1}^{n} \sum_{A_i\text{-type}} \hat{M}_j \Delta y_j \leq \sum_{i=1}^{n} M_i \left(\sum_{A_i\text{-type}} \Delta y_j \right)$$

$$= \sum_{i=1}^{n} M_i s_i = U(P^*, f) - \sum_{i=1}^{n} M_i (\Delta x_i - s_i)$$

$$\leq U(P^*, f) + K \sum_{i=1}^{n} (\Delta x_i - s_i)$$

$$= U(P^*, f) + K \left[(b-a) - \sum_{i=1}^{n} s_i \right]$$

$$\leq U(P^*, f) + K(n-1)\|P\|$$

Also

$$\sum_{\text{non-}A\text{-type}} \hat{M}_j \Delta y_j \leq K \sum_{\text{non-}A\text{-type}} \Delta y_j \leq K(n-1)\|P\|$$

Hence

$$U(P, f) \leq U(P^*, f) + 2K(n-1)\|P\| \leq U(P^*, f) + 2K(n-1)\delta_1$$

$$< \overline{\int} f + \frac{\varepsilon}{2} + 2K(n-1) \frac{\varepsilon}{4Kn} < \overline{\int} f + \varepsilon$$

Therefore, for each partition P with norm less than δ_1 we have

$$U(P, f) < \overline{\int} f + \varepsilon$$

Clearly $-f \in \mathcal{B}[a, b]$, and so by the first part of our proof there is a

$\delta_2 > 0$ such that

$$U(P, -f) < \overline{\int}(-f) + \varepsilon$$

whenever $\|P\| < \delta_2$. But by the discussion preceding this theorem

$$U(P, -f) = -L(P, f) \quad \text{and} \quad \overline{\int}(-f) = -\underline{\int} f$$

Hence for each P with $\|P\| < \delta_2$ we have

$$-L(P, f) < -\underline{\int} f + \varepsilon$$

and so

$$L(P, f) > \underline{\int} f - \varepsilon$$

Now if P is any partition with a norm less than $\delta = \min(\delta_1, \delta_2)$ then we have

$$\underline{\int} f - \varepsilon < L(P, f) \leq U(P, f) < \overline{\int} f + \varepsilon$$

and this concludes the proof.

In the following example we show how the above theorem can be used to evaluate the upper (or lower) integral.

Example 6.5 Let

$$f(x) = \begin{cases} x & \text{if } x \in \mathbf{Q} \cap [a, b] \\ 0 & \text{if } x \in (\mathbf{R} - \mathbf{Q}) \cap [a, b] \end{cases} \quad \text{where } a > 0$$

For any partition P, $L(P, f) = 0$, and so $\underline{\int} f = 0$. Let P_n be a regular partition of $[a, b]$ with n subintervals. Then $\Delta x_i = (b - a)/n$ for $i = 1, 2, \ldots, n$, and so

$$U(P_n, f) = \sum_{i=1}^{n} M_i \Delta x_i = \sum_{i=1}^{n} x_i \frac{b - a}{n}$$

$$= \frac{b - a}{n} \sum_{i=1}^{n} \left[a + i\left(\frac{b - a}{n}\right) \right]$$

$$= \frac{b - a}{n} \sum_{i=1}^{n} a + \left(\frac{b - a}{n}\right)^2 \sum_{i=1}^{n} i$$

$$= a(b - a) + \left(\frac{b - a}{n}\right)^2 \frac{n(n + 1)}{2}$$

$$= ab - a^2 + \frac{(b - a)^2}{2}\left(1 + \frac{1}{n}\right) = \frac{b^2 - a^2}{2} + \frac{(b - a)^2}{2n}$$

Since $\overline{\int} f \leq U(P, f)$ for every partition P, it follows that

$$\overline{\int} f \leq \frac{b^2 - a^2}{2} + \frac{(b-a)^2}{2n}$$

for every natural number n. Therefore

$$\overline{\int} f \leq \frac{b^2 - a^2}{2}$$

By Theorem 6.4, for every arbitrary $\varepsilon > 0$ there is a $\delta > 0$ such that $U(P, f) < \overline{\int} f + \varepsilon$ whenever $\|P\| < \delta$. Thus for n sufficiently large we have

$$U(P_n, f) < \overline{\int} f + \varepsilon$$

and so
$$\frac{b^2 - a^2}{2} + \frac{(b-a)^2}{2n} < \overline{\int} f + \varepsilon$$

Therefore
$$\frac{b^2 - a^2}{2} < \overline{\int} f + \varepsilon$$

and by the arbitrariness of ε it follows that $(b^2 - a^2)/2 \leq \overline{\int} f$. Hence $\overline{\int} f = (b^2 - a^2)/2$. Since $\underline{\int} f = 0$, we note that f is not Riemann-integrable on $[a, b]$, and so $\int_a^b f$ does not exist.

In the following we denote the set of all Riemann-integrable functions on $[a, b]$ by $\mathcal{R}[a, b]$. Clearly $\mathcal{R}[a, b] \subset \mathcal{B}[a, b]$; Theorem 6.2 says that $\mathcal{C}[a, b] \subseteq \mathcal{R}[a, b]$.

EXERCISES

6.1 Complete the proof of Theorem 6.3 by proving that a monotone decreasing function on $[a, b]$ is Riemann-integrable on $[a, b]$.

6.2 Let

$$f(x) = \begin{cases} 1 & \text{if } x \in [0, 1] \\ 0 & \text{if } x \in (1, 2] \end{cases}$$

For any partition P find $U(P, f)$ and $L(P, f)$. Determine the values of $\overline{\int} f$ and $\underline{\int} f$.

6.3 Give an example of a function f which is:
  (a) Bounded on $[a, b]$ but not Riemann-integrable
  (b) Riemann-integrable on $[a, b]$ but not in $\mathcal{C}[a, b]$
  (c) Riemann-integrable on $[a, b]$ but not monotone on $[a, b]$
  (d) Riemann-integrable on $[a, b]$ but neither continuous nor monotone on $[a, b]$

6.4 Determine whether f is Riemann-integrable on $[0, 1]$ and justify your answer.

(a) $f(x) = \dfrac{1}{x+2}$ (b) $f(x) = |x - \tfrac{1}{2}|$

(c) $f(x) = [\![x]\!]$ (d) $f(x) = \begin{cases} \dfrac{1}{x} & \text{if } x \neq 0 \\ 0 & \text{if } x = 0 \end{cases}$

(e) $f(x) = \begin{cases} x \sin \dfrac{1}{x} & \text{if } x \neq 0 \\ 0 & \text{if } x = 0 \end{cases}$ (f) $f(x) = \begin{cases} 0 & \text{if } x = 0 \text{ or } x = 1 \\ 1 & \text{otherwise} \end{cases}$

(g) $f(x) = \begin{cases} 0 & \text{if } x \text{ is rational} \\ \sqrt{x} & \text{if } x \text{ is irrational} \end{cases}$ (h) $f(x) = [\![10x]\!]$

(i) $f(x) = \begin{cases} \dfrac{1}{x - \tfrac{1}{2}} & \text{if } x \neq \tfrac{1}{2} \\ 0 & \text{if } x = \tfrac{1}{2} \end{cases}$ (j) $f(x) = \begin{cases} \sin \dfrac{1}{x} & \text{if } x \text{ is irrational} \\ 0 & \text{otherwise} \end{cases}$

6.5 Prove that if f is bounded on $[a, b]$ and has exactly one discontinuity in $[a, b]$ then f is Riemann-integrable on $[a, b]$.

6.6 Prove that if f is bounded on $[a, b]$ and f has only finitely many discontinuities in $[a, b]$ then f is Riemann-integrable on $[a, b]$.

6.7 Which of the following functions are Riemann-integrable on $[0, 1]$? Justify your answer.

(a) $f(x) = \begin{cases} 2x & \text{if } x \in [0, \tfrac{1}{4}] \\ 1 - x & \text{if } x \in (\tfrac{1}{4}, \tfrac{1}{2}] \\ 1 + x & \text{if } x \in (\tfrac{1}{2}, 1] \end{cases}$

(b) $f(x) = \begin{cases} \cos \dfrac{1}{x} & \text{if } x \neq 0 \\ 0 & \text{if } x = 0 \end{cases}$

6.8 Let $f(x) = x$ on $[a, b]$. Let P_n be a "regular partition" of $[a, b]$ with n subintervals. Then $\Delta x_i = (b - a)/n$ for $i = 1, 2, \ldots, n$. Find $L(P_n, f)$, $U(P_n, f)$, $\overline{\int} f$, $\underline{\int} f$, and $\int_a^b f$.

6.9 Let

$$f(x) = \begin{cases} x^2 & \text{if } x \in \mathbf{Q} \cap [a, b] \\ 0 & \text{if } x \in (\mathbf{R} - \mathbf{Q}) \cap [a, b] \end{cases} \quad \text{where } a > 0$$

Prove that $\overline{\int} f = \dfrac{b^3 - a^3}{3}$.

6.10 Let

$$f(x) = \begin{cases} x & \text{if } x \in \mathbf{Q} \cap [a, b] \\ a & \text{if } x \in (\mathbf{R} - \mathbf{Q}) \cap [a, b] \end{cases}$$

(a) For any partition P of $[a, b]$, find $L(P, f)$.
(b) Determine the value of $\underline{\int} f$.
(c) For a regular partition of P_n with n subintervals, find $U(P_n, f)$.
(d) Determine the value of $\overline{\int} f$. Prove your result.

6.11 Let

$$f(x) = \begin{cases} 4 & \text{if } x \in \mathbf{Q} \cap [0, 2] \\ 3 - x & \text{if } x \in (\mathbf{R} - \mathbf{Q}) \cap [0, 2] \end{cases}$$

(a) For any partition P of $[0, 2]$ find $U(P, f)$.
(b) Determine the value of $\overline{\int} f$.
(c) For a regular partition P_n with n subintervals, find $L(P_n, f)$.
(d) Determine the value of $\int f$. Prove your result.

6.12 Prove that the following function is Riemann-integrable on $[0, 1]$ even though it has infinitely many discontinuities on $[0, 1]$:

$$f(x) = \begin{cases} 1 & \text{if } x = \dfrac{1}{n} \text{ where } n = 1, 2, 3, \ldots \\ 0 & \text{otherwise} \end{cases}$$

6.2 PROPERTIES OF THE INTEGRAL

The reader may have seen the integral defined in terms of Riemann sums in elementary calculus. We define a Riemann sum and then in Theorem 6.5 state a condition which is sometimes given as the definition of a Riemann-integrable function.

Let $f \in \mathcal{B}[a, b]$ and let

$$P: a = x_0 < x_1 < \cdots < x_{i-1} < x_i < x_{i+1} < \cdots < x_{n-1} < x_n = b$$

be any partition of the interval $[a, b]$. We choose the intermediate points $\xi_i \in [x_{i-1}, x_i]$ ($i = 1, 2, \ldots, n$) arbitrarily but one from each of the n subintervals of P; for brevity, we denote the n intermediate points ξ_i by the single symbol ξ. The sum

$$S(P, f, \xi) = \sum_{i=1}^{n} f(\xi_i) \Delta x_i$$

is called a *Riemann sum*. Observe that for a given partition P of $[a, b]$ in general there will be many different Riemann sums depending upon the selection of the intermediate points ξ_i.

Definition 6.2 $\lim_{\|P\| \to 0} S(P, f, \xi) = \gamma$ ($\gamma \in \mathbf{R}$) if given any $\varepsilon > 0$ there exists a $\delta > 0$ such that $|S(P, f, \xi) - \gamma| < \varepsilon$ for all Riemann sums $S(P, f, \xi)$ for which $\|P\| < \delta$.

It is immediate from the above definition that for any partition P of $[a, b]$ and any selection of intermediate points $\xi_i \in [x_{i-1}, x_i]$ for $i = 1, 2, \ldots, n$ we have

$$L(P, f) \leq S(P, f, \xi) \leq U(P, f)$$

Theorem 6.5
(a) If $f \in \mathcal{R}[a, b]$ then $\lim_{||P|| \to 0} S(P, f, \xi) = \int_a^b f$.
(b) If $\lim_{||P|| \to 0} S(P, f, \xi) = \gamma$ then $f \in \mathcal{R}[a, b]$ and $\int_a^b f = \gamma$.

PROOF (a) Let $\varepsilon > 0$ be given. By Theorem 6.4 there is a $\delta > 0$ such that $U(P, f) < \int f + \varepsilon$ and $L(P, f) > \int f - \varepsilon$ whenever $||P|| < \delta$. Therefore, if $||P|| < \delta$, we have

$$\int_a^b f - \varepsilon < L(P, f) \le U(P, f) < \int_a^b f + \varepsilon$$

and so

$$\int_a^b f - \varepsilon < S(P, f, \xi) < \int_a^b f + \varepsilon$$

for all intermediate points ξ. Thus if $||P|| < \delta$ then $|S(P, f, \xi) - \int_a^b f| < \varepsilon$, and so $\lim_{||P|| \to 0} S(P, f, \xi) = \int_a^b f$.

(b) Assume that $\lim_{||P|| \to 0} S(P, f, \xi) = \gamma$ and let $\varepsilon > 0$ be arbitrary. There is a $\delta > 0$ such that if $||P|| < \delta$ then $|S(P, f, \xi) - \gamma| < \varepsilon/2$. Let P^* be a partition of $[a, b]$ with $||P^*|| < \delta$. Choose $\hat{\xi}_i \in [x_{i-1}, x_i]$ such that

$$f(\hat{\xi}_i) > M_i - \frac{\varepsilon}{2(b - a)}$$

and choose $\tilde{\xi}_i \in [x_{i-1}, x_i]$ such that

$$f(\tilde{\xi}_i) < m_i + \frac{\varepsilon}{2(b - a)} \qquad i = 1, 2, \ldots, n$$

Then

$$S(P^*, f, \hat{\xi}) = \sum_{i=1}^n f(\hat{\xi}_i) \Delta x_i$$

$$> \sum_{i=1}^n \left[M_i - \frac{\varepsilon}{2(b - a)} \right] \Delta x_i = U(P^*, f) - \frac{\varepsilon}{2}$$

and

$$S(P^*, f, \tilde{\xi}) = \sum_{i=1}^n f(\tilde{\xi}_i) \Delta x_i$$

$$< \sum_{i=1}^n \left[m_i + \frac{\varepsilon}{2(b - a)} \right] \Delta x_i = L(P^*, f) + \frac{\varepsilon}{2}$$

Consequently,

$$\overline{\int} f \le U(P^*, f) < S(P, f, \hat{\xi}) + \frac{\varepsilon}{2} < \left(\gamma + \frac{\varepsilon}{2} \right) + \frac{\varepsilon}{2} = \gamma + \varepsilon$$

and

$$\overline{\int} f \geq L(P^*, f) > S(P, f, \tilde{\xi}) - \frac{\varepsilon}{2} > \left(\gamma - \frac{\varepsilon}{2}\right) - \frac{\varepsilon}{2} = \gamma - \varepsilon$$

By the arbitrariness of ε, $\overline{\int} f \leq \gamma$ and $\underline{\int} f \geq \gamma$. Therefore $\overline{\int} f \leq \gamma \leq \underline{\int} f$. It follows that $\overline{\int} f = \underline{\int} f = \gamma$. Thus $f \in \mathcal{R}[a, b]$ and $\int_a^b f = \gamma$.

In the following example we show how Theorem 6.5 can be used to evaluate the integral provided we know that the function is Riemann-integrable.

Example 6.6 Evaluate $\int_a^b f$, where $f(x) = x$ and $[a, b]$ is any closed, bounded interval.

SOLUTION $f \in C[a, b]$, and so f is Riemann-integrable on $[a, b]$. Let P be any partition of $[a, b]$; choose the points $\xi_i = \frac{1}{2}(x_i + x_{i-1})$. Clearly $\xi_i \in [x_{i-1}, x_i]$ for $i = 1, 2, \ldots, n$. Now

$$S(P, f, \xi) = \sum_{i=1}^n f(\xi_i) \Delta x_i = \sum_{i=1}^n \frac{1}{2}(x_i + x_{i-1})(x_i - x_{i-1})$$

$$= \frac{1}{2} \sum_{i=1}^n (x_i^2 - x_{i-1}^2) = \tfrac{1}{2}(x_n^2 - x_0^2) = \tfrac{1}{2}(b^2 - a^2)$$

By Theorem 6.5, $\lim_{\|P\| \to 0} S(P, f, \xi)$ exists and equals $\int_a^b f$. The above shows that for each partition there is an appropriate choice of the intermediate points ξ_i for which $S(P, f, \xi) = \frac{1}{2}(b^2 - a^2)$. It follows that $\lim_{\|P\| \to 0} S(P, f, \xi) = \frac{1}{2}(b^2 - a^2)$, and so $\int_a^b f = \frac{1}{2}(b^2 - a^2)$.

The trick in the above example is to know how to choose the intermediate points ξ_i. It will be seen in the next section when we prove the fundamental theorem that the selection is based on the mean-value theorem of differential calculus.

Example 6.7 Evaluate $\int_0^b f$, where $f(x) = x^2$ and $b > 0$.

SOLUTION f is Riemann-integrable on $[0, b]$ since it is monotone on the interval. Let P be any partition of $[0, b]$ and this time choose the intermediate points $\xi_i = [\tfrac{1}{3}(x_i^2 + x_i x_{i-1} + x_{i-1}^2)]^{1/2}$. Then

$$0 \leq x_{i-1} = (x_{i-1}^2)^{1/2} < [\tfrac{1}{3}(x_i^2 + x_i x_{i-1} + x_{i-1}^2)]^{1/2} < (x_i^2)^{1/2} = x_i$$

for $i = 1, 2, \ldots, n$; that is, $\xi_i \in (x_{i-1}, x_i)$ for each i. Now

$$S(P, f, \xi) = \sum_{i=1}^n f(\xi_i) \Delta x_i = \sum_{i=1}^n \frac{1}{3}(x_i^2 + x_i x_{i-1} + x_{i-1}^2)(x_i - x_{i-1})$$

$$= \frac{1}{3} \sum_{i=1}^n (x_i^3 - x_{i-1}^3) = \tfrac{1}{3}(x_n^3 - x_0^3) = \tfrac{1}{3} b^3$$

Again, by Theorem 6.5, $\lim_{||P||\to 0} S(P, f, \xi)$ exists, and since for each P there is a selection of the intermediate points ξ_i for which $S(P, f, \xi) = \frac{1}{3}b^3$, it follows that $\lim_{||P||\to 0} S(P, f, \xi) = \frac{1}{3}b^3$. Hence $\int_0^b f = \frac{1}{3}b^3$.

The next theorem establishes the linearity properties of the integral. If we define $\varphi: \mathcal{R}[a, b] \to \mathbf{R}$ by the formula

$$\varphi(f) = \int_a^b f \quad \text{for each } f \in \mathcal{R}[a, b]$$

then φ satisfies the equation

$$\varphi(\alpha f + \beta g) = \alpha \varphi(f) + \beta \varphi(g)$$

for every pair $f, g \in \mathcal{R}[a, b]$ and every pair of real numbers α and β. A function which satisfies an equation of this form is called a *linear transformation*.

Theorem 6.6 Suppose $f, g \in \mathcal{R}[a, b]$ and $k \in \mathbf{R}$. Then
(a) $kf \in \mathcal{R}[a, b]$ and $\int_a^b kf = k \int_a^b f$.
(b) $(f + g) \in \mathcal{R}[a, b]$ and $\int_a^b (f + g) = \int_a^b f + \int_a^b g$.

PROOF (a) If $k = 0$, the result is immediate; thus we assume $k \neq 0$. Let $\varepsilon > 0$ be given; then there is a $\delta > 0$ such that

$$\left| S(P, f, \xi) - \int_a^b f \right| < \frac{\varepsilon}{|k|} \quad \text{whenever } ||P|| < \delta$$

Hence, if $||P|| < \delta$ then

$$\left| S(P, kf, \xi) - k \int_a^b f \right| = \left| kS(P, f, \xi) - k \int_a^b f \right|$$

$$= |k| \left| S(P, f, \xi) - \int_a^b f \right| < |k| \frac{\varepsilon}{|k|} = \varepsilon$$

Hence $\lim_{||P||\to 0} S(P, kf, \xi) = k \int_a^b f$. It follows from Theorem 6.5 that $kf \in \mathcal{R}[a, b]$ and

$$\int_a^b kf = k \int_a^b f$$

(b) Let $\varepsilon > 0$ be given. There exists $\delta_1 > 0$ such that

$$\left| S(P, f, \xi) - \int_a^b f \right| < \frac{\varepsilon}{2} \quad \text{whenever } ||P|| < \delta_1$$

and there exists $\delta_2 > 0$ such that

$$\left| S(P, g, \xi) - \int_a^b g \right| < \frac{\varepsilon}{2} \quad \text{whenever } ||P|| < \delta_2$$

Let $\delta = \min(\delta_1, \delta_2)$ and suppose that $||P|| < \delta$. Then we have

$$\left| S(P, f + g, \xi) - \left(\int_a^b f + \int_a^b g \right) \right|$$

$$= \left| S(P, f, \xi) + S(P, g, \xi) - \int_a^b f - \int_a^b g \right|$$

$$\leq \left| S(P, f, \xi) - \int_a^b f \right| + \left| S(P, g, \xi) - \int_a^b g \right| < \frac{\varepsilon}{2} + \frac{\varepsilon}{2} = \varepsilon$$

Hence $\lim_{||P|| \to 0} S(P, f + g, \xi) = \int_a^b f + \int_a^b g$; it follows that $f + g \in \mathcal{R}[a, b]$ and that

$$\int_a^b (f + g) = \int_a^b f + \int_a^b g$$

The following example shows how the linearity properties can be used to evaluate an integral.

Example 6.8 Evaluate $\int_a^b (cx + d)\, dx$, where c and d are arbitrary real numbers.

SOLUTION By Theorem 6.6,

$$\int_a^b (cx + d)\, dx = c \int_a^b x\, dx + d \int_a^b 1\, dx$$

By Examples 6.1 and 6.6 we know that

$$\int_a^b x\, dx = \tfrac{1}{2}(b^2 - a^2) \quad \text{and} \quad \int_a^b 1\, dx = b - a$$

Hence

$$\int_a^b (cx + d)\, dx = \frac{c}{2}(b^2 - a^2) + d(b - a) = \frac{b - a}{2}(ac + bc + 2d)$$

Definition 6.3 (a) If f is defined at the point $x = a$, we define

$$\int_a^a f(x)\, dx = 0$$

(b) If $f \in \mathcal{R}[a, b]$, we define

$$\int_b^a f(x)\, dx = -\int_a^b f(x)\, dx$$

When we deal with more than one interval in our discussion, additional notation is necessary to avoid ambiguity. For the interval $[a, b]$ we adopt the notation $U_a^b(P, f)$ and $L_a^b(P, f)$ for the upper and lower sums, respectively.

Theorem 6.7 If $f \in \mathcal{R}[a, b]$ and $[c, d] \subseteq [a, b]$ then $f \in \mathcal{R}[c, d]$.

PROOF Let $\varepsilon > 0$ be given. By Theorem 6.1 there is a partition P of $[a, b]$ such that $U_a^b(P, f) - L_a^b(P, f) < \varepsilon$. Refine P by choosing $P^* = P \cup \{c, d\}$. By Lemma 2 of Sec. 6.1, $U_a^b(P^*, f) \leq U_a^b(P, f)$ and $L_a^b(P^*, f) \geq L_a^b(P, f)$, and thus we have $U_a^b(P^*, f) - L_a^b(P^*, f) < \varepsilon$. Next we restrict P^* to the interval $[c, d]$ by defining $\hat{P} = P^* \cap [c, d]$. Then

$$U_c^d(\hat{P}, f) - L_c^d(\hat{P}, f) \leq U_a^b(P^*, f) - L_a^b(P^*, f)$$

since the left-hand side has fewer terms (all terms being nonnegative) than the right-hand side. Hence

$$U_c^d(\hat{P}, f) - L_c^d(\hat{P}, f) < \varepsilon$$

and again by Theorem 6.1, $f \in \mathcal{R}[c, d]$.

Theorem 6.8 If $f \in \mathcal{R}[a, b]$ and $c \in (a, b)$ then

$$\int_a^b f = \int_a^c f + \int_c^b f$$

PROOF By Theorem 6.7, $f \in \mathcal{R}[a, c]$ and $f \in \mathcal{R}[c, b]$. Let $\varepsilon > 0$ be given. By Theorem 6.5 there is a $\delta_1 > 0$ such that $|S(P, f, \xi) - \int_a^b f| < \varepsilon/3$ whenever P is a partition of $[a, b]$ with $||P|| < \delta_1$. Similarly, there is a $\delta_2 > 0$ such that $|S(P, f, \xi) - \int_a^c f| < \varepsilon/3$ whenever P is a partition of $[a, c]$ with $||P|| < \delta_2$, and there is a $\delta_3 > 0$ such that $|S(P, f, \xi) - \int_c^b f| < \varepsilon/3$ whenever P is a partition of $[c, b]$ for which $||P|| < \delta_3$. Now choose the partition P of $[a, b]$ such that P contains the point c and $||P|| < \min(\delta_1, \delta_2, \delta_3)$. Then $|S(P, f, \xi) - \int_a^b f| < \varepsilon/3$ for every choice of intermediate points ξ_i. Let $P_1 = P \cap [a, c]$ and $P_2 = P \cap [c, b]$. Then $S(P, f, \xi) = S(P_1, f, \hat{\xi}) + S(P_2, f, \xi^*)$, where $\hat{\xi} = \xi \cap [a, c]$ and $\xi^* = \xi \cap [c, b]$. Moreover, $||P_1|| \leq ||P|| < \delta_2$ and $||P_2|| \leq ||P|| <$

δ_3. Hence $|S(P_1, f, \hat{\xi}) - \int_a^c f| < \varepsilon/3$ and $|S(P_2, f, \xi^*) - \int_c^b f| < \varepsilon/3$. Finally we have that

$$\left| \int_a^b f - \left(\int_a^c f + \int_c^b f \right) \right|$$

$$= \left| \int_a^b f - S(P, f, \xi) + S(P_1, f, \hat{\xi}) - \int_a^c f + S(P_2, f, \xi^*) - \int_c^b f \right|$$

$$\leq \left| \int_a^b f - S(P, f, \xi) \right| + \left| \int_a^c f - S(P_1, f, \hat{\xi}) \right|$$

$$+ \left| \int_c^b f - S(P_2, f, \xi^*) \right| < \frac{\varepsilon}{3} + \frac{\varepsilon}{3} + \frac{\varepsilon}{3} = \varepsilon$$

Since $\varepsilon > 0$ was arbitrary, it follows that

$$\int_a^b f = \int_a^c f + \int_c^b f$$

Theorem 6.8 remains true if c is any real number provided, of course, that f is Riemann-integrable on the largest of the intervals $[a, b]$, $[c, b]$, and $[a, c]$. We leave the proof of this to the reader (see Exercise 6.15).

The following theorem establishes the *order-preserving property* of the integral. As the name suggests, if f is greater than or equal to g then the integral of f is greater than or equal to the integral of g.

Theorem 6.9 If $f, g \in \mathcal{R}[a, b]$ and $f(x) \geq g(x)$ for every $x \in [a, b]$ then

$$\int_a^b f \geq \int_a^b g$$

PROOF For each partition P of $[a, b]$ and each selection of intermediate points ξ

$$S(P, f, \xi) = \sum_{i=1}^n f(\xi_i) \Delta x_i \geq \sum_{i=1}^n g(\xi_i) \Delta x_i = S(P, g, \xi)$$

Therefore

$$\int_a^b f = \lim_{\|P\| \to 0} S(P, f, \xi) \geq \lim_{\|P\| \to 0} S(P, g, \xi) = \int_a^b g$$

Corollary If $f \in \mathcal{R}[a, b]$ and $f(x) \geq 0$ for all $x \in [a, b]$, then $\int_a^b f \geq 0$.

THE RIEMANN INTEGRAL **169**

The following theorem will prove to be a useful result; in some sense it indicates the effect of continuity on the value of the integral.

Theorem 6.10 If f is a nonnegative, integrable function on $[a, b]$ and f is positive and continuous at some point $x_0 \in [a, b]$ then $\int_a^b f > 0$.

PROOF By Lemma 3 in Sec. 4.2 there exists a $\delta > 0$ such that $f(x) > \frac{1}{2} f(x_0)$ for every $x \in (x_0 - \delta, x_0 + \delta) \cap [a, b]$. Now

$$\int_a^b f = \int_a^{x_0 - \delta} f + \int_{x_0 - \delta}^{x_0 + \delta} f + \int_{x_0 + \delta}^b f$$

which follows from Theorem 6.8 (obvious modifications are necessary when $x_0 - \delta < a$ or $x_0 + \delta > b$). By the order-preserving property of the integral we have that

$$\int_a^b f \geq 0 + \int_{x_0 - \delta}^{x_0 + \delta} \tfrac{1}{2} f(x_0) + 0 = 2\delta(\tfrac{1}{2} f(x_0)) = \delta \cdot f(x_0) > 0$$

With each function $f \in \mathcal{B}[a, b]$ we associate two nonnegative functions

$$f^+(x) = \begin{cases} f(x) & \text{if } f(x) \geq 0 \\ 0 & \text{otherwise} \end{cases} \qquad f^-(x) = \begin{cases} -f(x) & \text{if } f(x) \leq 0 \\ 0 & \text{otherwise} \end{cases}$$

(see Fig. 6.1). By definition f^+ and f^- are nonnegative and bounded on $[a, b]$; also, one can easily show that $f(x) = f^+(x) - f^-(x)$ and $|f(x)| = f^+(x) + f^-(x)$ for all $x \in [a, b]$. We write $f = f^+ - f^-$ and $|f| = f^+ + f^-$.

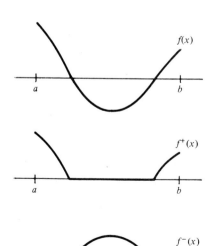

Figure 6.1

If P is a partition of $[a, b]$, then we define $M_i^+ = \sup_x f^+(x)$ and $m_i^+ = \inf_x f^+(x)$ for $x \in [x_{i-1}, x_i]$.

Lemma 1 $M_i - m_i \geq M_i^+ - m_i^+$ for each $i = 1, 2, \ldots, n$.

PROOF We consider three cases.

Case 1: $m_i \geq 0$ In this case $f(x) \geq 0$ on $[x_{i-1}, x_i]$, and so $f^+(x) = f(x)$ on $[x_{i-1}, x_i]$. Thus
$$M_i^+ = M_i \text{ and } m_i^+ = m_i$$

Case 2: $M_i \leq 0$ In this case $f(x) \leq 0$ on $[x_{i-1}, x_i]$, and so $f^+(x) = 0$ for all $x \in [x_{i-1}, x_i]$. Thus
$$M_i^+ = m_i^+ = 0$$

Case 3: $m_i < 0 < M_i$ Since f^+ is nonnegative, $m_i^+ \geq 0$, and so $m_i^+ > m_i$. Here $M_i^+ = M_i$, and again
$$M_i - m_i \geq M_i^+ - m_i^+$$

Theorem 6.11 If $f \in \mathcal{R}[a, b]$ then $f^+, f^- \in \mathcal{R}[a, b]$.

PROOF Let $\varepsilon > 0$ be given. By Theorem 6.1 there is a partition P of $[a, b]$ such that $U(P, f) - L(P, f) < \varepsilon$. Now

$$U(P, f^+) - L(P, f^+) = \sum_{i=1}^n M_i^+ \Delta x_i - \sum_{i=1}^n m_i^+ \Delta x_i = \sum_{i=1}^n (M_i^+ - m_i^+) \Delta x_i$$
$$\leq \sum_{i=1}^n (M_i - m_i) \Delta x_i = \sum_{i=1}^n M_i \Delta x_i - \sum_{i=1}^n m_i \Delta x_i$$
$$= U(P, f) - L(P, f) < \varepsilon$$

Again, by Theorem 6.1, $f^+ \in \mathcal{R}[a, b]$. Also, $f^- = f^+ - f$, and so $f^- \in \mathcal{R}[a, b]$ by Theorem 6.6.

The next result is extremely useful, as we shall see in the topics ahead.

Theorem 6.12 If $f \in \mathcal{R}[a, b]$ then $|f| \in \mathcal{R}[a, b]$ and

$$\int_a^b |f| \geq \left| \int_a^b f \right|$$

PROOF Let $f \in \mathcal{R}[a, b]$. Then $|f| = f^+ + f^- \in \mathcal{R}[a, b]$ by Theorems 6.11 and 6.6. Since f^+ and f^- are nonnegative on $[a, b]$,

$$\int_a^b f^+ \geq 0 \quad \text{and} \quad \int_a^b f^- \geq 0$$

(see the corollary to Theorem 6.9). Therefore,

$$\int_a^b |f| = \int_a^b f^+ + \int_a^b f^- \geq \int_a^b f^+ - \int_a^b f^- = \int_a^b (f^+ - f^-) = \int_a^b f$$

and

$$\int_a^b |f| = \int_a^b f^+ + \int_a^b f^- \geq \int_a^b f^- - \int_a^b f^+ = \int_a^b (f^- - f^+) = -\int_a^b f$$

It follows that

$$\int_a^b |f| \geq \left| \int_a^b f \right|$$

Theorem 6.13 If $f, g \in \mathcal{R}[a, b]$ then $fg \in \mathcal{R}[a, b]$.

We leave the proof of Theorem 6.13 to the reader (see Exercise 6.20).

EXERCISES

6.13 Prove that if $\lim_{\|P\| \to 0} S(P, f, \xi)$ exists and if for each partition P there is a choice of the intermediate points ξ_i such that $S(P, f, \xi) = \alpha$ then necessarily $\int_a^b f = \alpha$.

6.14 Using Theorem 6.5 and Exercise 6.13, evaluate the following integrals:

(a) $\int_a^b x^3 \, dx$ Hint: Consider the intermediate points

$$\xi_i = \left(\frac{x_i^3 + x_i^2 x_{i-1} + x_i x_{i-1}^2 + x_{i-1}^3}{4} \right)^{1/3}$$

(b) $\int_a^b \frac{1}{x^2} \, dx$ where $a > 0$ Hint: Consider the intermediate points $\xi_i = \sqrt{x_i x_{i-1}}$.

(c) $\int_a^b \frac{1}{\sqrt{x}} \, dx$ where $a > 0$ Hint: Consider the intermediate points

$$\xi_i = \left(\frac{\sqrt{x_i} + \sqrt{x_{i-1}}}{2} \right)^2$$

6.15 Using Definition 6.3 and Theorem 6.8, prove that if c is any real number then

$$\int_a^b f = \int_a^c f + \int_c^b f$$

provided f is Riemann-integrable on the largest of the intervals $[a, b]$, $[c, b]$, and $[a, c]$.

6.16 Prove that if f is nonnegative and continuous on $[a, b]$ then either $\int_a^b f > 0$ or else f is identically zero on $[a, b]$.

6.17 Give an example of a nonnegative function $f \in \mathcal{R}[a, b]$ such that $f(x_0) > 0$ for some $x_0 \in [a, b]$ but $\int_a^b f = 0$.

6.18 Prove that if $f \in \mathcal{C}[a, b]$ and $\int_a^b fg = 0$ for every $g \in \mathcal{C}[a, b]$ then $f(x) = 0$ for all $x \in [a, b]$.

6.19 Prove or give a counterexample: *Conjecture*: If $|f| \in \mathcal{R}[a, b]$ then $f \in \mathcal{R}[a, b]$.

6.20 Prove Theorem 6.13. *Hint*: First prove that the result holds when f and g are nonnegative integrable functions on $[a, b]$ and then use the identity

$$fg = (f^+ - f^-)(g^+ - g^-) = f^+g^+ - f^+g^- - f^-g^+ + f^-g^-$$

to establish Theorem 6.13.

6.3 FUNDAMENTAL THEOREM OF CALCULUS

In this section we obtain some results which will allow us to evaluate integrals without having to conside limits of Riemann sums as we did in the preceding section. We begin with a result which will be needed to establish the fundamental theorem of calculus.

Theorem 6.14: Mean-value theorem for integrals If f is continuous on $[a, b]$ then there is a point $c \in (a, b)$ such that

$$\int_a^b f = (b - a)f(c)$$

PROOF If f is constant on $[a, b]$, say $f(x) = k$ for all $x \in [a, b]$, then $\int_a^b f = k(b - a)$. Hence for any $c \in (a, b)$ we have

$$\int_a^b f = (b - a)f(c)$$

For the case where f is not constant on $[a, b]$ we shall use the intermediate-value theorem, but first there are some details we need in order to obtain the necessary inequalities. If $M = \sup_x f(x)$ and $m = \inf_x f(x)$ for all $x \in [a, b]$ then

$$g(x) = M - f(x) \geq 0 \quad \text{and} \quad h(x) = f(x) - m \geq 0 \quad \text{on } [a, b]$$

Moreover, our assumption that f is not constant on $[a, b]$ guarantees the existence of points $c_1, c_2 \in [a, b]$ for which $g(c_1) = M - f(c_1) > 0$

and $h(c_2) = f(c_2) - m > 0$. Since g and h are each continuous on $[a, b]$, it follows from Theorem 6.10 that

$$\int_a^b g > 0 \quad \text{and} \quad \int_a^b h > 0$$

Hence

$$\int_a^b M > \int_a^b f, \quad \text{and} \quad \int_a^b f > \int_a^b m$$

and so

$$m(b - a) < \int_a^b f < M(b - a)$$

Now, by the extreme-value theorem there are points $x_1, x_2 \in [a, b]$ with $f(x_1) = m$ and $f(x_2) = M$. Thus

$$f(x_1) < \frac{1}{b - a} \int_a^b f < f(x_2)$$

By the intermediate-value theorem there exists a point c between x_1 and x_2 (and therefore in the open interval (a, b)) such that

$$f(c) = \frac{1}{b - a} \int_a^b f$$

It follows that

$$\int_a^b f = (b - a) f(c)$$

The value $\int_a^b f/(b - a)$ is called the *mean value of the function on the interval* $[a, b]$. The preceding theorem says that a continuous function always assumes its mean value. To show the necessity of continuity in this result consider the function

$$f(x) = \begin{cases} 1 & \text{if } 1 \leq x < 2 \\ 2 & \text{if } 2 \leq x \leq 3 \end{cases}$$

$\int_1^3 f = 3$, and so the mean value of f on $[1, 3]$ is $\frac{3}{2}$, a value which the function fails to assume on the interval.

If f is Riemann-integrable on $[a, b]$ then, by Theorem 6.7, $f \in \mathcal{R}\,[a, x]$ for each $x \in (a, b]$. It follows that $\int_a^x f$ exists for each $x \in [a, b]$. Therefore the function $F(x) = \int_a^x f$ is a well-defined function on $[a, b]$. We observe that $F(a) = 0$ and $F(b) = \int_a^b f$.

✳ **Theorem 6.15: Indefinite-integral theorem** If f is continuous on $[a, b]$

then $F(x) = \int_a^x f$ is differentiable on $[a, b]$ and $F'(x) = f(x)$ for each $x \in [a, b]$.

PROOF Let x_0 be any point in $[a, b]$. Then

$$F'(x_0) = \lim_{h \to 0} \frac{F(x_0 + h) - F(x_0)}{h} = \lim_{h \to 0} \frac{1}{h}\left[\int_a^{x_0+h} f - \int_a^{x_0} f\right]$$

$$= \lim_{h \to 0} \frac{1}{h}\left[\int_a^{x_0+h} f + \int_{x_0}^{a} f\right] = \lim_{h \to 0} \frac{1}{h}\int_{x_0}^{x_0+h} f$$

(see Exercise 6.15). By the mean-value theorem for integrals there is a c between x_0 and $x_0 + h$ such that

$$\int_{x_0}^{x_0+h} f = h \cdot f(c)$$

Thus

$$F'(x_0) = \lim_{h \to 0} \frac{1}{h} [hf(c)] = \lim_{h \to 0} f(c)$$

Since f is continuous at x_0, $\lim_{h \to 0} f(c) = f(x_0)$. Therefore $F'(x_0) = f(x_0)$, and since $x_0 \in [a, b]$ was arbitrary, $F'(x) = f(x)$ for every $x \in [a, b]$.

We note that in the preceding proof we assumed that $x_0 + h$ is in the interval $[a, b]$ at all times. In particular, if x_0 is one of the endpoints a or b then h must be suitably restricted and the appropriate one-sided derivative is assumed.

The indefinite-integral theorem shows that every continuous function is a derivative of some function. We shall see shortly, however, that not every derivative is Riemann-integrable and that there are Riemann-integrable functions which are not derivatives. First we have the important fundamental theorem of calculus.

Theorem 6.16: Fundamental theorem of calculus If $f(x)$ is continuous on $[a, b]$ and $g(x)$ is any function for which $g'(x) = f(x)$ for all $x \in [a, b]$ then

$$\int_a^b f = g(b) - g(a)$$

PROOF By the indefinite-integral theorem the function $F(x) = \int_a^x f$ satisfies $F'(x) = f(x)$ for all $x \in [a, b]$. Therefore $g'(x) = F'(x)$ for every

$x \in [a, b]$, and so $g(x) = F(x) + k$ for some constant k. Now, $F(a) = 0$, and so $k = g(a)$. Thus $\int_a^b f = F(b) = g(b) - k = g(b) - g(a)$.

Example 6.9 Evaluate $\int_a^b x^n$, where $a > 0$ and $n > 0$.

SOLUTION The function $f(x) = x^n$ is continuous on $[a, b]$ and the function

$$g(x) = \frac{1}{n+1} x^{n+1}$$

satisfies $g'(x) = x^n$ for every $x \in [a, b]$. By the fundamental theorem of calculus

$$\int_a^b x^n = g(b) - g(a) = \frac{1}{n+1}(b^{n+1} - a^{n+1})$$

The fundamental theorem of calculus is an especially useful result; if f is continuous on $[a, b]$, the indefinite-integral theorem tells us that f is the derivative of some function F and the fundamental theorem of calculus then reduces the problem of evaluating $\int_a^b f$ to one of finding a function g which has f as its derivative.

We point out that if f is differentiable on $[a, b]$, the equation $\int_a^b f' = f(b) - f(a)$ is not always valid. For example, consider the function

$$f(x) = \begin{cases} x^2 \sin \dfrac{1}{x^2} & \text{if } x \in (0,1) \\ 0 & \text{if } x = 0 \end{cases}$$

f is differentiable on $[0, 1]$ and

$$f'(x) = \begin{cases} 2x \sin \dfrac{1}{x^2} - \dfrac{2}{x} \cos \dfrac{1}{x^2} & \text{if } x \in (0,1] \\ 0 & \text{if } x = 0 \end{cases}$$

f' is not bounded on $[0, 1]$ and therefore is not Riemann-integrable; that is, $\int_0^1 f'$ does not exist. In fact there are functions f which are differentiable on $[a, b]$ such that f' is bounded on $[a, b]$ but again not Riemann-integrable. Therefore the equation $\int_a^b f' = f(b) - f(a)$ fails to hold. The following extends the fundamental theorem of calculus and is a stronger form of that theorem.

Theorem 6.17 If f is differentiable on $[a, b]$ and f' is Riemann integrable on $[a, b]$ then

$$\int_a^b f' = f(b) - f(a)$$

Proof Let P be any partition of $[a, b]$. Since f' exists on $[a, b]$, f is continuous on $[x_{i-1}, x_i]$ and differentiable on (x_{i-1}, x_i) for $i = 1, 2, \ldots, n$. By the mean-value theorem of differential calculus, for each $i = 1, 2, \ldots, n$ there is a $\zeta_i \in (x_{i-1}, x_i)$ with $f(x_i) - f(x_{i-1}) = f'(\zeta_i)(x_i - x_{i-1})$. Now

$$S(P, f', \zeta) = \sum_{i=1}^{n} f'(\zeta_i) \Delta x_i = \sum_{i=1}^{n} f'(\zeta_i)(x_i - x_{i-1})$$

$$= \sum_{i=1}^{n} [f(x_i) - f(x_{i-1})] = f(x_n) - f(x_0) = f(b) - f(a)$$

Thus for every partition P there is a choice of ζ for which the Riemann sum $S(P, f', \zeta) = f(b) - f(a)$. Since $f' \in \mathcal{R}\,[a, b]$, $\lim_{\|P\| \to 0} S(P, f', \xi)$ exists. Necessarily $\lim_{\|P\| \to 0} S(P, f', \xi) = f(b) - f(a)$, and so

$$\int_a^b f' = f(b) - f(a)$$

Corollary If f is Riemann-integrable on $[a, b]$ and g is any function for which $g'(x) = f(x)$ for all $x \in [a, b]$, that is, any antiderivative of f, then

$$\int_a^b f = g(b) - g(a)$$

Example 6.10 Evaluate

$$\int_0^1 \left(2x \sin \frac{1}{x} - \cos \frac{1}{x}\right) dx$$

Solution

$$f(x) = \begin{cases} 2x \sin \dfrac{1}{x} - \cos \dfrac{1}{x} & \text{if } x \in (0, 1] \\ 0 & \text{if } x = 0 \end{cases}$$

is not continuous on $[0, 1]$ (it is discontinuous at $x = 0$), and hence the fundamental theorem of calculus does not apply. But f is bounded and continuous on $(0, 1]$, and thus it is Riemann-integrable on $[0, 1]$. (Why?) The function

$$g(x) = \begin{cases} x^2 \sin \dfrac{1}{x} & \text{if } x \in (0, 1] \\ 0 & \text{if } x = 0 \end{cases}$$

is differentiable on $[0, 1]$ and satisfies $g'(x) = f(x)$ for every $x \in [0, 1]$.

Therefore by the preceding corollary

$$\int_0^1 \left(2x \sin \frac{1}{x} - \cos \frac{1}{x}\right) dx = g(1) - g(0) = \sin 1$$

In noting above that not every derivative is Riemann-integrable, the question naturally arises: Is every Riemann-integrable function the derivative of some function? The function we considered earlier in this section

$$f(x) = \begin{cases} 1 & \text{if } 1 \le x < 2 \\ 2 & \text{if } 2 \le x \le 3 \end{cases}$$

was seen to be Riemann-integrable on $[1, 3]$ with $\int_1^3 f = 3$. Since this function has a simple discontinuity at the point $x = 2$, f is not the derivative of any function defined on $[1, 3]$ (recall that a derivative can have discontinuities only of the second kind). Therefore, there is a class of integrable functions for which the fundamental theorem is not directly applicable; other methods must be used to evaluate the integral of these functions.

EXERCISES

6.21 Use Theorem 6.8 and the fundamental theorem of calculus to evaluate the following integrals:

(a) $\int_0^5 [\![x]\!] dx$

(b) $\int_0^{2\pi} f(x) dx$, where f is the pie function of Chap. 3

(c) $\int_1^{5/2} \frac{|x|}{[\![2x - 5]\!]} dx$

6.22 Prove that if f and g are continuous on $[a, b]$ and if g does not change sign in $[a, b]$ then there is a point $c \in (a, b)$ such that

$$\int_a^b f(x)g(x) \, dx = f(c) \int_a^b g(x) \, dx$$

6.23 If $f \in \mathcal{R}[a, b]$ define $F(x) = \int_a^x f$ for every $x \in [a, b]$. Prove that F is continuous on $[a, b]$.

6.24 Find the error in the following:

(a) If $g(x) = -(1/x)$ then $g'(x) = 1/x^2$; hence

$$\int_{-1}^1 \frac{1}{x^2} dx = g(1) - g(-1) = (-1) - (1) = -2.$$

(b) If $g(x) = 2\sqrt{x}$ then $g'(x) = 1/\sqrt{x}$; hence

$$\int_0^1 \frac{1}{\sqrt{x}} dx = g(1) - g(0) = 2\sqrt{1} - 2\sqrt{0} = 2$$

6.25 Prove:
(a) If $f \in \mathcal{R}[0, b]$ and f is an even function then $f \in \mathcal{R}[-b, b]$ and
$$\int_{-b}^{b} f = 2 \int_{0}^{b} f$$
(b) If $f \in \mathcal{R}[0, b]$ and f is an odd function then $f \in \mathcal{R}[-b, b]$ and
$$\int_{-b}^{b} f = 0$$

6.26 Evaluate $\int_{a}^{b} |x|^{n} dx$, where:
(a) $a > 0, n = -1$ (b) $b < 0, n = -1$ (c) $n > 0$

6.27 Prove the *integration-by-parts formula*: if f, g are differentiable on $[a, b]$ and if f', g' are integrable on $[a, b]$ then
$$\int_{a}^{b} fg' = f(b)g(b) - f(a)g(a) - \int_{a}^{b} f'g$$

6.28 Consider $f(x) = 1/x^3$ on $[a, b]$, where $a > 0$. Prove that
$$\int_{a}^{b} f = \frac{b^2 - a^2}{2a^2 b^2}$$
without using the fundamental theorem of calculus but by selecting intermediate points ζ_i from any partition P in such a way that the resulting Riemann sum always has the value $(b^2 - a^2)/2a^2 b^2$.

6.4 NECESSARY AND SUFFICIENT CONDITIONS FOR RIEMANN INTEGRABILITY

We know that if a function f is Riemann-integrable on $[a, b]$ then it is necessarily bounded on $[a, b]$, but bounded functions are not necessarily Riemann-integrable. In Theorem 6.2 we saw that continuity of f on $[a, b]$ is sufficient to ensure Riemann integrability, but again we saw several examples of Riemann-integrable functions which are not continuous on all of $[a, b]$. Still, there is a close bond between integrability and the continuity properties of a bounded function defined on $[a, b]$. Theorem 6.3 shows that if f is monotone on $[a, b]$ then it is Riemann-integrable on $[a, b]$; monotone functions possess strong continuity properties (see Theorem 3.9). It is the purpose of this section to expose the close relationship between the notion of continuity of a function and the condition of Riemann integrability. In short, we shall see that Riemann-integrable functions must be "continuous almost everywhere."

We begin with a function f bounded on $[a, b]$. Recall that for a nonempty set $A \subseteq [a, b]$ the oscillation of f on A is given by
$$\varphi_f(A) = \sup_{x \in A} f(x) - \inf_{x \in A} f(x)$$
Since f is bounded on $[a, b]$, and hence on A, $\varphi_f(A)$ is a nonnegative real

number. It is clear that if B is a nonempty subset of A then $\varphi_f(B) \leq \varphi_f(A)$. If $x_0 \in [a, b]$, the oscillation of f at x_0 is defined by

$$\omega_f(x_0) = \inf_{\delta > 0} \varphi_f([x_0 - \delta, x_0 + \delta])$$

(see Definition 4.6), where the obvious restrictions are made if x_0 is an endpoint of the interval $[a, b]$. Again, for each $x_0 \in [a, b]$, $\omega_f(x_0)$ is a nonnegative real number. For example, if $f(x) = [\![x]\!]$, the greatest-integer function, then

$$\omega_f(x_0) = \begin{cases} 1 & \text{if } x_0 \text{ is an integer} \\ 0 & \text{otherwise} \end{cases}$$

For the function

$$g(x) = \begin{cases} \sin \dfrac{1}{x} & \text{if } x \neq 0 \\ 0 & \text{if } x = 0 \end{cases}$$

$\omega_g(0) = 2$.

Now f is continuous at x_0 if and only if $\omega_f(x_0) = 0$ (see Theorem 4.18). Thus a function is continuous at a point if and only if it has zero oscillation at that point. For each $\eta > 0$ we define

$$E_\eta = \{x \in [a, b] \mid \omega_f(x) \geq \eta\}$$

E_η is the subset of $[a, b]$ consisting of all points where the oscillation of f is not less than η. Clearly, f has a discontinuity at each point in E_η, and this holds for every positive η. Recall that

$$\mathcal{D}_f = \{x \in [a, b] \mid f \text{ is discontinuous at } x\}$$

It follows that

$$\mathcal{D}_f = \{x \in [a, b] \mid \omega_f(x) > 0\} = \bigcup_{\eta > 0} E_\eta$$

Since $0 < \eta_1 < \eta_2$ implies that $E_{\eta_2} \subseteq E_{\eta_1}$, we can write

$$\mathcal{D}_f = \bigcup_{n=k}^{\infty} E_{1/n} \qquad k = 1, 2, 3, \ldots$$

Finally, we recall Theorem 4.19: for each $\eta > 0$, E_η is a closed set; and Theorem 4.20: $\mathcal{D}_f$ is an F_σ set.

Definition 6.4: A set A is said to have *content zero* if given any $\varepsilon > 0$ there is a finite number of open intervals covering A with the sum of their lengths (*length sum*) less than ε.

Every finite set A has content zero; an interval does not have content zero, but there are infinite sets which do have content zero. For example, $A = \{1, \frac{1}{2}, \frac{1}{3}, \frac{1}{4}, \ldots\}$ has content zero. To show this let $\varepsilon > 0$ be arbitrary. Choose n_0 such that $1/n_0 < \varepsilon/5$ and consider the intervals $I_0 = (-\varepsilon/5, \varepsilon/5)$ and $I_k = (1/k - \varepsilon/4n_0, 1/k + \varepsilon/4n_0)$ for $k = 1, 2, \ldots, n_0 - 1$. Then

$$A \subset \bigcup_{k=0}^{n_0-1} I_k$$

and

$$\sum_{k=0}^{n_0-1} l(I_k) = \frac{2\varepsilon}{5} + \sum_{k=1}^{n_0-1} \frac{\varepsilon}{2n_0}$$
$$= \frac{2\varepsilon}{5} + \frac{\varepsilon(n_0 - 1)}{2n_0} < \frac{2\varepsilon}{5} + \frac{\varepsilon}{2} < \varepsilon$$

Every set of content zero is necessarily bounded, for it is a subset of a bounded set.

Definition 6.5 A set A is said to have *measure zero* if given any $\varepsilon > 0$ there is a countable set of open intervals covering A with length sum less than ε.

Intuitively, sets of measure zero are small in terms of length. It follows immediately from the definitions that every set of content zero must have measure zero. The set of all rational numbers has measure zero; to see this we order the rationals $r_1, r_2, r_3, \ldots$ and cover the set of rationals with intervals $I_k = (r_k - \varepsilon/2^{k+2}, r_k + \varepsilon/2^{k+2})$, $k = 1, 2, 3, \ldots$. Clearly

$$\sum_{k=1}^{\infty} l(I_k) = \sum_{k=1}^{\infty} \frac{\varepsilon}{2^{k+1}} = \frac{\varepsilon}{2} < \varepsilon$$

On the other hand, the set of all rationals is unbounded and so cannot have content zero. In fact, the set of rational numbers in a closed, bounded interval $[a, b]$ does not have content zero. Since in any subinterval there are rational points, any finite set of open intervals which covers the set of rationals in $[a, b]$ must have length sum greater than or equal to $b - a$.

Lemma If $\omega_f(x) < \varepsilon$ for every $x \in [a, b]$ then there is a $\delta > 0$ such that for every $[c, d] \subset [a, b]$ with $d - c < \delta$ we have $\varphi_f([c, d]) < \varepsilon$.

Proof Suppose $\omega_f(x) < \varepsilon$ for every $x \in [a, b]$. Then for each x there is a $\delta(x) > 0$ so that $\varphi_f([x - \delta(x), x + \delta(x)]) < \varepsilon$. The set of open intervals $(x - \frac{1}{2}\delta(x), x + \frac{1}{2}\delta(x))$ is an open covering of $[a, b]$. By the Heine-Borel theorem there are finitely many of these intervals, say

$$(x_j - \tfrac{1}{2}\delta(x_j), x_j + \tfrac{1}{2}\delta(x_j)) \quad \text{for } j = 1, 2, \ldots, n$$

which by themselves cover $[a, b]$. Let $\delta = \min_{1 \leq j \leq n} \frac{1}{2}\delta(x_j)$ and suppose $[c, d] \subset [a, b]$ with $d - c < \delta$. Since $c \in [a, b]$, there is an integer j_0 with $1 \leq j_0 \leq n$ such that $c \in (x_{j_0} - \frac{1}{2}\delta(x_{j_0}), x_{j_0} + \frac{1}{2}\delta(x_{j_0}))$. But $d - c < \delta < \frac{1}{2}\delta(x_{j_0})$ and so

$$[c, d] \subset (x_{j_0} - \delta(x_{j_0}), x_{j_0} + \delta(x_{j_0})) \subset [x_{j_0} - \delta(x_{j_0}), x_{j_0} + \delta(x_{j_0})]$$

Therefore

$$\varphi_f([c, d]) \leq \varphi_f([x_{j_0} - \delta(x_{j_0}), x_{j_0} + \delta(x_{j_0})]) < \varepsilon$$

Theorem 6.18 A bounded function f is Riemann-integrable on $[a, b]$ if and only if for every $\eta > 0$, E_η has content zero.

PROOF Suppose there is an $\eta > 0$ such that E_η does not have content zero. Then there is a $\delta > 0$ such that every finite set of open intervals which covers E_η has length sum greater than or equal to δ. Let P be any partition of $[a, b]$ and let $\mathcal{F}$ be the set of open subintervals of P which contain within them at least one point from E_η. The length sum of the intervals in $\mathcal{F}$ is greater than or equal to δ (otherwise we could get a finite set of open intervals covering E_η and with length sum less than δ). Therefore $U(P, f) - L(P, f) \geq \eta\delta > 0$. Since P was arbitrary, it follows from Theorem 6.1 that f is not Riemann-integrable on $[a, b]$.

Conversely, suppose that for each $\eta > 0$, E_η has content zero. Let $\varepsilon > 0$ be given and let $\mathcal{F}$ be a finite set of open intervals covering $E_{\varepsilon/2(b-a)}$ with length sum less than $\varepsilon/2(M - m)$. Recall that $M = \sup_x f(x)$, $m = \inf_x f(x)$ for all $x \in [a, b]$ and we are assuming that f is not constant on $[a, b]$. If $x \notin \bigcup_{F \in \mathcal{F}} F$ then $\omega_f(x) < \varepsilon/2(b - a)$. Now, $[a, b] - \bigcup_{F \in \mathcal{F}} F$ is the union of finitely many closed intervals and perhaps some finitely many singleton sets. By the previous lemma each of these closed intervals can be partitioned so that each subinterval $[c, d]$ in the partition satisfies $\varphi_f([c, d]) < \varepsilon/2(b - a)$. We define P to be the partition of $[a, b]$ formed from all points c, d obtained above together with the points from singleton sets in $[a, b] - \bigcup_{F \in \mathcal{F}} F$. There are two types of intervals which serve to make up P; those (from $\mathcal{F}$) which cover $E_{\varepsilon/2(b-a)}$ (they have length sum less than $\varepsilon/2(M - m)$) and those upon which the oscillation of the function does not exceed $\varepsilon/2(b - a)$. Hence

$$U(P, f) - L(P, f) < (M - m)\frac{\varepsilon}{2(M - m)} + (b - a)\frac{\varepsilon}{2(b - a)} = \frac{\varepsilon}{2} + \frac{\varepsilon}{2} = \varepsilon$$

It follows from Theorem 6.1 that f is Riemann-integrable on $[a, b]$.

We are now prepared to state the necessary and sufficient condition that a bounded function f defined on $[a, b]$ be Riemann-integrable on [a, b]. The condition shows the actual continuity character of Riemann-integrable functions.

Theorem 6.19 A bounded function f is Riemann-integrable on $[a, b]$ if and only if $\mathfrak{D}_f$, the set of points of discontinuity of f, has measure zero.

PROOF We first assume that f is Riemann-integrable on $[a, b]$. Let $\varepsilon > 0$ be given. $\mathfrak{D}_f = \bigcup_{n=1}^{\infty} E_{1/n}$; by Theorem 6.18 each set $E_{1/n}$ ($n = 1, 2, \ldots$) has content zero. Hence there exists a finite set of open intervals $\mathfrak{F}_n$ which covers $E_{1/n}$ and which has length sum less than $\varepsilon/2^n$. Let $\mathfrak{F} = \bigcup_{n=1}^{\infty} \mathfrak{F}_n$; $\mathfrak{F}$ is a countable collection of open intervals which covers $\mathfrak{D}_f$ and has length sum less than $\sum_{n=1}^{\infty} (\varepsilon/2^n) = \varepsilon$. It follows that $\mathfrak{D}_f$ has measure zero.

Conversely, suppose $\mathfrak{D}_f$ has measure zero and let $\varepsilon > 0$ be given. Consider any set E_η ($\eta > 0$ arbitrary). There is a countable set of open intervals $\mathfrak{F}$ such that $\mathfrak{F}$ covers $\mathfrak{D}_f$ and has length sum less than ε. Now $E_\eta \subseteq \mathfrak{D}_f$, and so $\mathfrak{F}$ covers E_η; also, E_η is a closed set (and hence compact). By the Heine-Borel theorem there is a finite subset $\hat{\mathfrak{F}} \subset \mathfrak{F}$ such that $\hat{\mathfrak{F}}$ covers E_η. Clearly the length sum of $\hat{\mathfrak{F}}$ is less than ε. Since $\varepsilon > 0$ was arbitrary, E_η has content zero. Due to the arbitrariness of E_η every such set must have content zero, and so, by Theorem 6.18, f is Riemann-integrable on $[a, b]$.

Example 6.11 Let $\{r_1, r_2, r_3, \ldots\}$ be an ordering of the rationals in $[a, b]$ and let

$$f(x) = \begin{cases} \dfrac{1}{n} & \text{if } x = r_n, \text{ where } n = 1, 2, 3, \ldots \\ 0 & \text{otherwise} \end{cases}$$

For each $x_0 \in [a, b]$ it is easy to show that $\lim_{x \to x_0} f(x) = 0$. Therefore $\mathfrak{D}_f = \{r_1, r_2, r_3, \ldots\}$, and by Theorem 6.19 the function f is Riemann-integrable on $[a, b]$.

EXERCISES

6.29 Find $\varphi_f([0, 1])$ for each of the following functions:

(a) $f(x) = [\![x]\!]$

(b) $f(x) = |x - \frac{1}{4}|$

(c) $f(x) = \begin{cases} \dfrac{\sin x}{x} & \text{if } x \neq 0 \\ 1 & \text{if } x = 0 \end{cases}$

(d) $f(x) = \begin{cases} \cos \dfrac{1}{x} & \text{if } x \neq 0 \\ 0 & \text{if } x = 0 \end{cases}$

6.30 Find $\omega_f(0)$ for each function f in Exercise 6.29.

6.31 Which of the following sets have content zero?
  (a) $J = \{0, \pm 1, \pm 2, \pm 3, \ldots\}$ (b) $\{n/(n+1) \mid n = 1, 2, 3, \ldots\}$
  (c) $[0, 1]$ (d) Set of all irrationals in $[0, 1]$

6.32 Which of the sets in Exercise 6.31 have measure zero?

6.33 Prove that every countable set has measure zero. Which countable sets have content zero?

6.34 Which of the following functions are Riemann-integrable on $[0, 1]$?

(a) $f(x) = [\![1000x]\!]$ (b) $f(x) = \begin{cases} x^2 & \text{if } x \text{ is rational} \\ 0 & \text{if } x \text{ is irrational} \end{cases}$

(c) $f(x) = \begin{cases} \dfrac{1}{q} & \text{if } x = \dfrac{p}{q} \text{ (properly reduced)} \\ 0 & \text{if } x \text{ is irrational} \end{cases}$

(d) $f(x) = \begin{cases} n & \text{if } x = \dfrac{1}{n} \text{ where } n = 1, 2, 3, \ldots \\ 0 & \text{otherwise} \end{cases}$

6.35 Let K be an arbitrary countably infinite subset of $[a, b]$. Find a function f defined and bounded on $[a, b]$ such that $\mathcal{D}_f = K$. Is $f \in \mathcal{R}[a, b]$?

6.36 Give an example of a function f which is differentiable on $[a, b]$ such that f' is bounded but not Riemann-integrable on $[a, b]$.

6.37 Let K be an arbitrary countably infinite subset of $[a, b]$. Define f on $[a, b]$ by
$$f(x) = \begin{cases} 1 & \text{if } x \in K \\ 0 & \text{otherwise} \end{cases}$$
Is $f \in \mathcal{R}[a, b]$? Justify your answer.

6.38 Suppose f and g are Riemann-integrable on $[a, b]$ and $f(x) \leq g(x)$ for all $x \in [a, b]$ except for those $x \in E$, where E is a subset of $[a, b]$ with measure zero. Prove that
$$\int_a^b f \leq \int_a^b g$$

6.39 Let $f : [a, b] \to \mathbf{R}$ be Riemann-integrable on $[a, b]$ and suppose $g(x) = f(x)$ for all $x \in [a, b]$ except for those $x \in E$ where E is a subset of $[a, b]$ with measure zero. Is g Riemann-integrable on $[a, b]$? Justify your answer.

6.40 Let $f : [a, b] \to \mathbf{R}$ and suppose that for any given $\varepsilon > 0$ there is a continuous function $g : [a, b] \to \mathbf{R}$ such that $f = g$ outside a union of finitely many open intervals with length sum less than ε. Is f Riemann-integrable on $[a, b]$? Justify your answer.

CHAPTER
SEVEN

SEQUENCES AND SERIES OF FUNCTIONS

7.1 INFINITE SERIES OF REAL NUMBERS

Infinite series are useful in a broad spectrum of mathematics: in elementary mathematics, for example, in defining a nonterminating decimal and in applied mathematics in defining special functions such as the Bessel functions. In mathematical analysis infinite series occur throughout the most advanced topics. This chapter should set a firm foundation upon which further study can be based.

Let $\{a_k\}$ be a sequence of real numbers. A *series*, denoted $\sum_{k=1}^{\infty} a_k$, is defined to be the sequence $\{S_n\}$, where $S_n = \sum_{k=1}^{n} a_k$. The numbers a_k are called the *terms* of the series, and the numbers S_n are called the *partial sums* of the series.

Definition 7.1 The series $\sum_{k=1}^{\infty} a_k$ is said to *converge* to S provided the sequence of partial sums $\{S_n\}$ converges to S. If $\{S_n\}$ diverges then we say the series *diverges*.

It is common to use the notation $\sum_{k=1}^{\infty} a_k$ both for the series and for the limit of the sequence of partial sums. We write $\sum_{k=1}^{\infty} a_k = S$ when the series converges to S.

Since the definition of convergence of a series is in terms of convergence of a sequence, many of the results in this section are based upon the corresponding results for sequences.

In Chap. 2 we had an example of a *geometric series*. Let $r \in \mathbf{R}$ and define $a_k = r^{k-1}$ for $k = 1, 2, 3, \ldots$. Then

$$S_n = \sum_{k=1}^{n} r^{k-1} = \begin{cases} \dfrac{1-r^n}{1-r} & \text{if } r \neq 1 \\ n & \text{if } r = 1 \end{cases}$$

If $|r| < 1$ then

$$\lim_{n \to \infty} S_n = \lim_{n \to \infty} \frac{1 - r^n}{1 - r} = \frac{1}{1 - r}$$

whereas if $|r| \geq 1$, $\{S_n\}$ diverges. Hence $\sum_{k=1}^{\infty} r^{k-1} = 1/(1 - r)$ if $|r| < 1$ and $\sum_{k=1}^{\infty} r^{k-1}$ diverges if $|r| \geq 1$.

Unfortunately, it is often not possible to obtain such a closed form for S_n, and so finding the value of $\sum_{k=1}^{\infty} a_k$ when the series converges can be extremely difficult. But even in these cases often one can readily determine whether or not the series converges. We develop several tests for convergence which allow us to say whether a series converges or diverges, and we put aside the question of finding the value of $\sum_{k=1}^{\infty} a_k$.

With each series two sequences are associated: The sequence $\{a_k\}$ of terms of the series and the sequence $\{S_n\}$ of partial sums. The reader must be careful not to confuse them. Our first theorem gives a fundamental relationship between the sequences.

Theorem 7.1 If $\sum_{k=1}^{\infty} a_k$ converges then $\lim_{k \to \infty} a_k = 0$.

PROOF If $\sum_{k=1}^{\infty} a_k$ converges, say to S, then $\lim_{n \to \infty} S_n = S$. Now $a_k = S_k - S_{k-1}$ for $k \geq 2$, and so

$$\lim_{k \to \infty} a_k = \lim_{k \to \infty} (S_k - S_{k-1}) = \lim_{k \to \infty} S_k - \lim_{k \to \infty} S_{k-1} = S - S = 0$$

The above theorem is useful in its contrapositive form: If $\lim_{k \to \infty} a_k \neq 0$ then $\sum_{k=1}^{\infty} a_k$ diverges. For example, the series $\sum_{k=1}^{\infty} \cos(\pi/k)$ diverges since $\lim_{k \to \infty} \cos(\pi/k) = 1$; the series $\sum_{k=1}^{\infty} (-1)^k$ diverges since $\lim_{k \to \infty} (-1)^k$ does not exist.

To show that the condition $\lim_{k \to \infty} a_k = 0$ is not sufficient to guarantee convergence of $\sum_{k=1}^{\infty} a_k$ we consider the *harmonic series* $\sum_{k=1}^{\infty} (1/k)$. For any $n \in \mathbb{N}$

$$S_{2^n} = 1 + \frac{1}{2} + \frac{1}{3} + \frac{1}{4} + \frac{1}{5} + \cdots + \frac{1}{2^n} = 1 + \frac{1}{2} + \left(\frac{1}{3} + \frac{1}{4}\right)$$

$$+ \left(\frac{1}{5} + \frac{1}{6} + \frac{1}{7} + \frac{1}{8}\right) + \cdots + \left(\frac{1}{2^{n-1} + 1} + \frac{1}{2^{n-1} + 2} + \cdots + \frac{1}{2^n}\right)$$

$$> 1 + \frac{1}{2} + \left(\frac{1}{4} + \frac{1}{4}\right) + \left(\frac{1}{8} + \frac{1}{8} + \frac{1}{8} + \frac{1}{8}\right) + \cdots + \left(\frac{1}{2^n} + \frac{1}{2^n} + \cdots + \frac{1}{2^n}\right)$$

$$= 1 + \left(\frac{1}{2} + \frac{1}{2} + \frac{1}{2} + \cdots + \frac{1}{2}\right) = 1 + \frac{n}{2}$$

It follows that the sequence $\{S_n\}$ is unbounded and hence diverges. Therefore $\sum_{k=1}^{\infty} (1/k)$ diverges, yet $\lim_{k \to \infty} (1/k) = 0$.

A necessary and sufficient condition for convergence of a series follows from the Cauchy criterion for sequences (see Definition 2.2 and Theorem 2.8).

Theorem 7.2: Cauchy criterion $\sum_{k=1}^{\infty} a_k$ converges if and only if given any $\varepsilon > 0$ there exists a natural number N such that
$$|a_{n+1} + a_{n+2} + \cdots + a_{n+p}| < \varepsilon$$
for every $n \geq N$ and every $p \in \mathbf{N}$.

Theorem 7.2 tells us that the size of the "first few terms" of an infinite series has no effect on whether or not the series converges. In other words, the convergence character of a series is completely determined in "the tail" of the series. This is made more precise in the following lemma, the proof of which is an immediate consequence of the Cauchy criterion.

Lemma If $\sum_{k=1}^{\infty} a_k$ converges then for any natural number N the series $\sum_{k=N}^{\infty} a_k$ converges. If $\sum_{k=N}^{\infty} a_k$ converges for some natural number N then $\sum_{k=1}^{\infty} a_k$ converges.

Theorem 7.3: Comparison test If $0 < a_k \leq b_k$ for every $k \geq N$, where $N \in \mathbf{N}$, then convergence of $\sum_{k=1}^{\infty} b_k$ implies that $\sum_{k=1}^{\infty} a_k$ converges.

PROOF By the preceding lemma it suffices to prove this theorem for the case $N = 1$. Let $A_n = \sum_{k=1}^{n} a_k$ and $B_n = \sum_{k=1}^{n} b_k$. Since a_k and b_k are positive terms, each of the sequences $\{A_n\}$, $\{B_n\}$ is strictly increasing. Moreover, since $a_k \leq b_k$ for $k = 1, 2, 3, \ldots$, we have that $0 < A_n \leq B_n$ for every $n \in \mathbf{N}$. Now if $\sum_{k=1}^{\infty} a_k$ diverges then $\{A_n\}$ diverges and so is an unbounded sequence. Hence $\{B_n\}$ is an unbounded sequence and therefore diverges. It follows that $\sum_{k=1}^{\infty} b_k$ diverges.

As an example of the use of the comparison test we note that the series $\sum_{k=1}^{\infty} (1/k!)$ converges by comparison with the geometric series, $r = \frac{1}{2}$. The reader should have little difficulty showing that $k! \geq 2^{k-1}$ holds for every natural number k (mathematical induction; Exercise 1.16).

As a second example we note that the series $\sum_{k=1}^{\infty} (1/\sqrt{k})$ diverges by comparison with the harmonic series.

Corollary 1 If $a_k > 0$, $b_k > 0$ for $k = 1, 2, 3, \ldots$ and $\{a_k/b_k\}$, $\{b_k/a_k\}$ are both bounded sequences then $\sum_{k=1}^{\infty} a_k$ and $\sum_{k=1}^{\infty} b_k$ either both converge or both diverge.

PROOF Suppose $a_k > 0$, $b_k > 0$, $a_k/b_k \leq M_1$, and $b_k/a_k \leq M_2$ for every natural number k, where $M_1 > 0$, $M_2 > 0$. Then $0 < \alpha \leq b_k/a_k \leq \beta$

holds for every $k \in \mathbf{N}$, where $\alpha = 1/M_1$ and $\beta = M_2$. Since $a_k > 0$ for every k,

$$0 < \alpha a_k \le b_k \le \beta a_k \qquad \text{holds for } k = 1, 2, 3, \ldots$$

Now if $\sum_{k=1}^{\infty} b_k$ converges then $\sum_{k=1}^{\infty} \alpha a_k$ converges by the comparison test, and so $\sum_{k=1}^{\infty} a_k$ converges (see Exercise 7.4). Conversely, if $\sum_{k=1}^{\infty} a_k$ converges then so does $\sum_{k=1}^{\infty} \beta a_k$, and hence by the comparison test $\sum_{k=1}^{\infty} b_k$ converges.

The following corollary is a special case of Corollary 1.

Corollary 2: Limit form of the comparison test If $a_k > 0$, $b_k > 0$ for $k = 1, 2, 3, \ldots$ and $\lim_{k \to \infty} (b_k/a_k)$ exists and is a positive real number then $\sum_{k=1}^{\infty} a_k$ and $\sum_{k=1}^{\infty} b_k$ either both converge or both diverge.

The next result gives a convergence test which is appropriate for a certain class of infinite series.

Theorem 7.4: Integral test If $\{a_k\}$ is a sequence of positive terms which is monotone decreasing to zero and if f is a monotone decreasing function defined on $[1, \infty)$ such that $f(k) = a_k$ for each natural number k then the series $\sum_{k=1}^{\infty} a_k$ and the sequence $\int_1^n f(x)\, dx$ either both converge or both diverge.

PROOF For each natural number k

$$a_{k+1} \le f(x) \le a_k \qquad \text{for every } x \in [k, k+1]$$

Hence

$$a_{k+1} \le \int_k^{k+1} f(x)\, dx \le a_k \qquad \text{for each } k \in \mathbf{N}$$

If $n \ge 2$

$$\sum_{k=1}^{n-1} a_{k+1} \le \sum_{k=1}^{n-1} \int_k^{k+1} f(x)\, dx \le \sum_{k=1}^{n-1} a_k$$

and so

$$S_n - a_1 \le \int_1^n f(x)\, dx \le S_{n-1}$$

Since $a_k > 0$ for each $k \in \mathbf{N}$, $\{S_n\}$ is a monotone increasing sequence; similarly, since $f(x) > 0$ on $[1, \infty)$, $\int_1^n f(x)\, dx$ is a monotone increasing sequence. Now if $\sum_{k=1}^{\infty} a_k$ converges, say $\{S_n'\}$ converges to S, then $\int_1^n f(x)\, dx$ is bounded by S and so converges. Conversely, if $\int_1^n f(x)\, dx$

converges to I then $\{S_n\}$ is bounded by $I + a_1$, and so $\{S_n\}$ converges. Therefore the series $\sum_{k=1}^{\infty} a_k$ converges.

The next example involves an application of the integral test.

Example 7.1 Let p be any real number. The series $\sum_{k=1}^{\infty} (1/k^p)$ is called a p series. We determine for which values of p the series converges.

SOLUTION If p is negative or zero, $\lim_{k \to \infty} (1/k^p) \neq 0$ and so the series diverges, by Theorem 7.1. If $p > 0$ then the sequence $\{1/k^p\}$ consists of positive terms which monotonically decrease to zero. $f(x) = 1/x^p$ is monotone decreasing on $[1, \infty)$, and $f(k) = 1/k^p$ for $k = 1, 2, 3, \ldots$. Thus by the integral test $\sum_{k=1}^{\infty} (1/k^p)$ converges if and only if the sequence $\int_1^n f(x)\, dx = \int_1^n (1/x^p)\, dx$ converges.

$$\int_1^n \frac{1}{x^p}\, dx = \begin{cases} \dfrac{n^{1-p} - 1}{1 - p} & \text{if } p \neq 1 \\ \ln n & \text{if } p = 1 \end{cases}$$

Hence $\int_1^n (1/x^p)\, dx$ converges if and only if $p > 1$. It follows that the p series $\sum_{k=1}^{\infty} (1/k^p)$ converges if and only if $p > 1$.

Example 7.1 provides us with a class of known series that can be used in conjunction with the comparison test in order to determine the convergence character of other series.

Example 7.2 Determine whether the series $\sum_{k=1}^{\infty} [1/\sqrt{k(k+1)}]$ converges or diverges.

SOLUTION We might observe that

$$\frac{1}{\sqrt{k(k+1)}} < \frac{1}{\sqrt{k^2}} = \frac{1}{k}$$

but the harmonic series diverges, and so a direct use of the comparison test here is not helpful. However, the limit form of the comparison test (Corollary 2 to Theorem 7.2) can be useful.

If we let $a_k = 1/\sqrt{k(k+1)}$ and $b_k = 1/k$, then

$$\frac{b_k}{a_k} = \frac{1/k}{1/\sqrt{k(k+1)}} = \frac{\sqrt{k(k+1)}}{k} = \sqrt{1 + \frac{1}{k}}$$

and

$$\lim_{k \to \infty} \frac{b_k}{a_k} = \lim_{k \to \infty} \sqrt{1 + \frac{1}{k}} = 1$$

Since
$$\sum_{k=1}^{\infty} b_k = \sum_{k=1}^{\infty} \frac{1}{k}$$
is known to diverge, the series
$$\sum_{k=1}^{\infty} a_k = \sum_{k=1}^{\infty} \frac{1}{\sqrt{k(k+1)}}$$
must also diverge.

An alternate approach to this example is to use the comparison test in conjunction with the inequalities
$$\frac{1}{\sqrt{k(k+1)}} > \frac{1}{\sqrt{(k+1)^2}} = \frac{1}{k+1}$$

Theorem 7.5: Ratio test If the terms $a_k > 0$, $\limsup (a_{k+1}/a_k) = R$, and $\liminf (a_{k+1}/a_k) = r$ then
(a) $R < 1$ implies that $\sum_{k=1}^{\infty} a_k$ converges.
(b) $r > 1$ implies that $\sum_{k=1}^{\infty} a_k$ diverges.

PROOF (a) If $R < 1$ then given ε, $0 < \varepsilon < 1 - R$, there is a natural number N such that $a_{k+1}/a_k < R + \varepsilon < 1$ whenever $k > N$. If $\eta = R + \varepsilon$ then $a_{k+1} < \eta a_k$ for all $k > N$, and hence $a_k < \eta^{k-N} a_N$ for all $k > N$. The series $\sum_{k=N+1}^{\infty} \eta^{k-N} a_N = a_N \sum_{j=1}^{\infty} \eta^j$ converges because $\eta < 1$, and it follows from the comparison test that $\sum_{k=N+1}^{\infty} a_k$ converges. Therefore $\sum_{k=1}^{\infty} a_k$ converges.
(b) If $r > 1$ then $a_{k+1}/a_k > 1$ for $k \geq N$, for some sufficiently large N. Thus $k > N$ implies $a_k > a_N > 0$, and so $\lim_{k \to \infty} a_k \neq 0$. Therefore $\sum_{k=1}^{\infty} a_k$ diverges.

Corollary If $a_k > 0$ for every $k \in \mathbf{N}$ and $\lim_{k \to \infty} (a_{k+1}/a_k) = l$, then
(a) $l < 1$ implies that $\sum_{k=1}^{\infty} a_k$ converges.
(b) $l > 1$ implies that $\sum_{k=1}^{\infty} a_k$ diverges.

Note that if $\lim_{k \to \infty} (a_{k+1}/a_k) = 1$, nothing can be concluded. For if $a_k = 1/k$ then $\lim_{k \to \infty} (a_{k+1}/a_k) = \lim_{k \to \infty} (k/(k+1)) = 1$ and the harmonic series $\sum_{k=1}^{\infty} 1/k$ diverges; on the other hand, if $a_k = 1/k^2$ then $\lim_{k \to \infty} (a_{k+1}/a_k) = \lim_{k \to \infty} [k^2/(k+1)^2] = 1$ and $\sum_{k=1}^{\infty} 1/k^2$ is a convergent p series.

Theorem 7.6: Root test If $a_k > 0$ and $\limsup \sqrt[k]{a_k} = R$ then
(a) $R < 1$ implies that $\sum_{k=1}^{\infty} a_k$ converges.
(b) $R > 1$ implies that $\sum_{k=1}^{\infty} a_k$ diverges.

The proof of Theorem 7.6 is similar to that of Theorem 7.5, and we leave it to the reader (see Exercise 7.9).

Corollary If $a_k > 0$ for every $k \in \mathbf{N}$ and $\lim_{k \to \infty} \sqrt[k]{a_k} = l$, then
(a) $l < 1$ implies that $\sum_{k=1}^{\infty} a_k$ converges.
(b) $l > 1$ implies that $\sum_{k=1}^{\infty} a_k$ diverges.

In practice the ratio test is generally easier to apply then the root test, but there are many examples where the ratio test fails and the root test yields the convergence character of the series. Consider the following.

Example 7.3 Test for convergence: $\sum_{k=1}^{\infty} a_k$, where $a_{2j} = 1/3^j$ and $a_{2j-1} = 1/(3^{j+1})$.

SOLUTION Equivalently,

$$a_k = \begin{cases} \dfrac{1}{3^{(k+3)/2}} & \text{if } k \text{ is odd} \\ \dfrac{1}{3^{k/2}} & \text{if } k \text{ is even} \end{cases}$$

Then

$$\frac{a_{k+1}}{a_k} = \begin{cases} 3 & \text{if } k \text{ is odd} \\ \frac{1}{9} & \text{if } k \text{ is even} \end{cases}$$

and so the ratio test yields no information. On the other hand

$$\sqrt[k]{a_k} = \begin{cases} \dfrac{1}{3^{(1/2) + 3/(2k)}} & \text{if } k \text{ is odd} \\ \dfrac{1}{3^{1/2}} & \text{if } k \text{ is even} \end{cases}$$

Hence $\lim_{k \to \infty} \sqrt[k]{a_k} = 1/\sqrt{3} < 1$, and it follows from the root test that $\sum_{k=1}^{\infty} a_k$ converges.

The previous tests have been for series with positive terms. If a series does not satisfy $a_k > 0$ for all $k \geq N$, where $N \in \mathbf{N}$, we still may be able to use the previous tests by virtue of the following theorem.

Theorem 7.7 If $\sum_{k=1}^{\infty} |a_k|$ converges then $\sum_{k=1}^{\infty} a_k$ converges.

PROOF If $\sum_{k=1}^{\infty} |a_k|$ converges then by the Cauchy criterion (Theorem 7.2) given any $\varepsilon > 0$ there is a natural number N such that

$$|a_{n+1}| + |a_{n+2}| + \cdots + |a_{n+p}| < \varepsilon$$

for every $n \geq N$ and for each $p \in \mathbf{N}$. Since

$$|a_{n+1} + a_{n+2} + \cdots + a_{n+p}| \leq |a_{n+1}| + |a_{n+2}| + \cdots + |a_{n+p}|$$

it follows again from Theorem 7.2 that the series $\sum_{k=1}^{\infty} a_k$ converges.

The converse of Theorem 7.7 does not hold; there exist series $\sum_{k=1}^{\infty} a_k$ which converge for which the series of absolute values $\sum_{k=1}^{\infty} |a_k|$ diverges. We shall show later (Theorem 7.9) that the alternating harmonic series $\sum_{k=1}^{\infty} [(-1)^{k+1}/k]$ converges, so this is an example of just such a series.

Definition 7.2 If $\sum_{k=1}^{\infty} |a_k|$ converges then we say the series $\sum_{k=1}^{\infty} a_k$ *converges absolutely*; if $\sum_{k=1}^{\infty} a_k$ converges but $\sum_{k=1}^{\infty} |a_k|$ diverges then we say the series $\sum_{k=1}^{\infty} a_k$ *converges conditionally*.

One of the consequences of absolute convergence is that it guarantees that any rearrangement of the terms of the series yields a series which converges to the same value as the original series. This is not the case with conditionally convergent series, as the following example demonstrates.

Example 7.4 We noted above that the alternating harmonic series $\sum_{k=1}^{\infty} [(-1)^{k+1}/k]$ is conditionally convergent. Writing a few terms of this series, we get

$$1 - \frac{1}{2} + \frac{1}{3} - \frac{1}{4} + \frac{1}{5} - \frac{1}{6} + \frac{1}{7} - \frac{1}{8} + \cdots = S \qquad (1)$$

We are calling the limit or sum of this series S; since $S_{2n} \geq \frac{1}{2}$, it is clear that $S \geq \frac{1}{2}$. The series $\sum_{k=1}^{\infty} [(-1)^{k+1}/2k]$ will have sum $S/2$ (see Exercise 7.4). Some terms of this series are

$$\frac{1}{2} - \frac{1}{4} + \frac{1}{6} - \frac{1}{8} + \frac{1}{10} - \frac{1}{12} + \frac{1}{14} - \frac{1}{16} + \cdots = \frac{S}{2} \qquad (2)$$

We consider a third series

$$0 + \frac{1}{2} + 0 - \frac{1}{4} + 0 + \frac{1}{6} + 0 - \frac{1}{8} + \cdots = \frac{S}{2} \qquad (3)$$

The sum of series (3) is $S/2$ since the sequence of partial sums for this series is

$$0, S'_1, S'_1, S'_2, S'_2, S'_3, S'_3, S'_4, S'_4, \ldots$$

where $\{S'_n\}$ is the sequence of partial sums of the series (2). If we add series (1) term by term to series (3) and use Exercise 7.5, we get

$$1 + 0 + \frac{1}{3} - \frac{1}{2} + \frac{1}{5} + 0 + \frac{1}{7} - \frac{1}{4} + \cdots = \frac{3S}{2} \qquad (4)$$

The series

$$1 + \frac{1}{3} - \frac{1}{2} + \frac{1}{5} + \frac{1}{7} - \frac{1}{4} + \frac{1}{9} + \frac{1}{11} - \frac{1}{6} + \cdots = \frac{3S}{2} \qquad (5)$$

has sum $3S/2$ since its sequence of partial sums is a subsequence of the sequence of partial sums of series (4). A close inspection of series (5) shows that it is a rearrangment of series (1), yet they converge to distinct values since $S \neq 0$.

Theorem 7.8 If $\sum_{k=1}^{\infty} a_k'$ is a rearrangement of the absolutely convergent series $\sum_{k=1}^{\infty} a_k$ and $\sum_{k=1}^{\infty} a_k = S$ then $\sum_{k=1}^{\infty} a_k' = S$.

PROOF If $\sum_{k=1}^{\infty} a_k'$ is a rearrangement of $\sum_{k=1}^{\infty} a_k$ then there is a one-to-one function φ from $\mathbf{N}$ onto $\mathbf{N}$ such that $a_k' = a_{\varphi(k)}$ for $k = 1, 2, 3, \ldots$. Let $\varepsilon > 0$ be given. Since the series $\sum_{k=1}^{\infty} |a_k|$ converges, it follows from the Cauchy criterion that there is a natural number N such that $\sum_{k=N+1}^{\infty} |a_k| < \varepsilon/2$. Choose the natural number K so large that $\{1, 2, 3, \ldots, N\} \subseteq \{\varphi(1), \varphi(2), \varphi(3), \ldots, \varphi(K)\}$. If $n \geq K$ then each term $a_1, a_2, a_3, \ldots, a_N$ occurs among the terms $a_{\varphi(1)}, a_{\varphi(2)}, a_{\varphi(3)}, \ldots, a_{\varphi(n)}$; that is, $\{a_1, a_2, a_3, \ldots, a_N\} \subseteq \{a_1', a_2', a_3', \ldots, a_n'\}$. Thus

$$\left| \sum_{k=1}^{N} a_k - \sum_{k=1}^{n} a_k' \right| \leq \sum_{k=N+1}^{\infty} |a_k| < \frac{\varepsilon}{2}$$

Therefore if $n \geq K$,

$$\left| S - \sum_{k=1}^{n} a_k' \right| = \left| S - \sum_{k=1}^{N} a_k + \sum_{k=1}^{N} a_k - \sum_{k=1}^{n} a_k' \right| \leq \left| S - \sum_{k=1}^{N} a_k \right|$$

$$+ \left| \sum_{k=1}^{N} a_k - \sum_{k=1}^{n} a_k' \right| = \left| \sum_{k=N+1}^{\infty} a_k \right| + \left| \sum_{k=1}^{N} a_k - \sum_{k=1}^{n} a_k' \right|$$

$$\leq \sum_{k=N+1}^{\infty} |a_k| + \frac{\varepsilon}{2} < \frac{\varepsilon}{2} + \frac{\varepsilon}{2} = \varepsilon$$

It follows that $\sum_{k=1}^{\infty} a_k' = S$.

We conclude this section with a final convergence test which is one of the easiest to apply.

Theorem 7.9 Alternating-series test If $\{a_k\}$ is a monotone decreasing sequence with $\lim_{k \to \infty} a_k = 0$ then the alternating series $\sum_{k=1}^{\infty} (-1)^k a_k$ converges.

PROOF We again let $S_n = \sum_{k=1}^{n} (-1)^k a_k$, and examine first the even-numbered partial sums

$$S_2 = a_2 - a_1 \leq 0$$
$$S_4 - S_2 = a_4 - a_3 \leq 0 \quad \text{and so} \quad S_4 \leq S_2$$
$$S_6 - S_4 = a_6 - a_5 \leq 0 \quad \text{and so} \quad S_6 \leq S_4$$

In general

$$S_{2n} - S_{2n-2} = a_{2n} - a_{2n-1} \leq 0 \quad \text{and so} \quad S_{2n} \leq S_{2n-2}$$

Consequently $\{S_{2n}\}$ is a monotone decreasing sequence. Moreover

$$S_{2n} = -a_1 + a_2 - a_3 + a_4 - a_5 + \cdots + a_{2n-2} - a_{2n-1} + a_{2n}$$
$$= -a_1 + (a_2 - a_3) + (a_4 - a_5) + \cdots + (a_{2n-2} - a_{2n-1}) + a_{2n} \geq -a_1$$

Hence $\{S_{2n}\}$ is bounded below and therefore must converge. We write $\lim_{n \to \infty} S_{2n} = S$. Now $S_{2n-1} = S_{2n} - a_{2n}$, and so

$$\lim_{n \to \infty} S_{2n-1} = \lim_{n \to \infty} S_{2n} - \lim_{n \to \infty} a_{2n} = S - 0 = S$$

Hence $\lim_{n \to \infty} S_n = S$ and therefore $\sum_{k=1}^{\infty} (-1)^k a_k$ converges.

We also note that

$$|S - S_n| = \left| \sum_{k=1}^{\infty} (-1)^k a_k - \sum_{k=1}^{n} (-1)^k a_k \right| = \left| \sum_{k=n+1}^{\infty} (-1)^k a_k \right|$$
$$= |a_{n+1} - a_{n+2} + a_{n+3} - a_{n+4} + \cdots|$$
$$= (a_{n+1} - a_{n+2}) + (a_{n+3} - a_{n+4}) + \cdots$$

since each term in parenthesis is nonnegative (see Exercise 7.14). It follows that

$$|S - S_n| = a_{n+1} - (a_{n+2} - a_{n+3}) - (a_{n+4} - a_{n+5}) - \cdots \leq a_{n+1}$$

since again each term in parenthesis is nonnegative. Therefore for each natural number n we have $|S - S_n| \leq a_{n+1}$.

Earlier we referred to the fact that the alternating harmonic series $\sum_{k=1}^{\infty} [(-1)^{k+1}/k]$ converges. This is an immediate consequence of the above alternating-series test. There are, of course, numerous other tests for convergence of series of constants; some of them, though important for their own sake, are rather subtle and have infrequent application. At this point we choose to go on to a discussion of series of functions and to develop the relationship these series have to the important properties of functions discussed in the previous chapters.

EXERCISES

7.1 The series $\sum_{k=1}^{\infty} [1/(k^2 + k)]$ converges; find its value. *Hint*:
$$\frac{1}{k^2 + k} = \frac{1}{k} - \frac{1}{k+1}$$

7.2 Prove the Cauchy criterion for series, Theorem 7.2.

7.3 Prove the lemma preceding Theorem 7.3.

7.4 Prove that if $\sum_{k=1}^{\infty} a_k$ converges to S and c is any constant then $\sum_{k=1}^{\infty} ca_k$ converges to cS.

7.5 Prove that if $\sum_{k=1}^{\infty} a_k$ converges to A and $\sum_{k=1}^{\infty} b_k$ converges to B then $\sum_{k=1}^{\infty} (a_k + b_k)$ converges to $A + B$.

7.6 Prove the limit form of the comparison test.

7.7 Determine which of the following series converge:

(a) $\sum_{k=1}^{\infty} \frac{1}{k^2 + 2k}$ (b) $\sum_{k=1}^{\infty} k^{-k}$ (c) $\sum_{k=2}^{\infty} \frac{1}{k \ln k}$

(d) $\sum_{k=1}^{\infty} \sin \frac{\pi}{k}$ (e) $\sum_{k=3}^{\infty} \frac{1}{k^2 - 3k + 2}$ (f) $\sum_{k=3}^{\infty} \frac{1}{k(k-2)}$

$(k-1)(k-2)$

7.8 Prove the corollary to Theorem 7.5.

7.9 Prove the root test, Theorem 7.6.

7.10 Prove the corollary to Theorem 7.6.

7.11 Determine which of the following series converge:

(a) $\sum_{k=1}^{\infty} \frac{1}{k!}$ (b) $\sum_{k=1}^{\infty} \frac{k!}{k^k}$ (c) $\sum_{k=1}^{\infty} \frac{3^k}{k!}$ (d) $\sum_{k=1}^{\infty} \frac{3^k + 4^k}{5^k}$

7.12 Determine which of the following series converge. Which converge absolutely?

(a) $\sum_{k=1}^{\infty} \frac{(-1)^k}{\sqrt{k}}$ (b) $\sum_{k=1}^{\infty} (-1)^k \frac{k}{k+1}$

(c) $\sum_{k=1}^{\infty} (-1)^k \frac{2k+1}{k^3 - 4}$ (d) $\sum_{k=1}^{\infty} (-1)^k \frac{2^k k^2}{k!}$

7.13 Examine the proof of Theorem 7.8 and indicate where the proof fails if the series $\sum_{k=1}^{\infty} a_k$ is conditionally convergent.

7.14 Prove that if $\sum_{k=1}^{\infty} a_k$ converges then any regrouping of the terms of the series converges to the same limit. *Note*: A regrouping of $\sum_{k=1}^{\infty} a_k$ is a series of the form
$$(a_1 + a_2 + \cdots + a_m) + (a_{m+1} + a_{m+2} + \cdots + a_n) + \cdots$$

7.15 The following extends the limit form of the comparison test.
(a) Prove that if $a_k > 0$, $b_k > 0$ for $k = 1, 2, 3, \ldots$ and $\lim_{k \to \infty} (b_k/a_k) = 0$ and $\sum_{k=1}^{\infty} a_k$ converges then $\sum_{k=1}^{\infty} b_k$ converges.
(b) Prove that if $a_k > 0$, $b_k > 0$ for $k = 1, 2, 3, \ldots$ and $\lim_{k \to \infty} (b_k/a_k) = \infty$ and $\sum_{k=1}^{\infty} a_k$ diverges then $\sum_{k=1}^{\infty} b_k$ diverges.

7.16 For the series in Exercises 7.12 which converge estimate the error in approximating the sum by the first five terms of the series (note the remark following the proof to Theorem 7.9).

7.17 Prove that if $\sum_{k=1}^{\infty} a_k$ converges to S and $\sum_{k=1}^{\infty} a'_k$ is a rearrangement of $\sum_{k=1}^{\infty} a_k$ such that $a'_k = a_k$ for $k \geq N$ then $\sum_{k=1}^{\infty} a'_k$ converges to S.

7.18 Prove that if $\sum_{k=1}^{\infty} a_k$ is conditionally convergent and p_j denotes the jth positive term of the series then the series $\sum_{j=1}^{\infty} p_j$ diverges.

7.19 Prove that if $\sum_{k=1}^{\infty} a_k$ is conditionally convergent and γ is any real number then there is some rearrangement of the series $\sum_{k=1}^{\infty} a_k$ which converges to γ.

7.20 Prove that if $\sum_{k=1}^{\infty} a_k$ is conditionally convergent then there exists some rearrangement of this series which diverges.

7.2 POINTWISE AND UNIFORM CONVERGENCE

Returning again to the geometric series

$$\sum_{k=1}^{\infty} x^{k-1} = \frac{1}{1-x} \quad \text{if } |x| < 1$$

we observe that the terms $f_k(x) = x^{k-1}$ ($k = 1, 2, 3, \ldots$) are functions, as is the limit $f(x) = 1/(1-x)$. The meaning of the equality above is that for each value of x satisfying $-1 < x < 1$ the series converges to the value $1/(1-x)$. Hence the series determines a function $f(x) = 1/(1-x)$ with domain $(-1, 1)$. We call this type of convergence *pointwise convergence*. We begin the discussion of convergence of functions by considering a sequence of functions $\{S_n(x)\}$ each defined on an interval I.

Definition 7.3 The sequence of functions $\{S_n(x)\}$ *converges pointwise* to the function $S(x)$ on the interval I if for each $x_0 \in I$ the sequence of real numbers $\{S_n(x_0)\}$ converges to the real number $S(x_0)$.

Thus $\{S_n(x)\}$ converges pointwise to the function $S(x)$ if, for every $x_0 \in I$, given any $\varepsilon > 0$ there is a natural number N (depending on ε and the point x_0) such that $|S_n(x_0) - S(x_0)| < \varepsilon$ whenever $n > N$.

Example 7.5 Consider the sequence of functions $S_n(x) = x^n$ ($n = 1, 2, 3, \ldots$) defined on $(0, 1)$. We show that $\{S_n(x)\}$ converges pointwise to the zero function $S(x) = 0$ on $(0, 1)$.

SOLUTION Let $x_0 \in (0, 1)$ be arbitrary and let $\varepsilon > 0$ be given. Choose the natural number $N \geq (\ln \varepsilon)/(\ln x_0)$; then $x_0^N \leq \varepsilon$. Now if $n > N$ then

$$|S_n(x_0) - S(x_0)| = |x_0^n - 0| = x_0^n < x_0^N \leq \varepsilon$$

It is clear that our choice of N in the above example depends not only on the given $\varepsilon > 0$ but also on the point x_0. If the point x_0 is closer to 1 then the N will have to be chosen appropriately larger. The question arises: Is there an n such that $|S_n(x) - S(x)| < \varepsilon$ for every $x \in (0, 1)$? To show that the answer to this question is no, let $\varepsilon = \frac{1}{2}$ and choose x_n such that

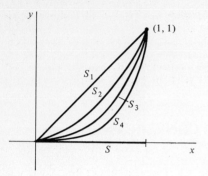

Figure 7.1

$1 > x_n \geq 1/\sqrt[n]{2}$. Then $|S_n(x_n) - S(x_n)| = x_n^n \geq \frac{1}{2} = \varepsilon$. Thus there is no natural number n for which $|S_n(x) - S(x)| < \frac{1}{2}$ for every $x \in (0, 1)$ (see Fig. 7.1).

Definition 7.4 The sequence of functions $\{S_n(x)\}$ *converges uniformly* to $S(x)$ on the interval I if for every $\varepsilon > 0$ there is a natural number N (depending only on ε) such that $|S_n(x) - S(x)| < \varepsilon$ for all $x \in I$ whenever $n > N$.

The geometric interpretation of uniform convergence is given in Fig. 7.2. Picture the limit function $S(x)$ and an ε band around the graph of $S(x)$; that is, $(S(x) - \varepsilon, S(x) + \varepsilon)$ for every x. If $\{S_n(x)\}$ converges uniformly to $S(x)$ then there must be a natural number N such that $S_n(x)$ lies in the ε band for every $n > N$.

Uniform convergence is stronger than pointwise convergence. If $\{S_n(x)\}$ converges uniformly to $S(x)$ on I then $\{S_n(x)\}$ converges pointwise to $S(x)$ on I (see Exercise 7.22). If $\{S_n(x)\}$ converges pointwise to $S(x)$ on I then $\{S_n(x)\}$ may or may not converge uniformly on I, but if it does then necessarily $S(x)$ is the uniform limit. To test a specific sequence of functions for uniform convergence first find the pointwise limit (if it exists) and then check to see whether the convergence is uniform.

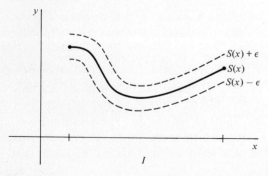

Figure 7.2

Example 7.6 Determine whether the sequence $\{S_n(x)\} = \{(1-x^n)/(1-x)\}$ converges uniformly on $(-1, 1)$.

SOLUTION For each $x \in (-1, 1)$ $\lim_{n \to \infty} S_n(x) = \lim_{n \to \infty} [(1-x^n)/(1-x)] = 1/(1-x)$; therefore $S(x) = 1/(1-x)$ is the pointwise limit of the sequence of functions. Consider

$$|S_n(x) - S(x)| = \left|\frac{1-x^n}{1-x} - \frac{1}{1-x}\right| = \left|\frac{x^n}{1-x}\right| = \frac{|x|^n}{1-x}$$

Since $|x|^n/(1-x)$ is unbounded on $(-1, 1)$ (it becomes large when x is very close to 1), it is clear that there is no natural number n for which $|S_n(x) - S(x)| < \varepsilon$ uniformly on $(-1, 1)$ (for all $x \in (-1, 1)$ at once). Therefore the sequence of functions $\{(1-x^n)/(1-x)\}$ converges pointwise to the function $1/(1-x)$ on the interval $(-1, 1)$, but the convergence is not uniform.

If we have a sequence of functions $\{S_n(x)\}$ with pointwise limit $S(x)$ on the interval I, define

$$T_n = \sup_{x \in I} |S_n(x) - S(x)|$$

It is clear that the sequence $\{S_n(x)\}$ converges uniformly to $S(x)$ if and only if $\lim_{n \to \infty} T_n = 0$. This follows from the definition of uniform convergence (we ask the reader to supply the details in Exercise 7.23). We use the sequence $\{T_n\}$ to determine whether we have uniform convergence in the following example. This technique is typical of the procedure that can be used to determine whether a sequence converges uniformly for a large class of sequences.

Example 7.7 Does $\{S_n(x)\} = \{x/(nx+1)\}$ converge uniformly on $[0, 1]$?

SOLUTION We first find the pointwise limit $S(x) = \lim_{n \to \infty} S_n(x) = \lim_{n \to \infty} [x/(nx+1)] = 0$. Therefore $\{x/(nx+1)\}$ converges pointwise to 0 on $[0, 1]$. Consider

$$T_n = \sup_x |S_n(x) - S(x)| = \sup_x \frac{x}{nx+1} = \max_x \frac{x}{nx+1} \qquad x \in [0, 1]$$

since $S_n(x) = x/(nx+1)$ is continuous on $[0, 1]$. To find the maximum value of $S_n(x)$ on $[0, 1]$ we note that $S_n'(x) = 1/(nx+1)^2 > 0$ for every $x \in [0, 1]$, and so $S_n(x)$ is strictly increasing on $[0, 1]$. Hence it assumes its maximum value at the right-hand endpoint $x = 1$. Therefore

$$\max_x \frac{x}{nx+1} = \max_x S_n(x) = S_n(1) = \frac{1}{n+1} \qquad x \in [0, 1]$$

and consequently $T_n = 1/(n + 1)$. By our remarks preceding this example we know that the sequence $\{x/(nx + 1)\}$ converges uniformly to 0 on $[0, 1]$ since clearly $\lim_{n \to \infty} T_n = 0$.

The above example could also have been done without explicitly finding the value T_n. Observe that for each $x \in (0, 1]$

$$|S_n(x) - S(x)| = \frac{x}{nx + 1} < \frac{x}{nx} = \frac{1}{n}$$

and also $|S_n(0) - S(0)| = 0 < 1/n$. Thus $0 \le T_n = \sup_x |S_n(x) - S(x)| \le 1/n$, where $x \in [0, 1]$. Again, the sequence $\{x/(nx + 1)\}$ converges uniformly on $[0, 1]$ since $\lim_{n \to \infty} T_n = 0$.

There are cases when the limit function $S(x)$ cannot be explicitly determined. This is especially apparent when the functions $S_n(x)$ are the partial sums of a series of functions. Uniform convergence can often be determined by the following version of the Cauchy criterion.

Theorem 7.10: Cauchy criterion $\{S_n(x)\}$ converges uniformly on I if and only if given any $\varepsilon > 0$ there is a natural number N such that $|S_n(x) - S_m(x)| < \varepsilon$ for all $x \in I$ whenever $m, n > N$.

PROOF Assume first that $\{S_n(x)\}$ converges uniformly, say to $S(x)$, on the interval I. Then given any $\varepsilon > 0$ there is a natural number N such that $|S_n(x) - S(x)| < \varepsilon/2$ for all $x \in I$ whenever $n > N$. Then if $m, n > N$, we have

$$|S_n(x) - S_m(x)| = |S_n(x) - S(x) + S(x) - S_m(x)|$$
$$\le |S_n(x) - S(x)| + |S(x) - S_m(x)|$$
$$< \frac{\varepsilon}{2} + \frac{\varepsilon}{2} = \varepsilon \quad \text{for all } x \in I$$

Conversely, suppose that given any $\varepsilon > 0$ there is a natural number N such that $|S_n(x) - S_m(x)| < \varepsilon$ for all $x \in I$ whenever $m, n > N$. Then for each $x_0 \in I$, $\{S_n(x_0)\}$ is a Cauchy sequence of real numbers and hence converges (see Theorem 2.8). We denote its limit by $S(x_0)$. In this way a function $S(x)$ is created which has domain I and is clearly the pointwise limit of the sequence $\{S_n(x)\}$. We show that $\{S_n(x)\}$ converges uniformly to $S(x)$ on I. Let $\varepsilon > 0$ be given; there is a natural number N such that $|S_n(x) - S_m(x)| < \varepsilon/2$ for every $x \in I$ whenever $m, n > N$. For each $x \in I$ there is a natural number $m_x > N$ for which

$$|S_{m_x}(x) - S(x)| < \varepsilon/2$$

If $n > N$ then

$$|S_n(x) - S(x)| = |S_n(x) - S_{m_x}(x) + S_{m_x}(x) - S(x)|$$

$$\leq |S_n(x) - S_{m_x}(x)| + |S_{m_x}(x) - S(x)| < \frac{\varepsilon}{2} + \frac{\varepsilon}{2} = \varepsilon$$

holds for every $x \in I$. Therefore $\{S_n(x)\}$ converges uniformly to $S(x)$ on I.

Next we consider pointwise and uniform convergence for series whose terms are functions.

Definition 7.5 The series $\sum_{k=1}^{\infty} f_k(x)$ is said to *converge pointwise* to the function $S(x)$ on I if $\sum_{k=1}^{\infty} f_k(x)$ converges to $S(x_0)$ for each $x_0 \in I$.

Hence the series $\sum_{k=1}^{\infty} f_k(x)$ converges pointwise to $S(x)$ if the sequence of partial sums $S_n(x) = \sum_{k=1}^{n} f_k(x)$ converges pointwise to $S(x)$.

Definition 7.6 The series $\sum_{k=1}^{\infty} f_k(x)$ *converges uniformly* to $S(x)$ on I if the sequence of partial sums $S_n(x) = \sum_{k=1}^{n} f_k(x)$ converges uniformly to $S(x)$ on I.

Consider the geometric series $\sum_{k=1}^{\infty} x^{k-1}$. Then the sequence of partial sums $S_n(x) = \sum_{k=1}^{n} x^{k-1} = (1 - x^n)/(1 - x)$ and, by Example 7.6, $\{S_n(x)\}$ converges pointwise to $S(x) = 1/(1 - x)$ on the interval $(-1, 1)$ but $\{S_n(x)\}$ does not converge uniformly on $(-1, 1)$. Therefore the geometric series $\sum_{k=1}^{\infty} x^{k-1}$ converges pointwise to the function $1/(1 - x)$ on $(-1, 1)$, but the convergence is not uniform.

The Cauchy criterion yields a test which is quite useful and applicable to a wide range of infinite series.

Theorem 7.11 Weierstrass M test If $|f_k(x)| \leq M_k$ for all $x \in I$ and for $k = 1, 2, 3, \ldots$ and if the series of nonnegative constants $\sum_{k=1}^{\infty} M_k$ converges then $\sum_{k=1}^{\infty} f_k(x)$ converges uniformly on I.

PROOF Let $S_n(x) = \sum_{k=1}^{n} f_k(x)$ and let $\varepsilon > 0$ be given. Since $\sum_{k=1}^{\infty} M_k$ converges there is a natural number N such that $\sum_{k=m+1}^{n} M_k < \varepsilon$ whenever $n > m > N$ (see Theorem 7.2). Therefore if $n > m > N$, we have

$$|S_n(x) - S_m(x)| = \left| \sum_{k=1}^{n} f_k(x) - \sum_{k=1}^{m} f_k(x) \right| = \left| \sum_{k=m+1}^{n} f_k(x) \right|$$

$$\leq \sum_{k=m+1}^{n} |f_k(x)| \leq \sum_{k=m+1}^{n} M_k < \varepsilon \quad \text{for every } x \in I$$

It follows from the Cauchy criterion (Theorem 7.10) that $\{S_n(x)\}$ converges uniformly on I; that is, $\sum_{k=1}^{\infty} f_k(x)$ converges uniformly on I.

Example 7.8 Show that $\sum_{k=1}^{\infty} (1/k^2) \sin kx$ is uniformly convergent on $(-\infty, \infty)$.

SOLUTION $|(1/k^2) \sin kx| \leq 1/k^2$ for all $x \in (-\infty, \infty)$, and $\sum_{k=1}^{\infty} (1/k^2)$ converges (it is a p series with $p = 2$). By the Weierstrass M test $\sum_{k=1}^{\infty} (1/k^2) \sin kx$ converges uniformly on $(-\infty, \infty)$.

Discussions of trigonometric series like these lead to a topic called *Fourier series*. The general theory of Fourier series involves special kinds of convergence tests which are applicable to these particular series.

Example 7.9 Does $\sum_{k=1}^{\infty} (xe^{-x})^k$ converge uniformly on $[0, 2]$?

SOLUTION We first bound $|f(x)|$ on $[0, 2]$, where $f(x) = xe^{-x}$. This function is continuous and differentiable on $[0, 2]$, and so we seek its maximum and minimum values. The derivative is $f'(x) = -xe^{-x} + e^{-x}$, which is zero only at $x = 1$. Hence the maximum and minimum values must be among the values

$$f(0) = 0$$
$$f(1) = e^{-1}$$
$$f(2) = 2e^{-2}$$

The largest of these values is e^{-1} and the smallest is 0. Hence $|f(x)| \leq e^{-1}$ on $[0, 2]$. It follows that $(xe^{-x})^k \leq e^{-k}$. The series $\sum_{k=1}^{\infty} e^{-k}$ converges (why?), and hence $\sum_{k=1}^{\infty} (xe^{-x})^k$ converges uniformly on $[0, 2]$.

EXERCISES

7.21 Find the pointwise limit $S(x)$ if it exists; then determine whether $S_n(x)$ converges uniformly to $S(x)$ on I.

(a) $S_n(x) = \dfrac{1}{nx + 1}$ $I = [0, 1]$ (b) $S_n(x) = \dfrac{x^{2n}}{1 + x^{2n}}$ $I = [-1, 1]$

(c) $S_n(x) = x^n$ $I = [-1, 1]$ (d) $S_n(x) = x^n$ $I = (-1, 1)$

(e) $S_n(x) = x^n(1 - x)$ $I = [0, 1]$ (f) $S_n(x) = n^2 x(1 - x^n)$ $I = [0, 1]$

7.22 (a) Prove that if $\{S_n(x)\}$ converges uniformly to $S(x)$ on I then $\{S_n(x)\}$ converges pointwise to $S(x)$ on I.

(b) Explain why it is impossible for $\{S_n(x)\}$ to converge pointwise to $S(x)$ on I and converge uniformly to $f(x)$ on I when $f(x) \neq S(x)$ on I.

7.23 Suppose $\{S_n(x)\}$ converges uniformly to $S(x)$ on I and $T_n = \sup_{x \in I} |S_n(x) - S(x)|$. Prove that $\{S_n(x)\}$ converges uniformly to $S(x)$ on I if and only if $\lim_{n \to \infty} T_n = 0$.

7.24 For what values of x do the infinite series converge (pointwise)?

(a) $\sum_{k=1}^{\infty} \left(\frac{1-x}{1+x}\right)^{k-1}$ (b) $\sum_{k=1}^{\infty} \frac{1}{k} x^k$ (c) $\sum_{k=1}^{\infty} \frac{1}{k^2} x^k$

(d) $\sum_{k=1}^{\infty} \frac{\cos kx}{k^2}$ (e) $\sum_{k=1}^{\infty} (\cos k\pi) x^{2k}$

7.25 Prove that if $\sum_{k=1}^{\infty} a_k$ converges absolutely then $\sum_{k=1}^{\infty} a_k \cos kx$ and $\sum_{k=1}^{\infty} a_k \sin kx$ converge uniformly on $(-\infty, \infty)$.

7.26 Suppose that all f_k are constant functions. What does it mean in this case to say that $\sum_{k=1}^{\infty} f_k(x)$ converges uniformly on an interval I?

7.27 Determine whether the geometric series $\sum_{k=1}^{\infty} x^{k-1}$ converges uniformly on $(-1, \frac{1}{2})$.

7.28 Prove that if $\sum_{k=1}^{\infty} f_k(x)$ converges uniformly on I and g is bounded on I then the series $\sum_{k=1}^{\infty} g(x) \cdot f_k(x)$ is uniformly convergent on I.

7.3 IMPORTANCE OF UNIFORM CONVERGENCE

If we have a sequence or series of functions, it is often important to know whether or not certain properties of these functions carry over to the limit function. For example, if each function in a sequence or series of functions is continuous or differentiable or integrable, can we assume then that the limit function will be continuous or differentiable or integrable?

Consider the sequence $\{x^n\}$ defined on $[0, 1]$. Each function in this sequence is continuous on $[0, 1]$, and the sequence converges pointwise to the function

$$S(x) = \begin{cases} 0 & \text{if } x \in [0, 1) \\ 1 & \text{if } x = 1 \end{cases}$$

on $[0, 1]$. Clearly $S(x)$ is not continuous on $[0, 1]$, and so continuity does not carry over to the pointwise limit function of a sequence of continuous functions. We do, however, have the following important result.

Theorem 7.12 If $\{S_n(x)\}$ is a sequence of continuous functions on the interval I which converges uniformly to $S(x)$ on I then $S(x)$ is continuous on I.

PROOF Let $x_0 \in I$ be arbitrary; we show that $S(x)$ is continuous at x_0 (if x_0 is an endpoint of I, we mean the appropriate one-sided continuity at x_0). Let $\varepsilon > 0$ be given. Choose n so large that $|S(x) - S_n(x)| < \varepsilon/3$ for every $x \in I$; this is possible by the uniform convergence of $\{S_n(x)\}$. Since $S_n(x)$ is continuous at x_0, there is a $\delta > 0$ such that $|S_n(x) - S_n(x_0)| < \varepsilon/3$

whenever $|x - x_0| < \delta$ (and $x \in I$ if x_0 is an endpoint of I). Now if $|x - x_0| < \delta$ then

$$|S(x) - S(x_0)| = |S(x) - S_n(x) + S_n(x) - S_n(x_0) + S_n(x_0) - S(x_0)|$$
$$< |S(x) - S_n(x)| + |S_n(x) - S_n(x_0)| + |S_n(x_0) - S(x_0)|$$
$$< \frac{\varepsilon}{3} + \frac{\varepsilon}{3} + \frac{\varepsilon}{3} = \varepsilon$$

Therefore $S(x)$ is continuous at x_0 and hence on I.

Of course, Theorem 7.12 does not apply to the sequence $\{x^n\}$ since this sequence does not converge uniformly on $[0, 1]$.

Corollary If $f_k(x)$ is continuous on I for $k = 1, 2, 3, \ldots$ and if $\sum_{k=1}^{\infty} f_k(x)$ converges uniformly to $S(x)$ on I then $S(x)$ is continuous on I.

Example 7.10 We show that $\sum_{k=0}^{\infty} (x^k/k!)$ converges to a continuous function on **R**.

SOLUTION For each $x \in \mathbf{R}$ the series $\sum_{k=0}^{\infty} (x^k/k!)$ converges by the ratio test. Hence $\sum_{k=0}^{\infty} (x^k/k!)$ converges pointwise to a function $S(x)$ on **R**, but the above corollary does not apply directly since the convergence is not uniform on **R** (see Exercise 7.29). We can circumvent the problem, however, by noting that $\sum_{k=0}^{\infty} (x^k/k!)$ converges uniformly to $S(x)$ on $[-M, M]$ for every $M > 0$. To see this note that

$$\left|\frac{x^k}{k!}\right| \leq \frac{M^k}{k!} \quad \text{on } [-M, M]$$

and $\sum_{k=0}^{\infty} (M^k/k!)$ converges (again, by the ratio test). Thus by the Weierstrass M test the series $\sum_{k=0}^{\infty} (x^k/k!)$ converges uniformly on $[-M, M]$. Since each function $f_k(x) = x^k/k!$ ($k = 0, 1, 2, 3, \ldots$) is continuous on $[-M, M]$, it now follows from the corollary that $S(x)$ is continuous on $[-M, M]$. But $M > 0$ was arbitrary. Therefore $S(x)$ is continuous on **R**. In the next section we shall show that $S(x) = e^x$.

Let $r_1, r_2, r_3, \ldots$ be an ordering of the countably infinite set of rationals in $[0, 1]$. If we define

$$S_n(x) = \begin{cases} 1 & \text{if } x \in \{r_1, r_2, \ldots, r_n\} \\ 0 & \text{otherwise} \end{cases}$$

then the set of points of discontinuity of S_n is $\mathfrak{D}_{S_n} = \{r_1, r_2, \ldots, r_n\}$. Moreover, it is clear that the sequence of $\{S_n(x)\}$ converges pointwise to the function

$$S(x) = \begin{cases} 1 & \text{if } x \in \mathbf{Q} \cap [0, 1] \\ 0 & \text{otherwise} \end{cases}$$

on the interval $[0, 1]$. Now, $\mathcal{D}_S = [0, 1]$, and it follows from Theorem 6.19 that each $S_n \in \mathcal{R}[0, 1]$ but $S \notin \mathcal{R}[0, 1]$. Therefore, the pointwise limit of a sequence of Riemann-integrable functions may well fail to be Riemann-integrable.

In the next example we show that even if the pointwise limit of a sequence of Riemann-integrable functions is Riemann-integrable, it may happen that the integral of the limit function is not equal to the limit of the sequence of integrals.

Example 7.11 Consider the sequence $\{S_n(x)\}$ defined on $[0, 1]$ by

$$S_n(x) = \begin{cases} n - n^2 x & \text{if } x \in (0, 1/n) \\ 0 & \text{otherwise} \end{cases}$$

$\{S_n(x)\}$ converges pointwise to $S(x) \equiv 0$ on $[0, 1]$, but the convergence is not uniform (see Exercise 7.30). For each natural number n, $S_n \in \mathcal{R}[0, 1]$ and $\int_0^1 S_n = \frac{1}{2}$. Thus $\lim_{n \to \infty} \int_0^1 S_n = \frac{1}{2}$. Now $S \in \mathcal{R}[0, 1]$ and $\int_0^1 S = 0$. Therefore

$$\lim_{n \to \infty} \int_0^1 S_n \neq \int_0^1 \lim_{n \to \infty} S_n$$

(see Figure 7.3).

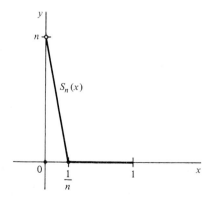

Figure 7.3

Theorem 7.13 If $\{S_n(x)\}$ is a sequence of Riemann-integrable functions on $[a, b]$ and if $\{S_n(x)\}$ converges uniformly to $S(x)$ on $[a, b]$ then $S(x) \in \mathcal{R}[a, b]$, and moreover

$$\int_a^b S(x)\, dx = \lim_{n \to \infty} \int_a^b S_n(x)\, dx$$

PROOF We first show that $S(x)$ is Riemann-integrable on $[a, b]$. Let $\varepsilon > 0$ be given. According to Theorem 6.1, it suffices to find a partition P of $[a, b]$ such that $U(P, S) - L(P, S) < \varepsilon$. Choose the natural number n so large that $|S_n(x) - S(x)| < \varepsilon/3(b - a)$ for all $x \in [a, b]$. Let P be a partition of $[a, b]$ for which $U(P, S_n) - L(P, S_n) < \varepsilon/3$; such a partition exists because $S_n \in \mathcal{R}[a, b]$. We define

$$M_i = \sup_x S(x) \qquad m_i = \inf_x S(x)$$

$$\bar{M}_i = \sup_x S_n(x) \qquad \bar{m}_i = \inf_x S_n(x) \qquad x \in [x_{i-1}, x_i]$$

Then $|S_n(x) - S(x)| < \varepsilon/3(b - a)$ for all $x \in [a, b]$ implies

$$\bar{m}_i - \frac{\varepsilon}{3(b - a)} \leq m_i \leq M_i \leq \bar{M}_i + \frac{\varepsilon}{3(b - a)} \qquad i = 1, 2, \ldots, n$$

Hence

$$U(P, S) - L(P, S) = \sum_{i=1}^n M_i\, \Delta x_i - \sum_{i=1}^n m_i\, \Delta x_i \leq \sum_{i=1}^n \left[\bar{M}_i + \frac{\varepsilon}{3(b - a)}\right] \Delta x_i$$

$$- \sum_{i=1}^n \left[\bar{m}_i - \frac{\varepsilon}{3(b - a)}\right] \Delta x_i = \sum_{i=1}^n \bar{M}_i\, \Delta x_i + \frac{\varepsilon}{3} - \sum_{i=1}^n \bar{m}_i\, \Delta x_i + \frac{\varepsilon}{3}$$

$$= U(P, S_n) - L(P, S_n) + \frac{2\varepsilon}{3} < \frac{\varepsilon}{3} + \frac{2\varepsilon}{3} = \varepsilon$$

It follows that $S(x)$ is Riemann-integrable on $[a, b]$.

We next show that $\lim_{n \to \infty} \int_a^b S_n = \int_a^b S$. Let $\varepsilon > 0$ be given and choose $N \in \mathbf{N}$ such that $|S_n(x) - S(x)| < \varepsilon/(b - a)$ for every $x \in [a, b]$ whenever $n > N$. Then

$$\left| \int_a^b S_n - \int_a^b S \right| = \left| \int_a^b (S_n - S) \right| \leq \int_a^b |S_n - S|$$

$$\leq \int_a^b \frac{\varepsilon}{b - a} = \frac{\varepsilon}{b - a}(b - a) = \varepsilon$$

whenever $n > N$ and so $\int_a^b S_n \to \int_a^b S$ as $n \to \infty$.

Corollary If $f_k(x) \in \mathcal{R}[a, b]$ for $k = 1, 2, 3, \ldots$ and if $\sum_{k=1}^\infty f_k(x)$ converges uniformly to $S(x)$ on $[a, b]$ then $S(x) \in \mathcal{R}[a, b]$ and

$$\int_a^b S(x)\,dx = \int_a^b \sum_{k=1}^\infty f_k(x)\,dx = \sum_{k=1}^\infty \int_a^b f_k(x)\,dx$$

The situation for differentiability is not quite the same as for continuity and integrability. The following example shows that the uniform limit of a sequence of differentiable function on I may fail to be differentiable on I.

Example 7.12 Let $S_n(x) = |x|^{1 + 1/n}$ on $[-1, 1]$. Each $S_n(x)$ is differentiable on $[-1, 1]$, and the sequence $\{S_n(x)\}$ converges uniformly to $S(x) = |x|$ on $[-1, 1]$ (see Exercise 7.31). But clearly $S(x)$ is not differentiable on $[-1, 1]$; $S'(0)$ does not exist (see Fig. 7.4).

Even if the uniform limit of a sequence of differentiable functions on I is again differentiable on the interval I, it can happen that the derivative of the limit is not the limit of the sequence of derivatives (see Exercise 7.35).

Theorem 7.14 Let $\{S_n(x)\}$ be a sequence of functions which have continuous derivatives $S_n'(x)$ on the interval $[a, b]$ ($n = 1, 2, 3, \ldots$). Also assume that $\{S_n(x)\}$ converges to $S(x)$ on $[a, b]$ and that $\{S_n'(x)\}$ converges uniformly to $g(x)$ on $[a, b]$. Then $S(x)$ is differentiable on $[a, b]$, and $S'(x) = g(x)$ for all $x \in [a, b]$.

PROOF For every $x \in [a, b]$,

$$S(x) = \lim_{n \to \infty} S_n(x) = S(a) + \lim_{n \to \infty} [S_n(x) - S_n(a)]$$

Now $S_n'(x)$ is continuous on $[a, b]$, and so by the fundamental theorem of calculus

$$\int_a^x S_n' = S_n(x) - S_n(a)$$

Therefore $S(x) = S(a) + \lim_{n \to \infty} \int_a^x S_n'$. But since $\{S_n'(x)\}$ converges uni-

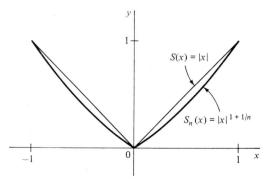

Figure 7.4

formly to $g(x)$ on $[a, b]$, it follows from Theorem 7.13 that $g \in \mathcal{R}[a, b]$ and

$$\lim_{n \to \infty} \int_a^x S_n' = \int_a^x g$$

Hence $S(x) = S(a) + \int_a^x g$. Now $g(x)$ is continuous on $[a, b]$ (Theorem 7.12), and so, by the indefinite-integral theorem, $F(x) = \int_a^x g$ is differentiable with derivative $F'(x) = g(x)$ for all $x \in [a, b]$. Thus $S'(x) = g(x)$ for every $x \in [a, b]$.

Corollary If $f_k(x)$ has a continuous derivative on $[a, b]$ for each $k = 1, 2, 3, \ldots$ and if $\sum_{k=1}^{\infty} f_k(x)$ converges to $S(x)$ on $[a, b]$ and if the series $\sum_{k=1}^{\infty} f_k'(x)$ converges uniformly to $g(x)$ on $[a, b]$ then $S'(x) = g(x)$ for every $x \in [a, b]$; equivalently

$$\frac{d}{dx} \sum_{k=1}^{\infty} f_k(x) = \sum_{k=1}^{\infty} \frac{d}{dx} f_k(x)$$

Under the conditions stated in the above corollary, the operation of term-by-term differentiation of an infinite series is valid. The reader may have seen the technique of solving linear second-order differential equations by assuming a series solution (of a particular type). With this technique the series is differentiated term by term and substituted into the differential equation. The corollary above justifies these steps.

Example 7.13 We define a sequence of functions graphically as shown in Fig. 7.5. Each function is defined on the closed interval $[0, 1]$ and has a graph consisting of straight-line segments; that is, a broken line. We continue the graphical construction indicated above to obtain a sequence $\{f_n(x)\}$ of continuous functions defined on $[0, 1]$. The sequence clearly converges uniformly to the zero function on $[0, 1]$. We consider the series $\sum_{n=1}^{\infty} f_n(x)$; since $|f_n(x)| \leq 1/2^n$ for each $n \in \mathbf{N}$, it follows from the Weierstrass M test that $\sum_{n=1}^{\infty} f_n(x)$ converges uniformly, say to $f(x)$, on the interval $[0, 1]$. Since each $f_n(x)$ is continuous on $[0, 1]$, it follows from Theorem 7.12 that $f(x)$ is continuous on $[0, 1]$. This function has the interesting property that, although continuous at each point in $[0, 1]$, $f(x)$ is not differentiable at a single $x \in [0, 1]$. To show this let a be an arbitrary point in $[0, 1]$. Then a has some binary representation, which can be obtained from the expansion

$$a = \sum_{n=1}^{\infty} \frac{a_n}{2^n}$$

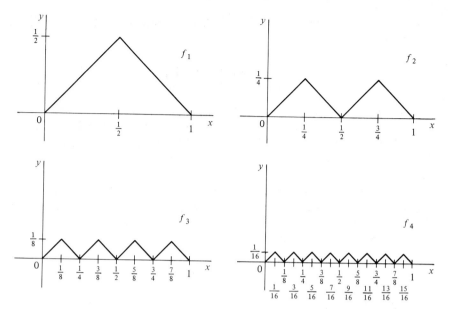

Figure 7.5

where the a_n's are 0 or 1. Define the sequence

$$x_n = \begin{cases} a - \dfrac{1}{2^n} & \text{if } a_n = 1 \\ a + \dfrac{1}{2^n} & \text{if } a_n = 0 \end{cases}$$

Clearly, $\lim_{n \to \infty} x_n = a$; however,

$$\lim_{n \to \infty} \frac{f(x_n) - f(a)}{x_n - a}$$

does not exist. We leave the details to the reader.

Mathematicians spent a good deal of time during the nineteenth century trying to prove that if a function is continuous on an interval then there must be some points in the interval where the derivative of the function exists. Karl Weierstrass put an end to these attempts late in 1887, when he exhibited the first known example of a continuous nowhere differentiable function.

EXERCISES

7.29 Prove that $\sum_{k=0}^{\infty} (x^k/k!)$ does not converge uniformly on **R**. *Hint:* Use the Cauchy criterion.

7.30 Prove that the sequence of function in Example 7.11 converges pointwise to the zero function on [0, 1] but that the convergence is not uniform.

7.31 Prove that each of the functions $S_n(x) = |x|^{1+1/n}$ ($n = 1, 2, 3, \ldots$) is differentiable on $[-1, 1]$ and that $\{S_n(x)\}$ converges uniformly to $S(x) = |x|$ on $[-1, 1]$.

7.32 (a) Prove that if $\{S_n(x)\}$ is a sequence of bounded functions on the interval I and $\{S_n(x)\}$ converges uniformly to $S(x)$ on I then $S(x)$ is bounded on I.

(b) Give an example of a sequence of bounded functions $\{S_n(x)\}$ on I such that the pointwise limit $S(x)$ is not bounded on I.

7.33 Prove or give a counterexample: *Conjecture*: If $\{S_n(x)\}$ is a sequence of Riemann-integrable functions on $[a, b]$ and if the pointwise limit $S(x) \in \mathcal{R}[a, b]$ and if $\{S_n(x)\}$ are uniformly bounded on $[a, b]$, that is, there is an $M > 0$ such that $|S_n(x)| \le M$ for every $x \in [a, b]$ and for all $n \in \mathbf{N}$, then

$$\int_a^b S = \lim_{n \to \infty} \int_a^b S_n$$

7.34 As an alternate proof that the uniform limit of a sequence of functions $S_n \in \mathcal{R}[a, b]$ is a function $S \in \mathcal{R}[a, b]$ prove that $S(x)$ is bounded on $[a, b]$ and that the set of points of discontinuity of $S(x)$ is a set of measure zero; then invoke Theorem 6.19.

7.35 Verify that the sequence of functions $S_n(x) = (1 - x)x^n$ converges uniformly to the zero function on $[0, 1]$ but that $\lim_{n \to \infty} S'_n(x) \ne 0$ on $[0, 1]$.

7.36 Let $S_n(x) = |x|^{1+1/n}$ ($n = 1, 2, 3, \ldots$). Show that the pointwise limit of $\{S'_n(x)\}$ is sgn (x) on $[-1, 1]$ and verify that the convergence is not uniform.

7.37 Show that each of the following sequences is not uniformly convergent on the indicated interval. *Hint*: Consider Theorem 7.12.

(a) $S_n(x) = \dfrac{1}{1 + nx}$ on $[0, 1]$ (b) $S_n(x) = \dfrac{1 - x^n}{1 + x^n}$ on $(0, 2)$

(c) $S_n(x) = \dfrac{x^{2n}}{1 + x^{2n}}$ on $[-2, 2]$

7.38 Let $S_n(x) = x/(1 + nx^2)$ on $[-1, 1]$. Prove that $\{S_n(x)\}$ converges uniformly to the zero function on $[-1, 1]$ but the sequence $\{S'_n(x)\}$ does not converge to zero. *Hint*: Consider the sequence $\{S'_n(0)\}$.

7.39 Let $S_n(x) = (2x/\pi) \tan^{-1} nx$. Prove:
(a) $\{S_n(x)\}$ converges uniformly to $S(x) = |x|$ on $(-\infty, \infty)$.
(b) $\{S'_n(x)\}$ converges pointwise to $T(x) = \text{sgn }(x)$.
(c) $S'(0) \ne T(0)$.

7.40 For each natural number n, define

$$S_n(x) = \begin{cases} \dfrac{1}{n} & \text{on } [n, 2n] \\ 0 & \text{otherwise} \end{cases}$$

Show that $\{S_n(x)\}$ converges uniformly to the zero function $S = 0$, $\int S_n = 1$ for each $n \in \mathbf{N}$ and $\int S = 0$. Indicate why this does not contradict Theorem 7.13.

7.4 POWER SERIES AND TAYLOR SERIES

One particular type of series, which has importance in applications as well as more advanced topics in analysis, is called *power series*. It has the general form $\sum_{k=0}^{\infty} a_k(x - x_0)^k$, where x_0 is a fixed real number and the coefficients a_k do not depend on x. Our results from previous sections of this chapter applied to this special series yield the following theorem.

Theorem 7.15 Associated with each power series $\sum_{k=0}^{\infty} a_k(x - x_0)^k$ is a *radius of convergence* R, $0 \leq R \leq \infty$, with the following properties:

(a) $1/R = \limsup |a_k|^{1/k}$. In the special cases when $\limsup |a_k|^{1/k} = 0$ define $R = \infty$, and when $\limsup |a_k|^{1/k} = \infty$ then $R = 0$.
(b) $\sum_{k=0}^{\infty} a_k(x - x_0)^k$ converges absolutely if $|x - x_0| < R$.
(c) $\sum_{k=0}^{\infty} a_k(x - x_0)^k$ diverges if $|x - x_0| > R$.
(d) $\sum_{k=0}^{\infty} a_k(x - x_0)^k$ converges uniformly in the interval $|x - x_0| \leq \rho$ for any ρ, $0 < \rho < R$.

PROOF (a) and (b). The series converges absolutely if

$$\limsup |a_k(x - x_0)^k|^{1/k} < 1$$

(see Theorem 7.6). Hence $\limsup |a_k|^{1/k} |x - x_0| < 1$, and so

$$|x - x_0| \limsup |a_k|^{1/k} < 1$$

Now substituting $1/R$ for $\limsup |a_k|^{1/k}$ gives $|x - x_0| 1/R < 1$, and therefore the series converges absolutely if $|x - x_0| < R$. In the special case where $\limsup |a_k|^{1/k} = 0$ the series converges for all x, and we say the radius of convergence $R = \infty$; if $\limsup |a_k|^{1/k} = \infty$ then the series converges only at $x = x_0$, and we say the radius of convergence $R = 0$.

(c) If $|x - x_0| > R$ then $|a_k(x - x_0)^k| > 1$ for an infinite number of values of k, and so $\lim_{k \to \infty} a_k(x - x_0)^k \neq 0$ and the series diverges.

(d) If $|x - x_0| \leq \rho$, where $0 < \rho < R$, then $|a_k(x - x_0)^k| \leq |a_k| \rho^k$. The series $\sum_{k=0}^{\infty} a_k \rho^k$ converges by the root test, and hence $\sum_{k=0}^{\infty} a_k(x - x_0)^k$ converges uniformly by the Weierstrass M test.

We make two observations on Theorem 7.15: (1) We could have used the ratio test instead of the root test. Therefore if $\lim_{k \to \infty} |a_{k+1}/a_k|$ exists, we have an easier alternate formula for R, namely $1/R = \lim_{k \to \infty} |a_{k+1}/a_k|$. Again if this limit is zero, define $R = \infty$; and if this limit is ∞, define $R = 0$. (2) Theorem 7.15 says nothing about convergence when $|x - x_0| = R$. The series may or may not converge at $x_0 - R$ and $x_0 + R$. If it is possible to determine whether or not the series converges at these points, we specify an *interval of convergence* as one of $[x_0 - R, x_0 + R]$, $[x_0 - R, x_0 + R)$, $(x_0 - R, x_0 + R]$, or $(x_0 - R, x_0 + R)$.

Example 7.14 Determine the interval of convergence for the series $\sum_{k=1}^{\infty} (x^k/k)$.

SOLUTION We use the ratio test here

$$\frac{1}{R} = \lim_{k \to \infty} \frac{1/(k+1)}{1/k} = \lim_{k \to \infty} \frac{k}{k+1} = 1$$

We know the series converges absolutely on the interval $(-1, 1)$. At $x = 1$, $\sum_{k=1}^{\infty} (1/k)$ diverges; at $x = -1$, $\sum_{k=1}^{\infty} [(-1)^k/k]$ converges. Hence the interval of convergence is $[-1, 1)$.

As a consequence of part (d) of Theorem 7.15, $f(x) = \sum_{k=0}^{\infty} a_k(x - x_0)^k$ is continuous at each point $x = x_1$ in the interval $(x_0 - R, x_0 + R)$, since we can always find a ρ such that

$$x_0 - R < x_0 - \rho < x_1 < x_0 + \rho < x_0 + R$$

The following theorem extends this result.

Theorem 7.16: Abel's theorem If $\sum_{k=0}^{\infty} a_k(x - x_0)^k$ has positive radius of convergence R and if $\sum_{k=0}^{\infty} a_k R^k$ converges then $\sum_{k=0}^{\infty} a_k(x - x_0)^k$ converges uniformly on $[x_0, x_0 + R]$. Similarly, if $\sum_{k=0}^{\infty} a_k(-R)^k$ converges then $\sum_{k=0}^{\infty} a_k(x - x_0)^k$ converges uniformly on $[x_0 - R, x_0]$.

PROOF We prove the first part of the theorem for the special case $x_0 = 0$ and $R = 1$. The general case is then obtained by a change of variables (see Exercise 7.45). Hence we want to use the fact that $\sum_{k=0}^{\infty} a_k$ converges to prove that $\sum_{k=0}^{\infty} a_k x^k$ converges uniformly on $[0, 1]$. Equivalently, we show that given any $\varepsilon > 0$ $|\sum_{k=n}^{\infty} a_k| < \varepsilon/2$ for all $n > N$ implies that $|\sum_{k=n}^{\infty} a_k x^k| < \varepsilon$ for all $n > N$ and all $x \in [0, 1]$. Denote $\sum_{k=n}^{\infty} a_k = A_n$ and assume $0 \leq x < 1$. Then

$$\sum_{k=n}^{\infty} a_k x^k = \sum_{k=n}^{\infty} (A_k - A_{k+1})x^k = \sum_{k=n}^{\infty} (A_k x^k - A_{k+1} x^k)$$

$$= A_n x^n + \sum_{k=n}^{\infty} A_{k+1}(-x^k + x^{k+1})$$

$$= A_n x^n + x^n(-1 + x) \sum_{k=n}^{\infty} A_{k+1} x^{k-n}$$

Now use the assumption $|A_k| < \varepsilon/2$, $k > N$

$$\left|\sum_{k=n}^{\infty} a_k x^k\right| \leq \frac{\varepsilon}{2} x^n + x^n(1-x)\frac{\varepsilon}{2} \sum_{k=n}^{\infty} x^{k-n} \qquad 0 \leq x < 1$$

$$< \frac{\varepsilon}{2} x^n + x^n(1-x)\frac{\varepsilon}{2}\frac{1}{1-x}$$

$$< \frac{\varepsilon}{2} x^n + x^n \frac{\varepsilon}{2} < \frac{\varepsilon}{2} + \frac{\varepsilon}{2} = \varepsilon$$

But this inequality holds for $x = 1$ also, and so

$$\left|\sum_{k=n}^{\infty} a_k x^k\right| < \varepsilon \qquad \text{for } 0 \leq x \leq 1 \text{ and } n > N$$

and the series converges uniformly on [0, 1].

Corollary If $f(x) = \sum_{k=0}^{\infty} a_k(x - x_0)^k$ has positive radius of convergence R and converges at $x_0 + R$ then $f(x)$ is left-continuous at $x_0 + R$; similarly if it converges at $x_0 - R$ then $f(x)$ is right-continuous at $x_0 - R$.

Example 7.15 Apply Abel's theorem to obtain a value for

$$\sum_{k=1}^{\infty} [(-1)^k/k]$$

SOLUTION First we enlist the aid of the geometric series to find $f(x)$

$$\frac{1}{1-t} = \sum_{k=1}^{\infty} t^{k-1} \qquad -1 < t < 1$$

$$\int_0^x \frac{1}{1-t} dt = \int_0^x \sum_{k=1}^{\infty} t^{k-1} dt \qquad -1 < x < 1$$

$$-\ln(1-x) = \sum_{k=1}^{\infty} \int_0^x t^{k-1} dt = \sum_{k=1}^{\infty} \frac{x^k}{k} \qquad -1 < x < 1$$

Since $\sum_{k=1}^{\infty}(x^k/k)$ has radius of convergence $R = 1$ and converges at $x = -1$, by the corollary above

$$\sum_{k=1}^{\infty} \frac{(-1)^k}{k} = \lim_{x \to -1^+}[-\ln(1-x)] = -\ln 2$$

Also by property (d) of Theorem 7.15 we can find the derivative of

$f(x) = \sum_{k=0}^{\infty} a_k(x - x_0)^k$ by differentiating the series term by term

$$f'(x) = \sum_{k=1}^{\infty} k a_k (x - x_0)^{k-1}$$

The new series is a power series with the same radius of convergence (see Exercise 7.43) and hence will converge uniformly in $|x - x_0| \leq \rho$, where $\rho < R$. By the same reasoning we used for continuity, the formula above for $f'(x)$ is valid for each x in the interval $(x_0 - R, x_0 + R)$.

We can continue to differentiate again and again, to obtain derivatives of all orders (as long as $R > 0$) and the formulas

$$f^{(n)}(x) = \sum_{k=n}^{\infty} a_k k(k-1) \cdots (k - n + 1)(x - x_0)^{k-n}$$

$$x_0 - R < x < x_0 + R$$

In particular, this formula with $x = x_0$ yields

$$f^{(n)}(x_0) = a_n \cdot n! \quad \text{or} \quad a_n = \frac{f^{(n)}(x_0)}{n!}$$

Theorem 7.17 If $f(x) = \sum_{k=0}^{\infty} a_k(x - x_0)^k$ has radius of convergence $R > 0$ then $f(x)$ has derivatives of all orders in $(x_0 - R, x_0 + R)$ and

$$a_k = \frac{f^{(k)}(x_0)}{k!}$$

Instead of starting with a power series, if we start with a function f with derivatives of all orders at x_0 and formally expand in a power series about x_0, that is,

$$\sum_{k=0}^{\infty} \frac{f^{(k)}(x_0)}{k!} (x - x_0)^k$$

we obtain the *Taylor series* for f at $x = x_0$. If this series converges to f in some neighborhood of x_0, we say that f is *analytic* at x_0. Specifying that a function f is analytic at x_0 is much stronger than specifying that f is continuous at x_0 or differentiable at x_0.

Some well-known Taylor series, which can be easily verified from the formula, are

$$e^x = \sum_{k=0}^{\infty} \frac{x^k}{k!} \qquad \sin x = \sum_{k=0}^{\infty} \frac{(-1)^k x^{2k+1}}{(2k+1)!}$$

$$\cos x = \sum_{k=0}^{\infty} \frac{(-1)^k x^{2k}}{(2k)!} \qquad \ln(1 + x) = \sum_{k=1}^{\infty} \frac{(-1)^{k+1} x^k}{k}$$

The reader should be able to determine the interval of convergence for each of these.

It follows from Theorem 7.17 that the Taylor series of a function which is analytic at x_0 is unique. Quite often the coefficients a_k are found by means other than the formula $f^{(k)}(x_0)/k!$. Integrating, differentiating, or substituting into the geometric series are popular techniques.

Example 7.16 Find the Taylor series for $f(x) = \tan^{-1} x$ at $x = 0$.

SOLUTION First recall that $f'(x) = 1/(1 + x^2)$.

$$\frac{1}{1-u} = \sum_{k=0}^{\infty} u^k \qquad -1 < u < 1$$

Substituting $u = -x^2$ gives

$$\frac{1}{1+x^2} = \sum_{k=0}^{\infty} (-1)^k x^{2k} \qquad -1 < x < 1$$

$$\tan^{-1} x - \tan^{-1} 0 = \int_0^x \frac{dt}{1+t^2} = \int_0^x \sum_{k=0}^{\infty} (-1)^k t^{2k} \, dt \qquad -1 < x < 1$$

$$\tan^{-1} x = \sum_{k=0}^{\infty} (-1)^k \int_0^x t^{2k} \, dt = \sum_{k=0}^{\infty} (-1)^k \frac{x^{2k+1}}{2k+1} \qquad -1 < x < 1$$

by uniform convergence.

It is interesting to note that $1/(1 + x^2)$ exists for all real numbers x but $\sum_{k=0}^{\infty} (-1)^k x^{2k}$ exists only if $-1 < x < 1$. Here there seems to be no way to predict the radius of convergence by examining the function. In complex analysis, where the variable assumes values in the complex plane, a relationship exists between the function and the radius of convergence of its Taylor series. In fact, a more satisfying and detailed study of analytic functions and Taylor series is found in complex analysis.

Most calculators and computers evaluate the transcendental functions e^x, $\sin x$, $\ln x$, etc., by truncating their Taylor series. Hence we close this section with a theorem which gives the remainder or error when the Taylor series is truncated.

Theorem 7.18 If f has continuous derivatives of order up to and including $n + 1$ in some neighborhood of $x = x_0$ then

$$f(x) = \sum_{k=0}^{n} \frac{f^{(k)}(x_0)}{k!} (x - x_0)^k + R_n(x)$$

where

$$R_n(x) = \frac{1}{n!} \int_{x_0}^{x} f^{(n+1)}(t)\,(x-t)^n\,dt$$

$R_n(x)$ is the remainder after the nth power of $x - x_0$.

PROOF We first change x_0 to a variable t and then define the function $g(t)$ by

$$g(t) = f(x) - \sum_{k=0}^{n} \frac{f^{(k)}(t)}{k!}(x-t)^k$$

Think of x and x_0 as constants; then $g(x) = 0$ and $g(x_0) = R_n(x)$. When $g'(t)$ is computed using the product rule, the sum telescopes to just one term

$$g'(t) = -\sum_{k=0}^{n} \frac{f^{(k+1)}(t)}{k!}(x-t)^k + \sum_{k=1}^{n} \frac{f^{(k)}(t)}{(k-1)!}(x-t)^{k-1}$$

$$= -\sum_{k=0}^{n} \frac{f^{(k+1)}(t)}{k!}(x-t)^k + \sum_{k=0}^{n-1} \frac{f^{(k+1)}(t)}{k!}(x-t)^k$$

$$= -\frac{f^{(n+1)}(t)}{n!}(x-t)^n$$

Now with the fundamental theorem of calculus

$$g(x) - g(x_0) = \int_{x_0}^{x} g'(t)\,dt$$

and

$$0 - R_n(x) = \int_{x_0}^{x} -\frac{f^{(n+1)}(t)}{n!}(x-t)^n\,dt$$

therefore

$$R_n(x) = \int_{x_0}^{x} \frac{f^{(n+1)}(t)}{n!}(x-t)^n\,dt$$

An alternate form of the remainder is obtained by using the mean value theorem for integrals (see Exercise 6.22). The functions $f^{(n+1)}(t)$ and $(x-t)^n$ are continuous functions of t, and $(x-t)^n$ does not change sign on $[x_0, x]$ if $x_0 < x$ or on $[x, x_0]$ if $x < x_0$. Hence

$$R_n(x) = \frac{1}{n!} \int_{x_0}^{x} f^{(n+1)}(t)\,(x-t)^n\,dt$$

$$= \frac{1}{n!} f^{(n+1)}(\xi) \int_{x_0}^{x} (x-t)^n\,dt \quad \text{for some } \xi \text{ between } x \text{ and } x_0$$

$$= \frac{1}{n!} f^{(n+1)}(\xi) \frac{(x-x_0)^{n+1}}{n+1}$$

Therefore
$$R_n(x) = \frac{f^{(n+1)}(\xi)(x-x_0)^{n+1}}{(n+1)!}$$
for some ξ between x and x_0.

Example 7.17 Estimate the remainder if
$$\cos x = 1 - \frac{x^2}{2} + \frac{x^4}{4} + R_5(x)$$

SOLUTION Since all the odd powers of x have zero coefficients, we can assume that we have terms up to the fifth power of x and so we have $R_5(x)$, not $R_4(x)$

$$R_5(x) = \frac{f^{(6)}(\xi)(x^6)}{6!} = \frac{(-\cos \xi)(x^6)}{6!}$$

Hence $|R_5(x)| \leq |x^6|/720$, and the closer x is to zero, the closer the remainder is to zero.

EXERCISES

7.41 Find the radius of convergence of each of the following power series:

(a) $\sum_{k=0}^{\infty} kx^k$ (b) $\sum_{k=1}^{\infty} \frac{k^k}{k!} x^k$ (c) $\sum_{k=0}^{\infty} 3^k x^k$ (d) $\sum_{k=0}^{\infty} (2^k + 3^k) x^k$

(e) $\sum_{k=0}^{\infty} \frac{x^k}{3^k + 5^k}$ (f) $\sum_{k=0}^{\infty} \frac{x^k}{k3^k + 5^k}$ (g) $\sum_{k=1}^{\infty} k! x^k$ (h) $\sum_{k=1}^{\infty} \frac{x^k}{k^k}$

7.42 Find the interval of convergence of each of the following power series:

(a) $\sum_{k=1}^{\infty} \frac{(x-2)^k}{k^2 3^k}$ (b) $\sum_{k=1}^{\infty} (-1)^k \frac{(x+3)^k}{k}$ (c) $\sum_{k=0}^{\infty} \frac{x^k}{k!}$ (d) $\sum_{k=1}^{\infty} \frac{(x-5)^k}{3^k + 4^k}$

7.43 Show that if $\sum_{k=0}^{\infty} a_k(x-x_0)^k$ has radius of convergence R then $\sum_{k=1}^{\infty} a_k k(x-x_0)^{k-1}$ has radius of convergence R.

7.44 Show that if $\sum_{k=0}^{\infty} a_k(x-x_0)^k$ has radius of convergence $R > 0$ then the series $\sum_{k=0}^{\infty} a_k u^k$, where $u = (x-x_0)/R$, has radius of convergence 1.

7.45 Using the substitution from Exercise 7.44, prove Abel's theorem in the general case.

7.46 Derive the Taylor series for $\ln(1+x)$ at $x = 0$ by first using the geometric series for $1/(1+t)$ and then integrating from 0 to x.

7.47 Estimate the error E in writing
$$\sin x = x - \frac{x^3}{3!} + \frac{x^5}{5!} + E$$
for x in the interval $[0, 0.5]$.

7.48 Estimate the error E in writing
$$\ln x = (x-1) - \frac{(x-1)^2}{2} + \frac{(x-1)^3}{3} + E$$
for x in the interval (0.5, 1.5).

CHAPTER
EIGHT

DIFFERENTIABLE FUNCTIONS OF SEVERAL VARIABLES

8.1 SETS AND FUNCTIONS IN $\mathbf{R}^n$

Functions used to model real-life situations often depend on more than one variable. For example, the pressure of a confined gas depends on both the temperature and the volume. The revenue realized from the manufacture and sales of a commodity depends on both the number of items sold and the price charged for each item. In these cases our general definition of function is still valid if we take as the domain some subset of the set of all ordered pairs of real numbers.

A pair of real numbers can be associated with a point in the plane; a triple of real numbers can be associated with a point in space. Algebraically, there is no reason why we should limit ourselves to pairs or triples of real numbers. For any fixed natural number n we define $\mathbf{R}^n$ to be the set of all ordered n-tuples of real numbers $\{(x_1, x_2, \ldots, x_n) \mid x_i \in \mathbf{R}\}$. Then for any natural number n, arithmetic operations can be defined and the algebra of n-tuples can be developed. But for our geometric intuition we shall often return to the cases $n = 2$ or $n = 3$. Geometric concepts such as length and angle can be generalized to higher dimensions. We shall refer to an n-tuple as a point or a vector.

Arithmetic operations are defined as follows: For all $(x_1, x_2, \ldots, x_n) \in \mathbf{R}^n$, $(y_1, y_2, \ldots, y_n) \in \mathbf{R}^n$ and $c \in \mathbf{R}$ we have

Definition 8.1 (a) $(x_1, x_2, \ldots, x_n) + (y_1, y_2, \ldots, y_n) = (x_1 + y_1, x_2 + y_2, \ldots, x_n + y_n)$ called *vector addition*.

(b) $c \cdot (x_1, x_2, \ldots, x_n) = (cx_1, cx_2, \ldots, cx_n)$ called *scalar multiplication*.

The operations of vector addition and scalar multiplication satisfy many of the familiar rules of algebra. These rules follow readily from the properties of real numbers. Collectively the rules say that the set of all vectors forms a *vector space*. For any vectors $U, V, W \in \mathbf{R}^n$ and scalars $c, d \in \mathbf{R}$

1. $U + V = V + U$.
2. $(U + V) + W = U + (V + W)$.
3. There is a vector $0 \in \mathbf{R}^n$ with the property that $0 + V = V$ for all $V \in \mathbf{R}^n$.
4. For each $V \in \mathbf{R}^n$ there is a vector $-V \in \mathbf{R}^n$ such that $V + (-V) = 0$.
5. $(cd)V = c(dV)$.
6. $(c + d)V = cV + dV$.
7. $c(U + V) = cU + cV$.
8. $1V = V$, where 1 is the real number 1.

In $\mathbf{R}^2$ we can visualize a vector $U = (a, b)$ as a directed line segment or *arrow* from the origin to the point with coordinates $x = a, y = b$. Any arrow in R^2 with the same length and direction as U is also a valid geometrical representation for this vector. Then the sum $U + V$ and the difference $U - V = U + (-V)$ correspond to the diagonals of the parallelogram with sides U and V, as in Fig. 8.1. This type of visualization is also appropriate for vectors in $\mathbf{R}^3$. Taking a scalar multiple c of the vector V corresponds to stretching V if $|c| > 1$, shrinking V if $|c| < 1$, and reversing its direction if $c < 0$. In fact, if V is any nonzero vector and if $c \neq 0$ then cV is a vector along the line through the point V and the origin.

Let U and V be distinct points in $\mathbf{R}^2$ or $\mathbf{R}^3$ and let P be any point on

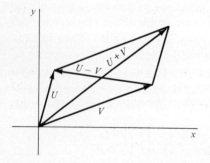

Figure 8.1

the line containing U and V. Then

$$P - U = t(V - U)$$

or equivalently $P = tV + (1 - t)U$ for some scalar t. The line segment connecting U and V is the set of all such P where $0 \leq t \leq 1$. This idea can be generalized to $\mathbf{R}^n$, where the *line segment* connecting points $U, V \in \mathbf{R}^n$ is defined to be the set of all points P which satisfy

$$P = tV + (1 - t)U \qquad 0 \leq t \leq 1$$

Suppose $x = x(t)$, $y = y(t)$, and $z = z(t)$ are each continuous functions of the parameter $t \in [a, b]$. Then the set of all points $P(t) = (x(t), y(t), z(t))$ for $a \leq t \leq b$ is called a *continuous curve* in $\mathbf{R}^3$. (Intuitively, just think of t as time; then as t changes, the point (x, y, z) moves through space, generating a curve.) More generally, if $x_j = x_j(t)$ is a continuous function defined on the closed, bounded interval $[a, b]$ for $j = 1, 2, \ldots, n$ then the set of points

$$P(t) = (x_1(t), x_2(t), \ldots, x_n(t))$$

is called a continuous curve in $\mathbf{R}^n$.

We define the *dot product* (also called *inner product*) of the vectors $U = (x_1, x_2, \ldots, x_n)$ and $V = (y_1, y_2, \ldots, y_n)$ to be

$$U \cdot V = \sum_{i=1}^{n} x_i y_i$$

Note that the dot product of two vectors is a scalar. This operation has some interesting algebraic and geometric properties. The following properties can be proved directly from the above definition.

1. $V \cdot V \geq 0$, and $V \cdot V = 0$ if and only if $V = 0$.
2. $U \cdot V = V \cdot U$ for all vectors U and V.
3. $(cU + dV) \cdot W = c(U \cdot W) + d(V \cdot W)$ for all vectors U, V, W and scalars c and d.

By virtue of these rules, the *length of a vector*, denoted $|V|$, can be defined as

$$|V| = \sqrt{V \cdot V}$$

This definition generalizes the notion of absolute value of a real number and corresponds to the usual notion of euclidean length of the arrow in $\mathbf{R}^2$ or $\mathbf{R}^3$. The *law of cosines* from trigonometry says that for nonzero vectors $U, V \in \mathbf{R}^2$ or $\mathbf{R}^3$

$$|U - V|^2 = |U|^2 + |V|^2 - 2|U||V| \cos \theta$$

where θ is the angle between U and V and satisfies $0 \leq \theta \leq \pi$. Since

$$|U - V|^2 = (U - V) \cdot (U - V) = U \cdot U - 2U \cdot V + V \cdot V$$
$$= |U|^2 + |V|^2 - 2U \cdot V$$

it follows that $U \cdot V = |U||V| \cos \theta$. It is clear that

$$\frac{|U \cdot V|}{|U||V|} \leq 1 \quad \text{for } U \neq 0 \text{ and } V \neq 0$$

Our first theorem generalizes this result to $\mathbf{R}^n$ and allows us to define the notion of angle in $\mathbf{R}^n$.

Theorem 8.1: Cauchy-Schwarz inequality Let U and V be any two vectors in $\mathbf{R}^n$. Then $|U \cdot V| \leq |U||V|$.

PROOF For any real number t

$$(U - tV) \cdot (U - tV) \geq 0$$

and so
$$(U \cdot V) - 2(U \cdot V)t + (V \cdot V)t^2 \geq 0$$

The quadratic in t on the left cannot have two distinct real roots because it would change sign. Hence the discriminant must be less than or equal to zero

$$4(U \cdot V)^2 - 4(U \cdot U)(V \cdot V) \leq 0$$
$$(U \cdot V)^2 \leq |U|^2 |V|^2$$
$$|U \cdot V| \leq |U||V|$$

We define the *angle* between two nonzero vectors $U, V \in \mathbf{R}^n$ to be the unique angle θ which satisfies

$$\cos \theta = \frac{U \cdot V}{|U||V|} \quad \text{and} \quad 0 \leq \theta \leq \pi$$

Whenever $U \cdot V = 0$, we say that U and V are *orthogonal*. In this case $\theta = \pi/2$, or else at least one of the vectors U, V is zero. Hence the zero vector is orthogonal to every vector.

Corollary: Triangle inequality For any two vectors $U, V \in \mathbf{R}^n$

$$|U + V| \leq |U| + |V|$$

Using the concept of length, we define the *distance* between $P_1, P_0 \in \mathbf{R}^n$ as $|P_1 - P_0|$. Then a *neighborhood* of P_0 is $N_\varepsilon(P_0) = \{P \,||\, P - P_0| < \varepsilon\}$. A set $S \subseteq \mathbf{R}^n$ is *open* if for each $P_0 \in S$ there is a neighborhood of P_0 which is completely contained in the set S. A set F is *closed* if its complement is open. The point P_0 is a *limit point* of the set S if each neighborhood of P_0 contains at least one $P \in S$ with $P \neq P_0$. As on the real line, a set F is closed if and only if it contains all its limit points. A point P_0 is an *interior point* of S if there is some $N_\varepsilon(P_0) \subseteq S$; P_0 is a *boundary point* of S if each $N_\varepsilon(P_0)$ contains points in S and points in the complement S' of S; P_0 is an *exterior point* of S if there is some $N_\varepsilon(P_0) \subseteq S'$. The *interior*, *boundary*, and *exterior* of S are the set of all interior points of S, the set of all boundary points of S, and the set of all exterior points of S, respectively.

A *polygonal line* is a set of line segments placed end to end. A subset S of $\mathbf{R}^n$ is called *polygonally connected* (or connected) if each pair of points in S can be joined by a polygonal line contained in S. (It should be noted here that there are other ways to define connected; another definition and its relationships to this one are given in the exercises.)

A *region* is any open connected set together with some, none, or all of its boundary points. Thus it may be open, closed, or neither. A region in $\mathbf{R}^n$ is the analog of an interval in $\mathbf{R}$.

A subset S of $\mathbf{R}^n$ is *bounded* (not to be confused with boundary) if there is a positive real number M such that $|P| < M$ for all $P \in S$. A bounded subset of $\mathbf{R}^2$ must be contained inside some circle centered at the origin. A bounded subset of $\mathbf{R}^3$ must be contained inside some sphere centered at the origin.

The sequence $\{P_k\}$, $P_k \in \mathbf{R}^n$ converges to P, written $P_k \to P$, if for any $\varepsilon > 0$ there is a natural number N such that $|P_k - P| < \varepsilon$ whenever $k > N$.

Lemma For each $P = (x_1, x_2, \ldots, x_n) \in \mathbf{R}^n$

$$|x_j| \leq |P| \leq \sum_{i=1}^n |x_i| \qquad j = 1, 2, \ldots, n$$

PROOF Since

$$|x_1|^2 + |x_2|^2 + \cdots + |x_n|^2 \leq (|x_1| + |x_2| + \cdots + |x_n|)^2$$

we have

$$|x_j| = \sqrt{x_j^2} \qquad \text{for each } j = 1, 2, \ldots, n$$

$$\leq \sqrt{\sum_{i=1}^n x_i^2} = |P| \leq \sum_{i=1}^n |x_i|$$

The relationship between convergence in $\mathbf{R}^n$ and convergence of real sequences is given in the following theorem.

Theorem 8.2 For each $k \in \mathbf{N}$ let $P_k = (x_{k1}, x_{k2}, \ldots, x_{kn})$ and let $P = (x_1, x_2, \ldots, x_n)$. Then $P_k \to P$ if and only if for each $j = 1, 2, \ldots, n$, $x_{kj} \to x_j$.

PROOF By the lemma, for each $j = 1, 2, \ldots, n$

$$|x_{kj} - x_j| \le |P_k - P| \le \sum_{i=1}^{n} |x_{ki} - x_i|$$

The result follows immediately.

The property of completeness was important in our study of limits in **R**. Fortunately $\mathbf{R}^n$ is complete for any natural number n. More precisely, a *Cauchy sequence* in $\mathbf{R}^n$ satisfies the property that for any $\varepsilon > 0$ there is a natural number K such that $|P_k - P_m| < \varepsilon$ whenever $k, m > K$. The completeness of $\mathbf{R}^n$ is given in the following theorem.

Theorem 8.3 A sequence $\{P_k\}$ in $\mathbf{R}^n$ converges if and only if it is a Cauchy sequence.

PROOF First assume that $\{P_k\}$ is a convergent sequence, say $P_k \to P$. Then for any $\varepsilon > 0$ there is a natural number N such that $|P_k - P| < \varepsilon/2$ when $k > N$. Let $m, k > N$. Then

$$|P_m - P_k| < |P_m - P| + |P - P_k| < \frac{\varepsilon}{2} + \frac{\varepsilon}{2} = \varepsilon$$

Next assume that $\{P_k\}$ is a Cauchy sequence. If $P_k = (x_{k1}, x_{k2}, \ldots, x_{kn})$ then

$$|x_{kj} - x_{mj}| \le |P_k - P_m| \quad \text{for each } j = 1, 2, \ldots, n$$

and hence $\{x_{kj}\}$ is a Cauchy sequence of real numbers for each fixed $j = 1, \ldots, n$. By the completeness of **R**, $\{x_{kj}\}$ converges to some real number x_j for each $j = 1, \ldots, n$. Finally since

$$|P_k - P| \le \sum_{i=1}^{n} |x_{ki} - x_i|$$

where $P = (x_1, x_2, \ldots, x_n)$, $P_k \to P$.

We note that two other familiar theorems are valid in $\mathbf{R}^n$: the Bolzano-Weierstrass theorem and the Heine-Borel theorem.

The main focus of this chapter is on functions, limits of functions, and derivatives. For the most part our functions will have as domain a region in $\mathbf{R}^2$ and as range a subset of **R**. Many of the definitions and theorems for these functions are easily generalized to functions defined on a region in $\mathbf{R}^n$. The graph of a function $f: D \to \mathbf{R}$, where D is a region in $\mathbf{R}^2$, is called a *surface* and is a subset of $\mathbf{R}^3$. Level curves are helpful in visualizing these surfaces. If $z = f(x, y)$, a *level curve* is obtained by setting one of the variables equal to a constant. Geometrically we are cutting the surface with a

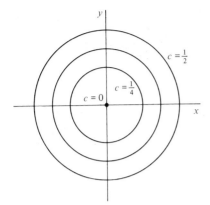

Figure 8.2

plane parallel to one of the coordinate planes. Then by assembling these level curves we get a picture of the surface.

Example 8.1 For $z = x^2 + y^2$ determine the level curves and sketch the surface represented by this function.

SOLUTION If $z = c$, we get a point when $c = 0$, namely $(0, 0)$. When $c > 0$, $x^2 + y^2 = c$ is a circle, which gets larger as c increases (see Fig. 8.2). If $x = c$ then $z = c^2 + y^2$ is a parabola in the $x = c$ plane (see Fig. 8.3). By symmetry the level curves for $y = c$ are the same as for $x = c$. Putting them together, we get the surface in Fig. 8.4, which is called a *paraboloid*. Large antennas are made in this shape to capture weak radio signals from great distances.

For more complicated surfaces, computer graphics can be used to trace

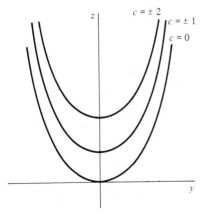

Figure 8.3

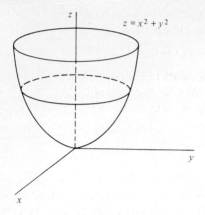

Figure 8.4

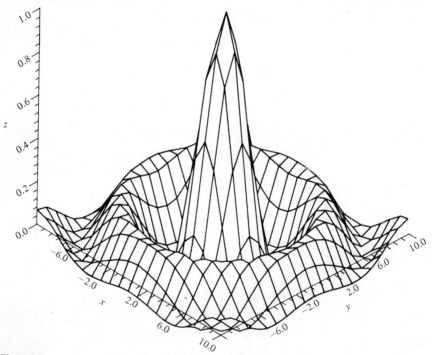

Figure 8.5

many of these level curves. Assembled together, they give us a good picture of the surface. Figure 8.5 shows the surface represented by

$$z = \frac{\sin(x^2 + y^2)^{1/2}}{(x^2 + y^2)^{1/2}}$$

We have seen that planes parallel to the coordinate planes have the form $x = c$, $y = c$, or $z = c$, where c is a constant. Generally the set of points P in a *plane* can be described by the equation

$$N \cdot (P - P_1) = 0$$

where N is a vector normal (perpendicular) to the plane and P_1 is a particular point in the plane. If we let $N = (a, b, c)$, $P = (x, y, z)$, and $P_1 = (x_1, y_1, z_1)$ then the equation of the plane takes the form

$$a(x - x_1) + b(y - y_1) + c(z - z_1) = 0$$

or

$$ax + by + cz = d$$

where d is the constant $ax_1 + by_1 + cz_1$.

Given three points in $\mathbf{R}^3$ which do not lie on a straight line, we should be able to find the equation of the plane containing those points. To accomplish this, it will be convenient to use a new operation called the *cross product*. If $V_1 = (a_1, a_2, a_3)$ and $V_2 = (b_1, b_2, b_3)$ then

$$V_1 \times V_2 = (a_2 b_3 - a_3 b_2, a_3 b_1 - a_1 b_3, a_1 b_2 - a_2 b_1)$$

Note that the components of $V_1 \times V_2$ are determinants

$$V_1 \times V_2 = \left(\begin{vmatrix} a_2 & a_3 \\ b_2 & b_3 \end{vmatrix}, -\begin{vmatrix} a_1 & a_3 \\ b_1 & b_3 \end{vmatrix}, \begin{vmatrix} a_1 & a_2 \\ b_1 & b_2 \end{vmatrix} \right)$$

In fact $V_1 \times V_2$ is often represented by the determinant

$$V_1 \times V_2 = \begin{vmatrix} i & j & k \\ a_1 & a_2 & a_3 \\ b_1 & b_2 & b_3 \end{vmatrix}$$

where i, j, k represent the unit vectors $(1, 0, 0)$, $(0, 1, 0)$, and $(0, 0, 1)$, respectively. The cross product of two vectors in $\mathbf{R}^3$ is a new vector in $\mathbf{R}^3$ which is orthogonal to each of the vectors V_1 and V_2

$$V_1 \cdot (V_1 \times V_2) = \begin{vmatrix} a_1 & a_2 & a_3 \\ a_1 & a_2 & a_3 \\ b_1 & b_2 & b_3 \end{vmatrix} = 0$$

and

$$V_2 \cdot (V_1 \times V_2) = \begin{vmatrix} b_1 & b_2 & b_3 \\ a_1 & a_2 & a_3 \\ b_1 & b_2 & b_3 \end{vmatrix} = 0$$

Furthermore the length of $V_1 \times V_2$ is given by

$$|V_1 \times V_2| = |V_1| |V_2| \sin \theta \qquad 0 \leq \theta \leq \pi$$

where θ is the angle between V_1 and V_2 (see Exercise 8.10). Thus the cross product of nonzero vectors is zero only when $\sin \theta = 0$, that is, when the vectors V_1 and V_2 lie along the same line (are collinear).

To find the equation of a plane through three points P_1, P_2, P_3 which are not collinear, first find two vectors $V_1 = P_2 - P_1$ and $V_2 = P_3 - P_1$ in the plane. Next use the cross product to find a vector $N = V_1 \times V_2$ normal to the plane. Finally, write $N \cdot (P - P_1) = 0$.

Example 8.2 Find the equation of the plane which contains the three points $P_1 = (1, -3, 2)$, $P_2 = (0, 1, 1)$, and $P_3 = (1, 0, 3)$.

SOLUTION Two vectors in the plane are

$$V_1 = P_2 - P_1 = (-1, 4, -1) \quad \text{and} \quad V_2 = P_3 - P_1 = (0, 3, 1)$$

A vector normal to the plane is

$$N = V_1 \times V_2 = (7, 1, -3)$$

The equation of the plane is

$$(7, 1, -3) \cdot (x - 1, y + 3, z - 2) = 0$$
$$7(x - 1) + (y + 3) - 3(z - 2) = 0$$
$$7x + y - 3z = -2$$

Throughout the remainder of this section we assume that the functions are defined on some region in $\mathbf{R}^2$. These results extend readily to the functions defined on regions in $\mathbf{R}^n$. Let $f: D \to \mathbf{R}$, where D is a region in $\mathbf{R}^2$, and assume that P_0 is a limit point of D.

Definition 8.2 $\lim_{P \to P_0} f(P) = L$ if for any $\varepsilon > 0$ there is a $\delta > 0$ such that $|f(P) - L| < \varepsilon$ whenever $0 < |P - P_0| < \delta$ and $P \in D$, the domain of f.

This is, of course, similar to the definition of limit for a function of one variable. As before, if the limit exists, it is unique; and f is bounded in some sets $S \cap (N_\delta(P_0) - \{P_0\})$, where $N_\delta(P_0) - \{P\}$ is some deleted neighborhood of P_0.

In the definition of limit $|P - P_0|$ is the magnitude (not absolute value) of the vector $P - P_0$. The multidimensional domain does make this limit more complicated than the one-dimensional limit, as we shall see in the examples below and the exercises at the end of this section.

Example 8.3 Evaluate and justify $\lim_{(x, y) \to (2, -1)} (y^2 - 3xy)$.

SOLUTION Intuitively, x approaches 2 and y approaches -1; so y^2 must approach 1 and $-3xy$ must approach 6. Hence the limit should be $1 + 6 = 7$. (*Caution*: This intuition may not always work.) The justification we give here is just short of an ε-δ proof. The idea is to write $|y^2 - 3xy - 7|$ in terms of $|y + 1|$ and $|x - 2|$, which go to zero because $|y + 1| \leq [(x - 2)^2 + (y + 1)^2]^{1/2} = |(x, y) - (2, -1)|$ and $|x - 2| \leq |(x, y) - (2, -1)|$. Now

$$|y^2 - 3xy - 7| = |y^2 - 1 - 3xy - 6| \leq |y^2 - 1| + 3|xy + 2|$$
$$\leq |y + 1||y - 1| + 3|xy + x - x + 2|$$
$$< |y + 1||y - 1| + 3|x||y + 1| + 3|x - 2|$$

The factors $|y - 1|$, $3|x|$, and 3 are bounded in any neighborhood of $(2, -1)$, while the factors $|y + 1|$ and $|x - 2|$ can be made arbitrarily small; hence we can make $|y^2 - 3xy - 7|$ arbitrarily small.

Example 8.4 Evaluate and justify $\lim_{(x, y) \to (0, 0)} [xy/(x^2 + y^2)]$.

SOLUTION Intuitively this is not as obvious as Example 8.3. Since both the numerator and denominator are close to zero, we cannot tell anything about the fraction. If (x, y) is close to zero, on the y axis $x = 0$, $y \neq 0$, and the fraction is zero. Similarly for the x axis. However, close to zero on the line $y = x$ the fraction is $\tfrac{1}{2}$. Since the limit cannot be 0 and $\tfrac{1}{2}$ simultaneously, we must say that the limit does not exist.

In problems similar to Example 8.4 we can use the technique of evaluating the limit as P approaches P_0 along various well-chosen paths in the domain of the function. For Example 8.4 we used the paths $y = x$ and $x = 0$ and got two different limits. Generally, if the limit exists, it must exist and be the same along any two paths to the point in the domain of the function. Of course, just being the same on two particular paths would not be sufficient reason to declare that the limit exists.

Example 8.5 Evaluate and justify $\lim_{(x, y) \to (0, 0)} [x^2 y/(x^2 + y^2)]$.

SOLUTION Along the path $y = 0$ the limit is zero; along the path $y = x$ the limit is $\lim_{x \to 0} (x^3/2x^2) = 0$, and so we suspect that the limit is zero. To prove it we use the inequality $|xy| \leq \tfrac{1}{2}(x^2 + y^2)$, which follows from $(x - y)^2 \geq 0$ and $(x + y)^2 \geq 0$

$$\left| \frac{x^2 y}{x^2 + y^2} \right| \leq \frac{|x||xy|}{x^2 + y^2} \leq \frac{1}{2} \frac{x^2 + y^2}{x^2 + y^2} |x| = \tfrac{1}{2}|x|$$

Since $|x|$ approaches zero, $|x^2 y/(x^2 + y^2)|$ must approach zero as $(x, y) \to (0, 0)$.

As in the one-dimensional case, we have various limit theorems which are summarized in the following theorem.

Theorem 8.4 Assume that P_0 is a limit point of the intersection of the domain of f and the domain of g. If $\lim_{P \to P_0} f(P) = L$ and $\lim_{P \to P_0} g(P) = M$ then
(a) $\lim_{P \to P_0} [f(P) + g(P)] = L + M$.
(b) $\lim_{P \to P_0} cf(P) = cL$, where $c = $ const.
(c) $\lim_{P \to P_0} [f(P) \cdot g(P)] = LM$.
(d) $\lim_{P \to P_0} [f(P)/g(P)] = L/M$ if $M \neq 0$.
(e) $L \leq M$ if $f(P) \leq g(P)$ for all P in the intersection of the domain of f and the domain of g with some deleted neighborhood of P_0.

Example 8.6 Evaluate and justify $\lim_{(x, y) \to (0, 0)} (1/xy) \sin(x^2y + xy^2)$.

SOLUTION Again the numerator and denominator both approach zero, but the following familiar limit should come to mind

$$\lim_{u \to 0} \frac{\sin u}{u} = 1$$

and

$$\lim_{(x, y) \to (0, 0)} \frac{1}{xy} \sin(x^2y + xy^2) = \lim_{(x, y) \to (0, 0)} \frac{x + y}{xy(x + y)} \sin(x^2y + xy^2)$$

$$= \lim_{(x, y) \to (0, 0)} (x + y) \lim_{(x, y) \to (0, 0)} \frac{\sin(x^2y + xy^2)}{(x^2y + xy^2)}$$

$$= 0 \cdot 1 = 0$$

We are able to identify the complicated quantity $x^2y + xy^2$ with u because $x^2y + xy^2$ approaches zero.

This limit concept is used to define the concept of continuity of functions of two variables.

Definition 8.3 Let f be defined in a region in $\mathbf{R}^2$ and let P_0 be a limit point of the region. Then f is continuous at P_0 if $\lim_{P \to P_0} f(P) = f(P_0)$.

If f is not continuous at P_0, we say that f is *discontinuous* at P_0. If $\lim_{P \to P_0} f(P)$ exists but f is discontinuous at P_0 then we say that f has a *removable discontinuity* at P_0. In this case, if f is defined or redefined at P_0 to be $\lim_{P \to P_0} f(P)$ then f will be continuous at P_0.

Returning to the preceding examples, we see that $f(x, y) = x^2 - 3xy$ is continuous at all points in $\mathbf{R}^2$, as is any polynomial in two variables; $g(x, y) = xy/(x^2 + y^2)$ is continuous at all points in $\mathbf{R}^2$ except at $(0, 0)$, where the limit does not exist; $h(x, y) = x^2y/(x^2 + y^2)$ is continuous at all points in $\mathbf{R}^2$ except at $(0, 0)$, where $h(x, y)$ has a removable discontinuity; and $k(x, y) = (1/xy) \sin(x^2y + xy^2)$ is continuous at all points in $\mathbf{R}^2$ except along the x axis and y axis. All the points of discontinuity of $k(x, y)$ are removable discontinuities.

The theory of continuity of a function of several variables is similar to that for a function of one variable. For the theorems that follow we include proofs that contain some new insights but generally omit proofs that are analogous to ones in Chap. 4. First, in Theorem 8.5 an equivalent form of continuity is given.

Theorem 8.5 Let $f: D \to \mathbf{R}$ where D is a region in $\mathbf{R}^2$. The function f is continuous at $P_0 \in D$ if and only if for each sequence $\{P_n\}$ in D which converges to P_0, $\{f(P_n)\}$ converges to $f(P_0)$.

The next theorem involves composite functions. The special cases $f(u, v) = u + v$, $f(u, v) = uv$, and $f(u, v) = u/v$, $v \neq 0$, yield the result that sums, products, and quotients (where denominator does not equal zero) of continuous functions are continuous.

Theorem 8.6 If $f(u, v)$ is continuous at $P_0 = (u_0, v_0)$ and if $u = g(x, y)$ and $v = h(x, y)$ are continuous at (x_0, y_0) with $u_0 = g(x_0, y_0)$ and $v_0 = h(x_0, y_0)$ then $f(g(x, y), h(x, y))$ is continuous at (x_0, y_0).

PROOF Given $\varepsilon > 0$ there is $\eta > 0$ such that $|f(u, v) - f(u_0, v_0)| < \varepsilon$ when $|(u, v) - (u_0, v_0)| < 2\eta$. Given $\eta > 0$ there is a $\delta_1 > 0$ such that $|g(x, y) - g(x_0, y_0)| < \eta$ whenever $|(x, y) - (x_0, y_0)| < \delta_1$; corresponding to $\eta > 0$ there is a $\delta_2 > 0$ such that $|h(x, y) - h(x_0, y_0)| < \eta$ whenever $|(x, y) - (x_0, y_0)| < \delta_2$. If $|(x, y) - (x_0, y_0)| < \delta = \min(\delta_1, \delta_2)$ then $|g(x, y) - g(x_0, y_0)| < \eta$ and $|h(x, y) - h(x_0, y_0)| < \eta$. Hence

$$|(g(x, y), h(x, y)) - (g(x_0, y_0), h(x_0, y_0))| < 2\eta$$

which implies

$$|f(g(x, y), h(x, y)) - f(g(x_0, y_0), h(x_0, y_0))| < \varepsilon$$

Corollary If $g(t)$ and $h(t)$ are continuous on $[\alpha, \beta]$ and if $f(x, y)$ is continuous on the range of $(g(t), h(t))$ then $f(g(t), h(t))$ is continuous on $[\alpha, \beta]$.

Theorem 8.7 If f is continuous at each point (x, y) in a region $D \subseteq \mathbf{R}^2$ and if $f(a, b) < v < f(c, d)$, where (a, b) and $(c, d) \in D$, then there is a point (x_0, y_0) in D such that $f(x_0, y_0) = v$.

PROOF Let $x = g(t)$, $y = h(t)$ be the equations of a polygonal line in D between (a, b) and (c, d), where t varies on some interval $[\alpha, \beta]$; then $g(t)$, $h(t)$ are continuous on $[\alpha, \beta]$ (see Exercise 8.15). By the corollary above, $f(g(t), h(t))$ is a continuous function of t with $f(g(\alpha), h(\alpha)) = f(a, b)$ and $f(g(\beta), h(\beta)) = f(c, d)$. By the intermediate-value theorem there is a point ξ between α, β such that $f(g(\xi), h(\xi)) = v$. But $(g(\xi), h(\xi))$ is a point on the polygonal line and hence in D. This is the point (x_0, y_0) we seek.

The above theorem is an intermediate-value theorem for functions of two variables. In the next two theorems we develop an extreme-value theorem guaranteeing that a function which is continuous on a closed, bounded region in $\mathbf{R}^2$ has an absolute maximum and absolute minimum. In Sec. 8.5 we shall use the concept of partial derivatives to find the extreme values of the function.

Theorem 8.8 If f is continuous on a closed bounded region D then f is bounded on D.

PROOF Assume that f is continuous on D but unbounded. Then there is a point $P_1 \in D$ with $|f(P_1)| > 1$, a point $P_2 \in D$ with $|f(P_2)| > 2$, and a point P_k (for each natural number k) with $|f(P_k)| > k$. The sequence $\{P_k\}$ is bounded because $P_k \in D$ and D is bounded. By the Bolzano-Weierstrass theorem $\{P_k\}$ has a convergent subsequence $\{P_{k_j}\}$, $\lim_{j \to \infty} P_{k_j} = P_0$, and since D is closed, $P_0 \in D$. But $|f(P_{k_j})| > k_j$ and so $\{f(P_{k_j})\}$ diverges. This contradicts Theorem 8.5, which says that $\lim_{j \to \infty} f(P_{k_j}) = f(P_0)$. Hence f is bounded on D.

Theorem 8.9 If f is continuous on a closed, bounded region D then there are points (a, b) and (c, d) in D where $f(a, b) \leq f(x, y) \leq f(c, d)$ for all $(x, y) \in D$.

PROOF By Theorem 8.8, f is bounded above on D. Define $M = \sup_{(x, y) \in D} f(x, y)$. If there were no point $(c, d) \in D$ with $f(c, d) = M$ then the function $g(x, y) = 1/[M - f(x, y)]$ would be continuous, positive, and bounded above on D. Hence there is a $B > 0$ such that

$$0 \leq \frac{1}{M - f(x, y)} \leq B$$

$$\frac{1}{B} \leq M - f(x, y)$$

$$f(x, y) \leq M - \frac{1}{B}$$

which contradicts the fact that $M = \sup_{(x, y) \in D} f(x, y)$; hence there is a point $(c, d) \in D$ where $f(c, d) = M$. A similar argument shows the existence of a point $(a, b) \in D$ which satisfies

$$f(a, b) = \inf_{(x, y) \in D} f(x, y)$$

Hence for all $(x, y) \in D$, $f(a, b) \leq f(x, y) \leq f(c, d)$. $f(a, b)$ is called the *absolute minimum* of f in D, and $f(c, d)$ is called the *absolute maximum* of f in D.

EXERCISES

8.1 Use the Cauchy-Schwarz inequality to prove the triangle inequality.

8.2 Prove that a neighborhood in $\mathbf{R}^n$ is an open set in $\mathbf{R}^n$.

8.3 Define the *closure* $\bar{S}$ of a set $S \subseteq \mathbf{R}^n$ as the union of S and the set of all limit points of S. Prove:
  (a) $\bar{S}$ is a closed set.
  (b) If F is any closed set which satisfies $S \subseteq F \subseteq \mathbf{R}^n$ then $S \subseteq \bar{S} \subseteq F$.

8.4 Find the closure of each of the following subsets of $\mathbf{R}^2$.
  (a) $\{(x, y) \mid x^2 + y^2 < 1\}$ (b) $\{(x, y) \mid x^2 + y^2 \leq 1\}$
  (c) $\left\{(x, y) \mid y = \sin \dfrac{1}{x}, x \neq 0\right\}$
  (d) $\{(x, y) \mid x \text{ rational and } y \text{ irrational}\}$
  (e) $\left\{\left(\dfrac{1}{n}, \dfrac{1}{m}\right) \mid n, m \text{ nonzero integers}\right\}$

8.5 Find the vector equation of the line through $(1, 7, -1)$ and $(2, 1, 0)$.

8.6 Sketch the curve represented by $P = (x, y, z) = (\cos t, \sin t, t)$ for $0 \leq t \leq 4\pi$.

8.7 Find the equation of the plane containing $(1, 2, 0)$, $(-1, 1, 0)$, and $(1, 1, -1)$.

8.8 Determine whether each of the following sets is open, closed, connected, or bounded and find its boundary:
  (a) $\{(x, y) \mid 2x^2 + 4y^2 < 1\}$ (b) $\{(x, y) \mid x^2 - y^2 \geq 1\}$
  (c) $\{(x, y) \mid |x| \leq 1 \text{ and } |y| \leq 1\}$ (d) $\{(x, y) \mid x \text{ and } y \text{ are rational}\}$

8.9 Find the level curves and sketch the surface:
  (a) $z = 4 - |p|$ for $p = (x, y)$ (b) $z = x^2 - y^2$
  (c) $z = 3x + y - 4$ (d) $z = xy$

8.10 Prove the following properties for the cross product of two vectors $V_1, V_2 \in \mathbf{R}^3$:
  (a) $V_2 \times V_1 = -(V_1 \times V_2)$
  (b) $|V_1 \times V_2| = |V_1||V_2| \sin \theta$, where θ is the angle between V_1 and V_2, $0 \leq \theta \leq \pi$.

8.11 Prove the Bolzano-Weierstrass theorem for $\mathbf{R}^2$: If $\{P_k\}$ is a bounded sequence in $\mathbf{R}^2$ then $\{P_k\}$ has a convergent subsequence.

8.12 Define the open set $G \subseteq \mathbf{R}^n$ as *connected* if G cannot be expressed as a disjoint union of nonempty open sets. Prove that if the open set G is connected then G is polygonally connected. *Hint*: Assume that G is not polygonally connected and show how G can be expressed as the disjoint union of nonempty open sets.

8.13 Evaluate the following limits and justify your answers:

(a) $\lim_{(x, y) \to (-1, 1)} (xy - 2x)$ (b) $\lim_{(x, y) \to (0, 0)} \dfrac{x^2 - y^2}{x^2 + y^2}$ (c) $\lim_{(x, y) \to (0, 0)} \dfrac{x^2}{x - y}$

(d) $\lim_{(x, y) \to (0, 0)} \dfrac{1 - \cos(x^2 + y^2)}{(x^2 + y^2)^2}$ (e) $\lim_{(x, y) \to (0, 0)} e^{-(x^2 + y^2)}$

(f) $\lim_{(x, y) \to (0, 0)} e^{-1/(x^2 + y^2)}$ (g) $\lim_{(x, y) \to (0, 0)} e^{-y/x}$

8.14 Prove for any pair of real numbers x, y that $|xy| \le \tfrac{1}{2}(x^2 + y^2)$. *Hint*: Start with $(x - y)^2 \ge 0$ and $(x + y)^2 \ge 0$.

8.15 Prove that any polygonal line in $\mathbf{R}^2$ can be written in the form $x = g(t)$, $y = h(t)$, $\alpha \le t \le \beta$, where g and h are continuous on $[\alpha, \beta]$.

8.16 Find the points of discontinuity of each of the following functions. Which are removable?

(a) $f(x, y) = \dfrac{x - y}{x^2 - y^2}$ (b) $f(x, y) = \sin \dfrac{y}{x}$

(c) $f(x, y) = \sqrt{|xy|}$ (d) $f(x, y) = \ln(x - y)$

8.17 Prove Theorem 8.5.

8.18 Define: f is uniformly continuous on a subset D of $\mathbf{R}^2$ if for any $\varepsilon > 0$ there is a $\delta > 0$ such that $|f(P) - f(P')| < \varepsilon$ whenever $|P - P'| < \delta$ and $P, P' \in D$. Prove that if f is continuous on a closed, bounded subset D of $\mathbf{R}^2$ then f is uniformly continuous on D. *Hint*: Assume that f is continuous but not uniformly continuous on D and show a contradiction. Under the assumption that f is not uniformly continuous on D, there are sequences $\{P_n\}$ and $\{P'_n\}$ with the property that $\lim_{n \to \infty} (P_n - P'_n) = 0$ but $|f(P_n) - f(P'_n)| \ge \varepsilon_0$ for some positive ε_0. Use the Bolzano-Weierstrass theorem to contradict the continuity of f on D.

8.2 DIFFERENTIABLE FUNCTIONS

Let f be a function from a region $D \subseteq \mathbf{R}^2$ into $\mathbf{R}$. Let β, a unit vector in $\mathbf{R}^2$, specify a direction. A measure of the rate of change of f as (x, y) changes in the direction β is the directional derivative.

Definition 8.4 The *directional derivative* of f in the direction $\beta = (b_1, b_2)$, $b_1^2 + b_2^2 = 1$, at the point (x_0, y_0) is

$$D_\beta f(x_0, y_0) = \lim_{t \to 0} \frac{f(x_0 + tb_1, y_0 + tb_2) - f(x_0, y_0)}{t}$$

Example 8.7 Find the directional derivative of $f(x, y) = x^2 - xy + 3$ at $(1, 1)$ in the direction $\beta = (\tfrac{3}{5}, -\tfrac{4}{5})$.

Solution

$$D_\beta f(1, 1) = \lim_{t \to 0} \frac{f(1 + \tfrac{3}{5}t, 1 - \tfrac{4}{5}t) - f(1, 1)}{t}$$

$$= \lim_{t \to 0} \frac{(1 + \tfrac{3}{5}t)^2 - (1 + \tfrac{3}{5}t)(1 - \tfrac{4}{5}t) + 3 - 3}{t}$$

$$= \lim_{t \to 0} \frac{t - \tfrac{3}{25}t^2}{t} = \lim_{t \to 0} (1 - \tfrac{3}{25}t) = 1$$

In the direction $\beta = (1, 0)$ the directional derivative is the *partial derivative of f with respect to x*

$$f_x(x_0, y_0) = \lim_{t \to 0} \frac{f(x_0 + t, y_0) - f(x_0, y_0)}{t}$$

Note that y is held constant at y_0 and the partial derivative $f_x(x_0, y_0)$ is obtained by considering the function $f(x, y_0)$ as a function of the single variable x and then finding the derivative of $f(x, y_0)$ at x_0. Similarly,

$$f_y(x_0, y_0) = \lim_{t \to 0} \frac{f(x_0, y_0 + t) - f(x_0, y_0)}{t}$$

is the derivative of $f(x_0, y)$ at y_0.

The following example shows that although both partial derivatives exist at a point the function may still be discontinuous at the point.

Example 8.8 Given

$$f(x, y) = \begin{cases} \dfrac{xy}{x^2 + y^2} & (x, y) \neq (0, 0) \\ 0 & (x, y) = (0, 0) \end{cases}$$

find $f_x(0, 0)$, $f_y(0, 0)$, and $D_\beta f(0, 0)$ for arbitrary β.

Solution Both partial derivatives exist

$$f_x(0, 0) = \lim_{t \to 0} \frac{0}{t} = 0 \qquad f_y(0, 0) = \lim_{t \to 0} \frac{0}{t} = 0$$

In fact the directional derivative

$$D_\beta f(0, 0) = \lim_{t \to 0} \frac{b_1 b_2}{b_1^2 + b_2^2} \frac{1}{t}$$

exists only when $\beta = (1, 0)$ or $(0, 1)$. In the previous section we showed that f is discontinuous at $(0, 0)$ (see Example 8.4).

Example 8.9 Let

$$f(x, y) = \begin{cases} \dfrac{x^2 y}{x^4 + y^2} & (x, y) \neq (0, 0) \\ 0 & (x, y) = (0, 0) \end{cases}$$

Show that f has a directional derivative at $(0, 0)$ in any direction $\beta = (b_1, b_2)$ but f is discontinuous at $(0, 0)$.

SOLUTION

$$\begin{aligned} D_\beta f(0, 0) &= \lim_{t \to 0} \frac{f(tb_1, tb_2) - f(0, 0)}{t} \\ &= \lim_{t \to 0} \frac{t^3 b_1^2 b_2 / (t^4 b_1^4 + t^2 b_2^2)}{t} \\ &= \begin{cases} \dfrac{b_1^2}{b_2} & \text{if } b_2 \neq 0 \\ 0 & \text{if } b_2 = 0 \end{cases} \end{aligned}$$

To show that f is discontinuous at $(0, 0)$ consider $\lim_{(x, y) \to (0, 0)} f(x, y)$. Along the path $y = 0$ the limit is zero, and along the path $y = x^2$ the limit is $\tfrac{1}{2}$. Hence $\lim_{(x, y) \to (0, 0)} f(x, y)$ does not exist and f is discontinuous at $(0, 0)$.

We are about to define the notion of f being differentiable at the point (x_0, y_0). Considering Example 8.8, it would not be enough to say that $f_x(x_0, y_0)$ and $f_y(x_0, y_0)$ exist because we want all the directional derivatives to exist at (x_0, y_0). Considering Example 8.9 it would not be enough to say $D_\beta f(x_0, y_0)$ exists for any direction β because f differentiable at (x_0, y_0) should certainly imply f continuous at (x_0, y_0), as in the case of functions of one variable.

How, then, shall we define f to be differentiable at (x_0, y_0)? First we require that $f_x(x_0, y_0)$ and $f_y(x_0, y_0)$ exist. Next consider the plane

$$\pi: z - z_0 = f_x(x_0, y_0)(x - x_0) + f_y(x_0, y_0)(y - y_0)$$

where $z_0 = f(x_0, y_0)$. This plane contains the lines

$$L_1: z - z_0 = f_x(x_0, y_0)(x - x_0) \quad \text{in the plane } y = y_0$$

and

$$L_2: z - z_0 = f_y(x_0, y_0)(y - y_0) \quad \text{in the plane } x = x_0$$

The line L_1 is the tangent line to the level curve $z = f(x, y_0)$ in the plane $y = y_0$ at (x_0, y_0, z_0). Similarly the line L_2 is tangent to the level curve $z = f(x_0, y)$ in the plane $x = x_0$ at the point (x_0, y_0, z_0). We would like the definition of $f(x, y)$ differentiable at (x_0, y_0) to be strong enough for curves cut by other planes through this point, for example, $y - y_0 = x - x_0$, to have tangent lines in the plane π also. To accomplish this we require that

$$\lim_{(x, y) \to (x_0, y_0)} \frac{z_s - z_\pi}{\sqrt{(x - x_0)^2 + (y - y_0)^2}} = 0$$

where z_s is the value of z on the surface: $z_s = f(x, y)$ and z_π is the corresponding value of z on the plane π

$$z_\pi = f(x_0, y_0) + f_x(x_0, y_0)(x - x_0) + f_y(x_0, y_0)(y - y_0)$$

The difference $z_s - z_\pi$ can be thought of as the error in approximating $f(x, y)$ by the value of z on the plane π.

Definition 8.5 The function f is differentiable at (x_0, y_0) if $f_x(x_0, y_0)$ and $f_y(x_0, y_0)$ exist and if

$$\lim_{(x, y) \to (x_0, y_0)} \eta(x, y) = 0$$

where $\eta(x, y)$ is defined by

$$f(x, y) = f(x_0, y_0) + f_x(x_0, y_0)(x - x_0) + f_y(x_0, y_0)(y - y_0)$$
$$+ \eta(x, y) \sqrt{(x - x_0)^2 + (y - y_0)^2}$$

If f is differentiable at (x_0, y_0) then the plane π is called the *tangent plane* to the surface $z = f(x, y)$ at (x_0, y_0, z_0).

Theorem 8.10 If f is differentiable at (x_0, y_0) then $D_\beta f(x_0, y_0)$ exists for any unit vector β and

$$D_\beta f(x_0, y_0) = f_x(x_0, y_0)b_1 + f_y(x_0, y_0)b_2 \qquad \text{where } \beta = (b_1, b_2)$$

PROOF $\lim_{t \to 0} \dfrac{f(x_0 + tb_1, y_0 + tb_2) - f(x_0, y_0)}{t}$

$= \lim_{t \to 0} \dfrac{f_x(x_0, y_0)tb_1 + f_y(x_0, y_0)tb_2 + \eta(x_0 + tb_1, y_0 + tb_2)\sqrt{(tb_1)^2 + (tb_2)^2}}{t}$

$= \lim_{t \to 0} \left[f_x(x_0, y_0)b_1 + f_y(x_0, y_0)b_2 + \eta(x_0 + tb_1, y_0 + tb_2) \dfrac{|t|}{t} \right]$

but $\lim_{t \to 0} \eta(x_0 + tb_1, y_0 + tb_2) = 0$

and so $\quad D_\beta f(x_0, y_0) = f_x(x_0, y_0)b_1 + f_y(x_0, y_0)b_2$

The formula given by Theorem 8.10 is a dot product of the vector $(f_x(x_0, y_0), f_y(x_0, y_0))$ with β. It is convenient to give this vector a name:

$$\nabla f(x_0, y_0) = (f_x(x_0, y_0), f_y(x_0, y_0))$$

is called the *gradient* of f at (x_0, y_0) (or sometimes the derivative of f at (x_0, y_0)). Since

$$D_\beta f(x_0, y_0) = |\nabla f| |\beta| \cos \theta = |\nabla f| \cos \theta$$

where θ is the angle between β and ∇f, the maximum rate of change of f is in the direction $\theta = 0$, or β is in the same direction as the gradient. Similarly the direction for the greatest rate of decrease in $f(x, y)$ at (x_0, y_0) is $\theta = \pi$, or β is in the direction of $-\nabla f(x_0, y_0)$. One popular method for locating the minimum of a function of several variables is the *method of steepest descent*. We move a small distance from (x_0, y_0) in the direction of $-\nabla f(x_0, y_0)$ to (x_1, y_1), some point where $f(x_1, y_1) < f(x_0, y_0)$. (What if there is no such point?) Again evaluate the direction of $-\nabla f(x_1, y_1)$ of the steepest descent from (x_1, y_1) and move to a new point (x_2, y_2), where $f(x_2, y_2) < f(x_1, y_1)$. Continue this process until (x_n, y_n) is sufficiently close to the minimum.

Theorem 8.11 If f is differentiable at (x_0, y_0) and if $f(x_0, y_0)$ is a local minimum or local maximum of $f(x, y)$ then

$$\nabla f(x_0, y_0) = 0$$

PROOF By Theorem 8.10, $D_\beta f(x_0, y_0)$ exists for any unit vector β. Assume that $f(x_0, y_0)$ is a local minimum; then

$$D_\beta f(x_0, y_0) = \lim_{t \to 0^-} \frac{f(x_0 + tb_1, y_0 + tb_2) - f(x_0, y_0)}{t} \leq 0$$

$$= \lim_{t \to 0^+} \frac{f(x_0 + tb_1, y_0 + tb_2) - f(x_0, y_0)}{t} \geq 0$$

Therefore $D_\beta f(x_0, y_0) = 0$ for any unit vector β. Choosing $\beta = (1, 0)$ implies $f_x(x_0, y_0) = 0$ and $\beta = (0, 1)$ implies $f_y(x_0, y_0) = 0$. Hence $\nabla f(x_0, y_0) = 0$. The proof is similar for $f(x_0, y_0)$ a local maximum.

Up to now we have seen some consequences of differentiability (necessary conditions for differentiability). The following theorem gives a sufficient condition for differentiability which is usually much easier to check for a given function than the definition.

Theorem 8.12 If $f_x(x, y)$ and $f_y(x, y)$ exist in some neighborhood of (x_0, y_0) and if $f_x(x, y)$ and $f_y(x, y)$ are continuous at (x_0, y_0) then $f(x, y)$ is differentiable at (x_0, y_0).

PROOF We wish to show that $\lim_{(x, y) \to (x_0, y_0)} \eta(x, y) = 0$, where

$$f(x, y) - f(x_0, y_0) = f_x(x_0, y_0)(x - x_0) + f_y(x_0, y_0)(y - y_0)$$
$$+ \eta(x, y) \sqrt{(x - x_0)^2 + (y - y_0)^2}$$

Now

$$f(x, y) - f(x_0, y_0) = f(x, y) - f(x_0, y) + f(x_0, y) - f(x_0, y_0)$$
$$= f_x(\bar{x}, y)(x - x_0) + f_y(x_0, \bar{y})(y - y_0)$$

by the mean-value theorem for some $\bar{x}$ between x and x_0 and for some $\bar{y}$ between y and y_0. Then

$$f(x, y) - f(x_0, y_0) = f_x(x_0, y_0)(x - x_0) + f_y(x_0, y_0)(y - y_0)$$
$$+ E_1(x - x_0) + E_2(y - y_0)$$

where

$$E_1 = f_x(\bar{x}, y) - f_x(x_0, y_0) \quad \text{and} \quad E_2 = f_y(x_0, \bar{y}) - f_y(x_0, y_0)$$

By the continuity of f_x and f_y at (x_0, y_0)

$$\lim_{(x, y) \to (x_0, y_0)} E_1 = 0 \quad \text{and} \quad \lim_{(x, y) \to (x_0, y_0)} E_2 = 0$$

Then

$$\lim_{(x, y) \to (x_0, y_0)} \eta(x, y) = \lim_{(x, y) \to (x_0, y_0)} \frac{E_1(x - x_0) + E_2(y - y_0)}{\sqrt{(x - x_0)^2 + (y - y_0)^2}}$$
$$= \lim_{(x, y) \to (x_0, y_0)} E_1 \frac{x - x_0}{\sqrt{(x - x_0)^2 + (y - y_0)^2}}$$
$$+ E_2 \frac{y - y_0}{\sqrt{(x - x_0)^2 + (y - y_0)^2}} = 0$$

If $f_x(x, y)$ and $f_y(x, y)$ are continuous in a region D of the (x, y) plane, we say that $f \in \mathcal{C}^1(D)$; that is, $\mathcal{C}^1(D)$ is the set of all functions which have continuous (first-order) partial derivatives in D. The functions $f_{xx}(x, y)$ and $f_{xy}(x, y)$ are the partial derivatives of $f_x(x, y)$, and $f_{yx}(x, y)$ and $f_{yy}(x, y)$ are the partial derivatives of $f_y(x, y)$. If all these second-order partial derivatives are continuous at each point in D, we say that $f \in \mathcal{C}^2(D)$. For any natural number k, $\mathcal{C}^k(D)$ is the set of all functions whose kth-order partial derivatives are continuous in D. The set $\mathcal{C}^\infty(D)$ contains those functions having continuous partial derivatives of all orders in D. We note that for any natural number k, $\mathcal{C}^\infty(D) \subset \mathcal{C}^{k+1}(D) \subset \mathcal{C}^k(D)$ (see Exercise 8.29).

The definitions of this section are easily extended to higher derivatives. Let f be defined on some open region $D \subseteq \mathbf{R}^n$. The partial derivative of f

with respect to the ith variable x_i is given by

$$f_{x_i}(x_1, \ldots, x_n) = \lim_{t \to 0} \frac{f(x_1, x_2, \ldots, x_{i-1}, x_i + t, x_{i+1}, \ldots, x_n) - f(x_1, \ldots, x_n)}{t}$$

Alternate notation for f_{x_i} is f_i or $\partial f/\partial x_i$. The function f is differentiable at a point $P_0 = (x_1^0, x_2^0, \ldots, x_n^0)$ if all partial derivatives exist at P_0 and if $\lim_{P \to P_0} \eta(P) = 0$, where

$$f(P) = f(P_0) + \sum_{i=1}^{n} f_i(P_0)(x_i - x_i^0) + \eta(P)|P - P_0|$$

and $P = (x_1, \ldots, x_n)$.

The extension of Theorem 8.12 to higher dimensions says that if f has partial derivatives in some neighborhood of P_0 and if the partial derivatives are continuous at P_0, that is, $\lim_{P \to P_0} f_i(P) = f_i(P_0)$ for $i = 1, \ldots, n$, then f is differentiable at P_0. Again it is useful to define $C^k(D)$, $D \subseteq \mathbf{R}^n$, as the set of all functions $f: D \to \mathbf{R}$ whose kth order partial derivatives are continuous on the region D.

EXERCISES

8.19 Prove that if f is differentiable at (x_0, y_0) then f is continuous at (x_0, y_0).

8.20 For each of the following functions find $f_x(x, y)$ and $f_y(x, y)$ and give the domain of each partial derivative:

(a) $f(x, y) = \dfrac{xy}{x^2 + y^2}$ (b) $f(x, y) = \ln(x - y)$ (c) $f(x, y) = \sin \dfrac{y}{x}$

8.21 Given $f(x, y) = \sqrt{|xy|}$ find $f_x(0, 0)$ and $f_y(0, 0)$. Show that these are the only directional derivatives of f that exist at $(0, 0)$.

8.22 Which of the following functions are differentiable at $(0, 0)$? Carefully give your reasons.

(a) $f(x, y) = \sqrt{|xy|}$

(b) $f(x, y) = \begin{cases} \dfrac{xy}{x^2 + y^2} & (x, y) \neq (0, 0) \\ 0 & (x, y) = (0, 0) \end{cases}$

(c) $f(x, y) = \begin{cases} \dfrac{x^2 y}{x^2 + y^2} & (x, y) \neq (0, 0) \\ 0 & (x, y) = (0, 0) \end{cases}$

(d) $f(x, y) = \begin{cases} \dfrac{x^2 y^2}{x^2 + y^2} & (x, y) \neq (0, 0) \\ 0 & (x, y) = (0, 0) \end{cases}$

8.23 In each step for the method of steepest descent we move a certain distance d in the direction of $-\nabla f$. What is one way of selecting an appropriate d at each step?

8.24 Let $f(x, y)$ be differentiable at (x_0, y_0) and let $\beta = (b_1, b_2)$ be a fixed unit vector. Let $F(t) = f((x_0, y_0) + t(b_1, b_2))$. Prove that dF/dt at $t = 0$ is equal to $D_\beta f(x_0, y_0)$. Interpret F and dF/dt geometrically.

8.25 State a theorem for a function defined on region $D \subseteq \mathbf{R}^n$ which is analogous to Theorem 8.11.

8.26 For the function in Example 8.9

$$f(x, y) = \begin{cases} \dfrac{x^2 y}{x^4 + y^2} & (x, y) \neq (0, 0) \\ 0 & (x, y) = (0, 0) \end{cases}$$

show that along any straight-line path to $(0, 0)$ the limit is zero. (Yet the function is discontinuous at $(0, 0)$.)

8.27 Let

$$f(x, y) = \begin{cases} \dfrac{x^3 y - y^3 x}{x^2 + y^2} & (x, y) \neq (0, 0) \\ 0 & (x, y) = (0, 0) \end{cases}$$

Find $f_x(0, 0), f_y(0, 0), f_{xy}(0, 0)$, and $f_{yx}(0, 0)$; note that $f_{xy}(0, 0) \neq f_{yx}(0, 0)$.

8.28 Is the function in Exercise 8.27 in the set $\mathcal{C}(\mathbf{R}^2)$? In $\mathcal{C}^1(\mathbf{R}^2)$? In $\mathcal{C}^2(\mathbf{R}^2)$?

8.29 Show that for any natural number k and region $D \subseteq \mathbf{R}^2$

$$\mathcal{C}^\infty(D) \subset \mathcal{C}^{k+1}(D) \subset \mathcal{C}^k(D)$$

8.30 Prove that the converse of Theorem 8.12 is false.

8.3 CHAIN RULES AND TAYLOR'S FORMULA

Quite often in mathematics a change of variables for a function is necessary. The appropriate change of variable in an integral or differential equation may suffice to solve the problem. We use a chain rule to substitute for the derivatives in such cases.

Consider changing the variables of the function $z = F(x, y)$ by the substitution $x = f(u, v)$, $y = g(u, v)$. The following theorem relates the new partial derivative to the old ones.

Theorem 8.13 Assume that $z = F(x, y)$, $x = f(u, v)$, and $y = g(u, v)$, where f and g are in $\mathcal{C}^1(D)$ for some open region $D \subseteq \mathbf{R}^2$ and $F \in \mathcal{C}^1(D_1)$, where D_1 is an open region in $\mathbf{R}^2$ which contains the set $\{(x, y) = (f(u, v), g(u, v)) \mid (u, v) \in D\}$. Then at each $(u_0, v_0) \in D$

$$\frac{\partial z}{\partial u} = \frac{\partial z}{\partial x}\frac{\partial x}{\partial u} + \frac{\partial z}{\partial y}\frac{\partial y}{\partial u} \qquad \frac{\partial z}{\partial v} = \frac{\partial z}{\partial x}\frac{\partial x}{\partial v} + \frac{\partial z}{\partial y}\frac{\partial y}{\partial v}$$

Proof In order to find $\partial z/\partial u$ at a point $(u_0, v_0) \in D$ let

$$x_0 = f(u_0, v_0) \qquad y_0 = g(u_0, v_0) \qquad x = f(u, v_0) \qquad y = g(u, v_0)$$

Since F has continuous first-order partial derivatives, F is differentiable at (x_0, y_0) and

$$F(x, y) - F(x_0, y_0) = F_x(x_0, y_0)(x - x_0) + F_y(x_0, y_0)(y - y_0) + \eta(x, y)\sqrt{(x - x_0)^2 + (y - y_0)^2}$$

where $\lim_{(x, y) \to (x_0, y_0)} \eta(x, y) = 0$. For $u \neq u_0$

$$\frac{F(x, y) - F(x_0, y_0)}{u - u_0} = F_x(x_0, y_0) \frac{f(u, v_0) - f(u_0, v_0)}{u - u_0}$$

$$+ F_y(x_0, y_0) \frac{g(u, v_0) - g(u_0, v_0)}{u - u_0}$$

$$+ \frac{\eta(x, y)}{u - u_0} [(f(u, v_0) - f(u_0, v_0))^2 + (g(u, v_0) - g(u_0, v_0))^2]^{1/2}$$

To show that the last term vanishes as u approaches u_0 consider

$$\left| \frac{\eta(x, y)}{u - u_0} [(f(u, v_0) - f(u_0, v_0))^2 + (g(u, v_0) - g(u_0, v_0))^2]^{1/2} \right|$$

$$= |\eta(x, y)| \left[\left(\frac{f(u, v_0) - f(u_0, v_0)}{u - u_0} \right)^2 + \left(\frac{g(u, v_0) - g(u_0, v_0)}{u - u_0} \right)^2 \right]^{1/2}$$

By the continuity of f and g, as u approaches u_0, we have that (x, y) approaches (x_0, y_0) and $\eta(x, y)$ approaches zero. The radical above approaches

$$[(f_u(u_0, v_0))^2 + (g_u(u_0, v_0))^2]^{1/2}$$

and hence the product approaches zero. Therefore

$$\lim_{u \to u_0} \frac{F(x, y) - F(x_0, y_0)}{u - u_0} = F_x(x_0, y_0)f_u(u_0, v_0) + F_y(x_0, y_0)g_u(u_0, v_0)$$

or

$$\frac{\partial z}{\partial u} = \frac{\partial z}{\partial x} \frac{\partial x}{\partial u} + \frac{\partial z}{\partial y} \frac{\partial y}{\partial u} \qquad \text{at } (u_0, v_0)$$

A similar argument shows that

$$\frac{\partial z}{\partial v} = \frac{\partial z}{\partial x} \frac{\partial x}{\partial v} + \frac{\partial z}{\partial y} \frac{\partial y}{\partial v} \qquad \text{at } (u_0, v_0)$$

Example 8.10 Given $z = x^2 - y^2$, $x = u \cos v$, and $y = u \sin v$, use Theorem 8.13 to find $\partial z/\partial u$ and $\partial z/\partial v$.

SOLUTION

$$\frac{\partial z}{\partial u} = \frac{\partial z}{\partial x}\frac{\partial x}{\partial u} + \frac{\partial z}{\partial y}\frac{\partial y}{\partial u} = 2x \cos v - 2y \sin v = 2u \cos^2 v - 2u \sin^2 v$$

$$\frac{\partial z}{\partial v} = \frac{\partial z}{\partial x}\frac{\partial x}{\partial v} + \frac{\partial z}{\partial y}\frac{\partial y}{\partial v} = -2xu \sin v - 2yu \cos v$$

$$= -2u^2 \cos v \sin v - 2u^2 \cos v \sin v = -4u^2 \cos v \sin v$$

Note that these can be verified by first substituting for x and y in $z = x^2 - y^2$ and then differentiating.

A useful way to remember this chain rule is by writing a chain of dependence as follows:

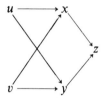

where an arrow is drawn from one variable to another; for example, $u \to y$ if the second variable depends on the first, that is, y depends on u. Then the chain rule for $\partial z/\partial u$ is found by tracing the arrows in reverse from z to u along all possible paths. A product of partial derivatives is formed for each path, and then these products are added. The path $z \leftarrow x \leftarrow u$ yields the product $(\partial z/\partial x)(\partial x/\partial u)$, and the path $z \leftarrow y \leftarrow u$ yields the product $(\partial z/\partial y)(\partial y/\partial u)$. Adding, we get

$$\frac{\partial z}{\partial u} = \frac{\partial z}{\partial x}\frac{\partial x}{\partial u} + \frac{\partial z}{\partial y}\frac{\partial y}{\partial u}$$

Other dependence realtionships yield other chain rules. The chain of dependence can be used to derive these rules. Of course, a derivation of this type does not constitute a proof. Suitable continuity requirements have to be imposed on the partial derivatives and a proof similar to that of Theorem 8.13 supplied.

Example 8.11 Derive the chain rule for the relationships

$$z = F(x, y) \quad \text{and} \quad y = g(x, t)$$

SOLUTION The chain of dependence is

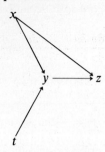

Hence z is ultimately a function of x and t. To find $\partial z/\partial x$ we use the path $z \leftarrow x$ and $z \leftarrow y \leftarrow x$. From these paths

$$\frac{\partial z}{\partial x} = F_1 + F_2 \frac{\partial y}{\partial x}$$

where F_1 is the partial of z with respect to x before the substitution for y in the formula for z and $\partial z/\partial x$ is the partial of z with respect to x after the substitution. F_2, of course, is $\partial z/\partial y$.

The partial derivative $\partial z/\partial t$ is found from the single path $z \leftarrow y \leftarrow t$ and hence is

$$\frac{\partial z}{\partial t} = F_2 \frac{\partial y}{\partial t}$$

The next example is an application of Theorem 8.13 to solving a partial differential equation. Since second partial derivatives are used, we shall look for solutions which are in $\mathcal{C}^2(D)$ for some open region $D \subseteq \mathbf{R}^2$.

Example 8.12 An important partial differential equation in mathematical physics is the wave equation

$$\frac{\partial^2 y}{\partial x^2} = \frac{1}{c^2} \frac{\partial^2 y}{\partial t^2}$$

where $y = F(x, t)$ and where c is a positive constant whose value is determined by the nature of the physical application. Using the substitution $u = x + ct$ and $v = x - ct$, we reduce the equation to one which can be integrated twice, and then we obtain the general solution. The chain of dependence we use is

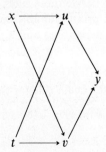

The second partial derivatives are obtained by two applications of the chain rule.

$$\frac{\partial y}{\partial x} = \frac{\partial y}{\partial u}\frac{\partial u}{\partial x} + \frac{\partial y}{\partial v}\frac{\partial v}{\partial x} = \frac{\partial y}{\partial u}(1) + \frac{\partial y}{\partial v}(1)$$

$$\frac{\partial^2 y}{\partial x^2} = \frac{\partial}{\partial x}\left(\frac{\partial y}{\partial x}\right) = \frac{\partial}{\partial x}\left(\frac{\partial y}{\partial u} + \frac{\partial y}{\partial v}\right)$$

$$= \frac{\partial}{\partial u}\left(\frac{\partial y}{\partial u} + \frac{\partial y}{\partial v}\right)\frac{\partial u}{\partial x} + \frac{\partial}{\partial v}\left(\frac{\partial y}{\partial u} + \frac{\partial y}{\partial v}\right)\frac{\partial v}{\partial x}$$

$$= \frac{\partial^2 y}{\partial u^2} + \frac{\partial^2 y}{\partial u\, \partial v} + \frac{\partial^2 y}{\partial v\, \partial u} + \frac{\partial^2 y}{\partial v^2}$$

$$= \frac{\partial^2 y}{\partial u^2} + 2\frac{\partial^2 y}{\partial u\, \partial v} + \frac{\partial^2 y}{\partial v^2}$$

(In Theorem 9.10 we prove that if the mixed partials: $\partial^2 y/(\partial u\, \partial v)$ and $\partial^2 y/(\partial v\, \partial u)$ are continuous then they are equal.)

$$\frac{\partial y}{\partial t} = \frac{\partial y}{\partial u}\frac{\partial u}{\partial t} + \frac{\partial y}{\partial v}\frac{\partial v}{\partial t} = \frac{\partial y}{\partial u}c + \frac{\partial y}{\partial v}(-c) = c\left(\frac{\partial y}{\partial u} - \frac{\partial y}{\partial v}\right)$$

$$\frac{\partial^2 y}{\partial t^2} = \frac{\partial}{\partial t}\left(\frac{\partial y}{\partial t}\right) = \frac{\partial}{\partial t}\left[c\left(\frac{\partial y}{\partial u} - \frac{\partial y}{\partial v}\right)\right]$$

$$= \frac{\partial}{\partial u}\left[c\left(\frac{\partial y}{\partial u} - \frac{\partial y}{\partial v}\right)\right]\frac{\partial u}{\partial t} + \frac{\partial}{\partial v}\left[c\left(\frac{\partial y}{\partial u} - \frac{\partial y}{\partial v}\right)\right]\frac{\partial v}{\partial t}$$

$$= \left[c\left(\frac{\partial^2 y}{\partial u^2} - \frac{\partial^2 y}{\partial u\, \partial v}\right)\right]c + \left[c\left(\frac{\partial^2 y}{\partial v\, \partial u} - \frac{\partial^2 y}{\partial v^2}\right)\right](-c)$$

$$= c^2\left(\frac{\partial^2 y}{\partial u^2} - 2\frac{\partial^2 y}{\partial u\, \partial v} + \frac{\partial^2 y}{\partial v^2}\right)$$

again assuming that the mixed parts are equal. The equation

$$\frac{\partial^2 y}{\partial x^2} = \frac{1}{c^2}\frac{\partial^2 y}{\partial t^2} \qquad \text{becomes} \qquad \frac{\partial^2 y}{\partial u\, \partial v} = 0$$

Integrating once while holding v constant, we obtain

$$\frac{\partial y}{\partial v} = f(v)$$

where $f(v)$ is an arbitrary function of v. Integrating again, holding u constant we have $y = g(v) + h(u)$, where g and h are arbitrary functions with $g'(v) = f(v)$. Substituting back for u and v, we obtain

$$y = g(x - ct) + h(x + ct)$$

In a physical application of the wave equation auxiliary conditions usually exist which determine the functions g and h (see Exercise 8.37).

A form of the chain rule can be used to derive *Taylor's formula* for functions of two variables. For convenience we return to the vector notation. Let $P = (x, y)$ and $P_0 = (x_0, y_0)$. To find the Taylor polynomial of degree n for $f(x, y)$ at (x_0, y_0) let $F(t) = f(P_0 + t(P - P_0))$. Think of P_0 and P as fixed points and t as the only variable. Then Taylor's formula for a function of one variable is

$$F(t) = F(0) + F'(0)t + \frac{F''(0)}{2!} t^2 + \cdots + F^{(n)}(0) \frac{t^n}{n!} + F^{(n+1)}(\xi) \frac{t^{n+1}}{(n+1)!}$$

where ξ is between 0 and t. By the chain rule (now allowing P to vary)

$$F'(t) = f_x(P_0 + t(P - P_0))(x - x_0) + f_y(P_0 + t(P - P_0))(y - y_0)$$
$$F'(0) = f_x(x_0, y_0)(x - x_0) + f_y(x_0, y_0)(y - y_0)$$
$$F''(t) = f_{xx}(P_0 + t(P - P_0))(x - x_0)^2 + 2f_{xy}(P_0 + t(P - P_0))(x - x_0)(y - y_0)$$
$$+ f_{yy}(P_0 + t(P - P_0))(y - y_0)^2$$
$$F''(0) = f_{xx}(x_0, y_0)(x - x_0)^2 + 2f_{xy}(x_0, y_0)(x - x_0)(y - y_0)$$
$$+ f_{yy}(x_0, y_0)(y - y_0)^2$$

$F^{(k)}(0)$ involves kth-order partials of f at (x_0, y_0), the corresponding powers of $x - x_0$ and $y - y_0$, and the binomial coefficients.

$$F^{(3)}(0) = f_{xxx}(x_0, y_0)(x - x_0)^3 + 3f_{xxy}(x_0, y_0)(x - x_0)^2(y - y_0)$$
$$+ 3f_{xyy}(x_0, y_0)(x - x_0)(y - y_0)^2 + f_{yyy}(x_0, y_0)(y - y_0)^3$$

The last term is the error term R_n, where the partials are evaluated at some point close to (x_0, y_0). Hence

$$f(x, y) = F(1) = f(x_0, y_0) + f_x(x_0, y_0)(x - x_0) + f_y(x_0, y_0)(y - y_0)$$
$$+ f_{xx}(x_0, y_0)(x - x_0)^2 + 2f_{xy}(x_0, y_0)(x - x_0)(y - y_0)$$
$$+ f_{yy}(x_0, y_0)(y - y_0)^2 + f_{xxx}(x_0, y_0)(x - x_0)^3$$
$$+ 3f_{xxy}(x_0, y_0)(x - x_0)^2(y - y_0) + 3f_{xyy}(x_0, y_0)(x - x_0)(y - y_0)^2$$
$$+ f_{yyy}(x_0, y_0)(y - y_0)^3 + \cdots + f_{x \cdots x}(x_0, y_0)(x - x_0)^n$$
$$+ nf_{x \cdots xy}(x_0, y_0)(x - x_0)^{n-1}(y - y_0) + \cdots$$
$$+ f_{y \cdots y}(x_0, y_0)(y - y_0)^n + R_n$$

This is Taylor's formula for two variables for a function f having partial derivatives up to order $n+1$ that are continuous in some neighborhood of (x_0, y_0). This formula without R_n is called the *Taylor polynomial* of degree n. If we increase n without bound then this becomes the Taylor series of $f(x, y)$ at (x_0, y_0). If $\lim_{n\to\infty} R_n = 0$ in some neighborhood of (x_0, y_0) then the Taylor series converges to the function in this neighborhood. As in the case of functions of one variable, the Taylor series represents the function in this neighborhood and can be used in approximating the function.

EXERCISES

8.31 Given $z = 2xy$, $x = e^u \cos v$, and $y = e^u \sin v$, find $\partial z/\partial u$ and $\partial z/\partial v$:
  (a) Using the chain rule.
  (b) By first substituting for x and y in $z = 2xy$.

8.32 Given that $z = F(x, y)$ has continuous second-order partial derivatives and $x = au + bv$, $y = cu + dv$, where a, b, c, d are constants, find

$$\frac{\partial z}{\partial u} \quad \frac{\partial z}{\partial v} \quad \frac{\partial^2 z}{\partial u^2} \quad \frac{\partial^2 z}{\partial u \, \partial v} \quad \text{and} \quad \frac{\partial^2 z}{\partial v^2}$$

8.33 (a) Derive a chain rule for the following chain of dependence; that is, find

$$\frac{\partial w}{\partial x} \quad \frac{\partial w}{\partial y} \quad \frac{\partial w}{\partial z}$$

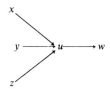

  (b) Assume $f(x, y, z) = g(x^2 + y^2 + z^2)$. Find an expression for $f_{xx} + f_{yy} + f_{zz}$ in terms of x, y, z and the function g.

8.34 Find the general solution of

$$\frac{\partial^2 y}{\partial x^2} + 2\frac{\partial^2 y}{\partial x \, \partial t} + \frac{\partial^2 y}{\partial t^2} = 0$$

by first changing the variables (x, t) to (u, v) by

$$u = x + t \qquad v = x - t$$

8.35 A function $f \in C^2(\mathbf{R}^2)$ is *homogeneous of degree n*, a nonnegative integer, if $f(tx, ty) = t^n f(x, y)$ for all x, y, t. Prove:
  (a) $xf_1(x, y) + yf_2(x, y) = nf(x, y)$ for all (x, y).
  (b) $x^2 f_{11}(x, y) + 2xyf_{12}(x, y) + y^2 f_{22}(x, y) = n(n-1)f(x, y)$ for all (x, y).

8.36 Laplace's equation is given by

$$\frac{\partial^2 z}{\partial x^2} + \frac{\partial^2 z}{\partial y^2} = 0$$

Find Laplace's equation in polar coordinates (r, θ), where $x = r \cos \theta$ and $y = r \sin \theta$.

8.37 Find a solution of the wave equation (Example 8.12) which satisfies the initial conditions

$$F(x, 0) = \begin{cases} x + 1 & -1 < x < 0 \\ -x + 1 & 0 < x < 1 \\ 0 & \text{otherwise} \end{cases}$$

$$F_t(x, 0) = 0$$

Sketch the graph of the solution in the (x, y) plane for $t = 0$, $t = 1/2c$, $t = 1/c$, and $t = 2/c$.

8.38 Use a chain of dependence for each of the following to obtain appropriate chain rules (assume that all partials are continuous).

(a) $z = F(x, y)$, $x = g(t)$, $y = h(t)$
(b) $z = F(x, y, u)$, $u = g(x, y)$
(c) $z = F(x, y, u, v)$, $u = f(x, y)$, $v = g(x, y)$
(d) $z = F(x, y, t)$, $x = f(y, t)$, $y = g(t)$

8.39 Write an expression for the remainder R_n in Taylor's formula for a function $f(x, y)$ with continuous partial derivatives of order $n + 1$.

8.40 Find the Taylor polynomial of degree 3 of each of the following functions at the indicated point:

(a) $f(x, y) = e^x \cos y$ $(0, \pi)$ (b) $f(x, y) = \ln xy$ $(1, 1)$
(c) $f(x, y) = x^2 + 2xy + y^3$ $(1, 0)$

8.41 Prove that if $z = F(x, y)$, $x = g(t)$, and $y = h(t)$, where g and h are in $\mathcal{C}^1(I)$, I an open interval, and if $F \in \mathcal{C}^1(D)$, where D contains the set $\{(x, y) \mid x = g(t), y = h(t) \text{ for some } t \in I\}$, then

$$\frac{dz}{dt} = \frac{\partial z}{\partial x}\frac{dx}{dt} + \frac{\partial z}{\partial y}\frac{dy}{dt}$$

8.42 Use Exercise 8.41 to prove the mean-value theorem: If $F \in \mathcal{C}^1(D)$, where D is an open region containing P_0 and P_1, and if the line segment between P_0 and P_1 is contained in D then there is a point P^* in D where

$$F(P_1) - F(P_0) = \nabla F(P^*) \cdot (P_1 - P_0)$$

8.4 IMPLICIT- AND INVERSE-FUNCTION THEOREMS

In this section we give sufficient conditions for being able to solve an equation for one variable as a function of the others in the equation. Except for the simplest of equations (linear equations, quadratic equations, etc.) this is a very difficult task. Next we consider two equations in two unknowns and then finally the inverse of a transformation. For the proofs in this section a linear-algebra approach† is more elegant but requires a deeper background than is needed for the proofs given here.

Theorem 8.14: Implicit-function theorem Let $F \in \mathcal{C}^1(D)$ for some open region $D \subseteq \mathbf{R}^3$. Let $P_0 = (x_0, y_0, z_0)$ be a point in D where

$$F(x_0, y_0, z_0) = 0 \quad \text{and} \quad F_z(x_0, y_0, z_0) \neq 0$$

† See R. C. Buck, *Advanced Calculus*, 3d ed., McGraw-Hill, New York, 1978.

Then there is neighborhood $N_c(z_0) \subseteq \mathbf{R}$, a neighborhood $N_\delta(x_0, y_0) \subseteq \mathbf{R}^2$, and a unique function $z = g(x, y)$ defined for $(x, y) \in N_\delta(x_0, y_0)$ with values $z \in N_c(z_0)$ such that

$$z_0 = g(x_0, y_0) \quad \text{and} \quad F(x, y, g(x, y)) = 0$$

for all $(x, y) \in N_\delta(x_0, y_0)$; that is, in this neighborhood $z = g(x, y)$ is the solution of the equation $F(x, y, z) = 0$ for z. Also, $g \in C^1(N_\delta(x_0, y_0))$, and

$$g_x(x, y) = -\frac{F_x(x, y, z)}{F_z(x, y, z)} \qquad g_y(x, y) = -\frac{F_y(x, y, z)}{F_z(x, y, z)}$$

for all $(x, y) \in N_\delta(x_0, y_0)$ and where $z = g(x, y)$.

PROOF Assume that $F_z(x_0, y_0, z_0) > 0$. (If $F_z(x_0, y_0, z_0) < 0$, the same proof applies to $-F$.) By the continuity of F_z in D there is a neighborhood $N_\varepsilon(x_0, y_0, z_0) \subset D$ in which F_z is continuous and positive. Therefore for fixed x and y, F is a strictly increasing function of z with $F(x_0, y_0, z_0) = 0$. So there is $c > 0$ such that $F(x_0, y_0, z_0 - c) < 0$ and $F(x_0, y_0, z_0 + c) > 0$ with $(x_0, y_0, z_0 - c) \in N_\varepsilon(x_0, y_0, z_0)$ and $(x_0, y_0, z_0 + c) \in N_\varepsilon(x_0, y_0, z_0)$.

By the continuity of F there is a $\delta > 0$ such that $F(x, y, z_0 - c) < 0$ and $F(x, y, z_0 + c) > 0$ for $(x, y) \in N_\delta(x_0, y_0)$ and where δ is small enough to ensure that $(x, y, z_0 - c) \in N_\delta(x_0, y_0, z_0)$ and $(x, y, z_0 + c) \in N_\delta(x_0, y_0, z_0)$ for all $(x, y) \in N_\delta(x_0, y_0)$.

Consider now $(x, y) \in N_\delta(x_0, y_0)$ arbitrary but fixed. Since $F(x, y, z)$ is continuous and strictly increasing for $z_0 - c < z < z_0 + c$ with $F(x, y, z_0 - c) < 0$ and $F(x, y, z_0 + c) > 0$, there is a unique z, $z_0 - c < z < z_0 + c$ such that $F(x, y, z) = 0$. Thus z is a function of (x, y), say $z = g(x, y)$, where $(x, y) \in N_\delta(x_0, y_0)$ and $z_0 - c < z < z_0 + c$.

To show that $g(x, y)$ is continuous in $N_\delta(x_0, y_0)$, let (x_1, y_1) be a point in $N_\delta(x_0, y_0)$ and let $z_1 = g(x_1, y_1)$. Then $F(x_1, y_1, z_1) = 0$. Given any $\varepsilon_1 > 0$ consider $|z - z_1| < \varepsilon_1$. By an argument similar to that used in defining g there is a c_1 (choose it small enough to ensure that $c_1 < \varepsilon_1$ and $N_{c_1}(z_1) \subseteq N_c(z_0)$) and $\delta_1 > 0$ (choose it small enough to ensure that $N_{\delta_1}(x_1, y_1) \subseteq N_\delta(x_0, y_0)$) such that if $(x, y) \in N_{\delta_1}(x_1, y_1)$ then the unique z which satisfies $F(x, y, z) = 0$ is between $z_1 - c_1$ and $z_1 + c_1$ and hence $|z - z_1| < \varepsilon_1$. Thus $f(x, y)$ is continuous in $N_\delta(x_0, y_0)$.

To establish the formula for $g_x(x, y)$ consider $F(x, y, z)$ a function of x and z with y fixed. Let h be a small change in x and $z + k = g(x + h, y)$, where $(x, y) \in N_\delta(x_0, y_0)$, $(x + h, y) \in N_\delta(x_0, y_0)$, and $z + k \in N_c(z_0)$. This is possible by the continuity of $g(x, y)$ in $N_\delta(x_0, y_0)$. Both $F(x, y, z) = 0$ and $F(x + h, y, z + k) = 0$, and so

$$0 = F(x + h, y, z + k) - F(x, y, z)$$
$$= F_x(x^*, y, z^*)h + F_z(x^*, y, z^*)k$$

where x^* is between x and $x + h$ and z^* is between z and $z + k$ by the mean-value theorem (see Exercise 8.42). Then

$$g_x(x, y) = \lim_{h \to 0} \frac{g(x + h, y) - g(x, y)}{h} = \lim_{h \to 0} \frac{k}{h}$$

$$= \lim_{h \to 0} -\frac{F_x(x^*, y, z^*)}{F_z(x^*, y, z^*)} = -\frac{F_x(x, y, z)}{F_z(x, y, z)}$$

using the continuity of g, F_x and F_z. The formula for g_y can be obtained in a similar fashion. Since F_x, F_y, F_z are continuous and $F_z \neq 0$ in $N_\varepsilon(x_0, y_0, z_0)$, it follows that $g_x(x, y)$ and $g_y(x, y)$ are continuous in $N_\delta(x_0, y_0)$, and so $g \in \mathcal{C}^1(N_\delta(x_0, y_0))$.

Example 8.13 Show that $F(x, y, z) = x^2 + y^2 - z^2 - 1 = 0$ defines z as a function $g(x, y)$ in a neighborhood of any point (x_0, y_0, z_0) with $F(x_0, y_0, z_0) = 0$ and $z_0 \neq 0$. Also find the first-order partial derivatives of $g(x, y)$.

SOLUTION The first-order partials of F are

$$F_x(x, y, z) = 2x \qquad F_y(x, y, z) = 2y \qquad F_z(x, y, z) = -2z$$

Clearly $F \in \mathcal{C}^1(\mathbf{R}^3)$. Assume (x_0, y_0, z_0) is a point which satisfies $F(x_0, y_0, z_0) = 0$. Then $F_z(x_0, y_0, z_0) \neq 0$ if $z_0 \neq 0$. By Theorem 8.14 there is a function g with continuous partial derivatives such that $z = g(x, y)$ in some $N_\delta(x_0, y_0)$. The partial derivatives of g are given by

$$g_x(x, y) = -\frac{x}{z} \qquad g_y(x, y) = -\frac{y}{z} \qquad \text{where } z = g(x, y)$$

The particular function F in this example is so simple that $g(x, y)$ can be found explicitly

$$g(x, y) = \begin{cases} \sqrt{x^2 + y^2 - 1} & \text{if } z_0 > 0 \\ -\sqrt{x^2 + y^2 - 1} & \text{if } z_0 < 0 \end{cases}$$

where the size of the neighborhood $N_\delta(x_0, y_0)$ is restricted by $x^2 + y^2 - 1 \geq 0$.

The general implicit-function theorem is stated below. Its proof is conceptually the same as the proof of Theorem 8.14 but more complicated in notation.

Theorem 8.15 Let $F \in \mathcal{C}^1(D)$ for some open region $D \subseteq \mathbf{R}^{n+1}$. Let $(x_1^0, x_2^0, \ldots, x_n^0, z_0)$ be a point in D where

$$F(x_1^0, x_2^0, \ldots, x_n^0, z_0) = 0 \qquad F_z(x_1^0, x_2^0, \ldots, x_n^0, z_0) \neq 0$$

Then there is a neighborhood $N_\epsilon(z_0) \subset \mathbf{R}$, a neighborhood $N_\delta(x_1^0, \ldots, x_n^0) \subset \mathbf{R}^n$, and unique function $z = g(x_1, \ldots, x_n)$ defined for $(x_1, \ldots, x_n) \in N_\delta(x_1^0, \ldots, x_n^0)$ with values $z \in N_\epsilon(z_0)$ such that

$$z_0 = g(x_1^0, \ldots, x_n^0) \quad \text{and} \quad F(x_1, \ldots, x_n, g(x_1, \ldots, x_n)) = 0$$

for all $(x_1, \ldots, x_n) \in N_\delta(x_1^0, \ldots, x_n^0)$. Moreover $g \in C^1(N_\delta(x_1^0, \ldots, x_n^0))$ and

$$g_{x_i}(x_1, \ldots, x_n) = -\frac{F_{x_i}(x_1, \ldots, x_n, z)}{F_z(x_1, \ldots, x_n, z)} \qquad i = 1, \ldots, n$$

for all $(x_1, \ldots, x_n) \in N_\delta(x_1^0, \ldots, x_n^0)$ and where $z = g(x_1, \ldots, x_n)$.

Next we consider the case of two simultaneous equations with two unknowns. We wish to solve

$$F(x, y, u, v) = 0$$
$$G(x, y, u, v) = 0$$

for u and v. The sufficient condition for this is given in terms of a determinant which is called a *Jacobian*:

$$\frac{\partial(F, G)}{\partial(u, v)} = \begin{vmatrix} F_u & F_v \\ G_u & G_v \end{vmatrix}$$

There are other Jacobians involved in the next theorem but for brevity we shall call this one J.

Theorem 8.16 Let $F, G \in C^1(D)$, where D is an open region in $\mathbf{R}^4$, and let $(x_0, y_0, u_0, v_0) \in D$ satisfy

$$F(x_0, y_0, u_0, v_0) = 0$$
$$G(x_0, y_0, u_0, v_0) = 0$$

and $J = \partial(F, G)/\partial(u, v) \neq 0$ at (x_0, y_0, u_0, v_0). Then there are neighborhoods $N_\delta(x_0, y_0)$, $N_{\epsilon_1}(u_0)$, and $N_{\epsilon_2}(v_0)$ such that for any $(x, y) \in N_\delta(x_0, y_0)$ there is a unique pair (u, v), $u \in N_{\epsilon_1}(u_0)$, $v \in N_{\epsilon_2}(v_0)$, such that

$$F(x, y, u, v) = 0$$
$$G(x, y, u, v) = 0$$

This defines unique functions

$$u = f(x, y) \quad \text{and} \quad v = g(x, y)$$

with f and g in $C^1(N_\delta(x_0, y_0))$. The first-order partial derivatives of f and g are given by

$$f_x(x, y) = -\frac{1}{J} \frac{\partial(F, G)}{\partial(x, v)} \qquad f_y(x, y) = -\frac{1}{J} \frac{\partial(F, G)}{\partial(y, v)}$$

$$g_x(x, y) = -\frac{1}{J} \frac{\partial(F, G)}{\partial(u, x)} \qquad g_y(x, y) = -\frac{1}{J} \frac{\partial(F, G)}{\partial(u, y)}$$

PROOF F_v and G_v cannot both vanish at $P_0 = (x_0, y_0, u_0, v_0)$; otherwise $J = 0$ at P_0. Assume $F_v \neq 0$. (The proof is similar for $G_v \neq 0$.) By Theorem 8.15 there is an $N_{\delta_1}(x_0, y_0, u_0)$ and an $N_{c_2}(v_0)$ and a function $\varphi(x, y, u)$ such that $v = \varphi(x, y, u)$ for $(x, y, u) \in N_{\delta_1}(x_0, y_0, u_0)$ with $v \in N_{c_2}(v_0)$. $v = \varphi(x, y, u)$ is a solution of $F(x, y, u, v) = 0$. Next consider the function

$$H(x, y, u) = G(x, y, u, \varphi(x, y, u))$$

By the chain rule

$$H_u = G_u + G_v \varphi_u \qquad \text{where} \qquad \varphi_u = -\frac{F_u}{F_v}$$

and it follows that

$$H_u = \frac{G_u F_v - G_v F_u}{F_v} = -\frac{J}{F_v} \neq 0$$

By Theorem 8.15 again, there is a neighborhood $N_\delta(x, y)$, $\delta < \delta_1$, a neighborhood $N_{c_1}(u_0)$, and a function f such that $u = f(x, y)$ is a solution of $H(x, y, u) = 0$. Finally if we define g by

$$g(x, y) = \varphi(x, y, f(x, y))$$

then $u = f(x, y)$ and $v = g(x, y)$ are solutions of

$$F(x, y, u, v) = 0$$
$$G(x, y, u, v) = 0$$

for $(x, y) \in N_\delta(x_0, y_0)$, $u \in N_{c_1}(u_0)$, $v \in N_{c_2}(v_0)$. By Theorem 8.15 f and g have continuous first-order partial derivatives in $N_\delta(x_0, y_0)$. We derive the formula for f_x and leave the other formulas to the exercises.

$$f_x(x, y) = -\frac{H_x}{H_u} = -\frac{G_x + G_v(-F_x/F_v)}{(G_u F_v - G_v F_u)/F_v}$$

$$= -\frac{F_x G_v - G_x F_v}{F_u G_v - G_u F_v} = -\frac{1}{J} \frac{\partial(F, G)}{\partial(x, v)}$$

We next apply Theorem 8.16 to the two equations

$$x = f(u, v) \qquad y = g(u, v)$$

In the preceding section we thought of these as a change of variables. Another interpretation is that they represent a transformation or mapping from $\mathbf{R}^2$ in the (u, v) plane into $\mathbf{R}^2$ in the (x, y) plane. Indeed the ordered pair (x, y) is a vector function of the ordered pair (u, v). For each pair (u, v) in the common domain of f and g there is a unique pair (x, y) associated. We say $(x, y) = T(u, v)$, where T denotes this vector function. If this function is one-to-one then for each (x, y) in the range of T, there is a unique (u, v) in the domain of T such that $(x, y) = T(u, v)$. In this case an inverse function T^{-1} exists such that $(u, v) = T^{-1}(x, y)$, and we say that T is *invertible*.

A special class of transformations is the set of all linear transformations. A linear transformation has the form

$$x = au + bv \qquad y = cu + dv \qquad \text{where } a, b, c, d = \text{const}$$

A transformation of this form is invertible if and only if $ad - bc \neq 0$ (see Exercise 8.52). The reader may recognize $ad - bc$ as the determinant of the coefficient matrix

$$\begin{bmatrix} a & b \\ c & d \end{bmatrix}$$

The criterion for a linear transformation from $\mathbf{R}^n$ to $\mathbf{R}^n$ is the same: $\det(A) \neq 0$, where A is the coefficient matrix.

The reader should also recognize $ad - bc \neq 0$ as $\partial(f, g)/\partial(u, v) \neq 0$, where

$$f(u, v) = au + bv \qquad g(u, v) = cu + dv$$

A sufficient condition for invertibility of more general transformations is given in the following *inverse-function* theorem.

Theorem 8.17 Inverse-function theorem Let $x = f(u, v)$ and $y = g(u, v)$, where $f, g \in \mathcal{C}^1(D)$, D being an open region in $\mathbf{R}^2$. Let $(u_0, v_0) \in D$ satisfy $x_0 = f(u_0, v_0)$, $y_0 = g(u_0, v_0)$ and

$$J = \frac{\partial(f, g)}{\partial(u, v)} \neq 0 \qquad \text{at } (u_0, v_0)$$

Then there are neighborhoods $N_\delta(x_0, y_0)$, $N_{c_1}(u_0)$ and $N_{c_2}(v_0)$ such that for each pair $(x, y) \in N_\delta(x_0, y_0)$ there is a unique pair (u, v), $u \in N_{c_1}(u_0)$, $v \in N_{c_2}(v_0)$ such that

$$f(u, v) = x \qquad \text{and} \qquad g(u, v) = y$$

Hence this defines unique functions

$$u = \varphi(x, y) \quad \text{and} \quad v = \psi(x, y)$$

with $\varphi, \psi \in C^1(N_\delta(x_0, y_0))$. The first-order partial derivatives of φ and ψ are given by

$$\varphi_x = \frac{1}{J} g_v \qquad \varphi_y = -\frac{1}{J} f_v \qquad \psi_x = -\frac{1}{J} g_u \qquad \psi_y = \frac{1}{J} f_u$$

PROOF This theorem is an immediate consequence of Theorem 8.16 with

$$F(x, y, u, v) = f(u, v) - x = 0$$
$$G(x, y, u, v) = g(u, v) - y = 0$$

We show the derivation of the formula for one of the partial derivatives φ_x:

$$\varphi_x = -\frac{1}{J} \begin{vmatrix} -1 & f_v \\ 0 & g_v \end{vmatrix} \quad \text{by Theorem 8.16}$$

$$= \frac{1}{J} g_v$$

Example 8.14 Find the Jacobian of the transformation

$$x = u \cos v \qquad y = u \sin v$$

and determine where there is an inverse transformation.

SOLUTION Clearly these functions have continuous partial derivatives and

$$J = \begin{vmatrix} \cos v & -u \sin v \\ \sin v & u \cos v \end{vmatrix} = u$$

Hence if (u_0, v_0) is a point where $u_0 \neq 0$ then there is a neighborhood of (x_0, y_0) where $x_0 = u_0 \cos v_0$, $y_0 = u_0 \sin v_0$ on which an inverse transformation exists.

The inverse-function theorem only guarantees an inverse transformation on a neighborhood, that is, a *local inverse*. The hypothesis of Theorem 8.17 is not strong enough to guarantee the existence of an inverse for an entire region, that is, a *global inverse*. The transformation of Example 8.14 is not invertible throughout a region which includes the points $(u, 0)$, $(u, 2\pi)$ $u \neq 0$, for example, even though $J \neq 0$, because the transformation is not one-to-one in such a region. A necessary and sufficient condition for a transformation T to have a global inverse is that the transformation be one-to-one.

EXERCISES

8.43 Which of the following equations have solutions for z in a neighborhood of the given point? Compute $\partial z/\partial x$ and $\partial z/\partial y$.
(a) $e^{x^2+y^2+z^2} - 1 = 0$, $(0, 0, 0)$ (b) $z + 2 \cos(x + y + z) = 2$, $(0, 0, 0)$
(c) $x^2 y^2 z^2 = 1$, $(1, 1, 1)$ (d) $x^2 + y^2 - z^2 = 1$, $(0, 0, 1)$

8.44 Given that $F(x, y, z) = 0$, where F has continuous nonzero first-order partial derivatives in an open region $D \subseteq \mathbf{R}^3$, prove that

$$\frac{\partial x}{\partial y} \frac{\partial y}{\partial z} \frac{\partial z}{\partial x} = -1 \quad \text{in } D$$

8.45 State and prove an implicit-function theorem for when $F(x, y) = 0$ defines y as a function of x (include a formula for dy/dx).

8.46 If y is eliminated from the two equations $z = f(x, y)$ and $g(x, y) = 0$, the result is expressible in the form $z = h(x)$. Express $h'(x)$ in terms of f_1, f_2, g_1, g_2.

8.47 Given sufficient conditions for when

$$xy - uv + v^2 = 0 \quad \text{and} \quad xv + 2u^2 - y^3 = 0$$

define u, v as a function of x, y. Find $\partial u/\partial x$, $\partial u/\partial y$, $\partial v/\partial x$, and $\partial v/\partial y$.

8.48 Complete the proof of Theorem 8.16 by deriving the formulas for $f_y, g_x,$ and g_y.

8.49 Complete the details of the proof of Theorem 8.17.

8.50 State and prove an inverse-function theorem for $y = f(x)$.

8.51 Find the image of $R = \{(u, v) \mid 0 \le u \le 1, 0 \le v \le 1\}$ under each of the following transformations:

(a) $x = u \cos \pi v$
$y = u \sin \pi v$

(b) $x = u - v$
$y = u + v$

(c) $x = u^2 - v^2$
$y = 2uv$

(d) $x = 2u - 5v$
$y = -4u + 10v$

8.52 Prove directly (without using Theorem 8.17) that the linear transformation

$$x = au + bv \qquad y = cu + dv$$

is invertible if and only if $ad - bc \ne 0$.

8.53 Prove that as a consequence of Theorem 8.17

$$\frac{\partial(u, v)}{\partial(x, y)} = \frac{1}{\partial(x, y)/\partial(u, v)}$$

8.54 Compute the Jacobian of each of the following transformations. Determine where local inverses exist.

(a) $x = e^u \cos v$
$y = e^u \sin v$

(b) $x = u^2 - v^2$
$y = 2uv$

(c) $x = u^2 - uv$
$y = v - u$

(d) $x = \sin(u + v)$
$y = \cos(u + v)$

8.5 EXTREMA OF FUNCTIONS OF SEVERAL VARIABLES

Recall that a function f has a *local maximum* at P_0 if there is some neighborhood $N_\delta(P_0)$ such that $f(P) \leq f(P_0)$ for all $P \in N_\delta(P_0)$; f has a *local minimum* at P_0 if there is some neighborhood $N_\delta(P_0)$ such that $f(P) \geq f(P_0)$ for all $P \in N_\delta(P_0)$. If f has a local maximum or local minimum at P_0 then we say that f has an *extremum* at P_0.

By Theorem 8.11 a necessary condition for an extremum of a differentiable function f at P_0 is $\nabla f(P_0) = 0$. For a particular function this is a set of simultaneous equations for P_0. Simultaneous equations are generally difficult to solve, and it may be necessary to use an iterative method to approximate a solution. Newton's method, covered in most numerical analysis texts, is one of the most popular.

The necessary condition $\nabla f(P_0) = 0$ is not sufficient for an extremum. The function $f(x, y) = -xy$ satisfies this condition at $(0, 0)$ but clearly does not have an extremum at $(0, 0)$ because $f(x, y)$ is positive when $x > 0$ and $y < 0$ but negative when $x > 0$ and $y > 0$. A point of this type is called a *saddle point* (see Fig. 8.6). Generally the point $(x_0, y_0, f(x_0, y_0))$ is a saddle point of the function f if $\nabla f(x_0, y_0) = 0$ but $f(x_0, y_0)$ is neither a local maximum nor a local minimum.

The purpose of this section is to provide sufficient conditions for a local maximum (and for a local minimum), to discuss extrema of functions under constraints, and to present a method for finding the absolute maximum and the absolute minimum of a differentiable (and therefore continuous) function on a closed, bounded region.

Theorem 8.18 If $f \in C^3(D)$, where $D \subseteq \mathbf{R}^2$ is an open region, and if $(x_0, y_0) \in D$ such that $\nabla f(x_0, y_0) = 0$ then

(a) $f(x_0, y_0)$ is a local maximum if $f_{xx}(x_0, y_0) < 0$ and

$$f_{xy}(x_0, y_0)^2 - f_{xx}(x_0, y_0)f_{yy}(x_0, y_0) < 0$$

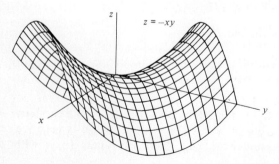

Figure 8.6

(b) $f(x_0, y_0)$ is a local minimum if $f_{xx}(x_0, y_0) > 0$ and
$$f_{xy}(x_0, y_0)^2 - f_{xx}(x_0, y_0)f_{yy}(x_0, y_0) < 0$$
(c) $(x_0, y_0, f(x_0, y_0))$ is a saddle point if
$$f_{xy}(x_0, y_0)^2 - f_{xx}(x_0, y_0)f_{yy}(x_0, y_0) > 0$$

PROOF By Taylor's formula
$$f(x, y) - f(x_0, y_0) = f_x(x_0, y_0)(x - x_0) + f_y(x_0, y_0)(y - y_0)$$
$$+ f_{xx}(x_0, y_0)(x - x_0)^2 + 2f_{xy}(x_0, y_0)(x - x_0)(y - y_0)$$
$$+ f_{yy}(x_0, y_0)(y - y_0)^2 + R_2$$

$f_x(x_0, y_0)$ and $f_y(x_0, y_0)$ are both zero, and since R_2 contains cubic terms, it is negligible for (x, y) sufficiently close to (x_0, y_0). Hence the sign of $f(x, y) - f(x_0, y_0)$ is the same as the sign of the quadratic
$$f_{xx}(x_0, y_0)(x - x_0)^2 + 2f_{xy}(x_0, y_0)(x - x_0)(y - y_0)$$
$$+ f_{yy}(x_0, y_0)(y - y_0)^2$$

It has one sign for all (x, y) close to (x_0, y_0) if the discriminant is negative
$$f_{xy}(x_0, y_0)^2 - f_{xx}(x_0, y_0)f_{yy}(x_0, y_0) < 0$$
If in addition $f_{xx}(x_0, y_0) < 0$ then the quadratic is negative and
$$f(x, y) - f(x_0, y_0) < 0$$
that is, $\qquad f(x, y) < f(x_0, y_0)$
for all (x, y) sufficiently close to (x_0, y_0), and so $f(x_0, y_0)$ is a local maximum. Similarly if
$$f_{xy}(x_0, y_0)^2 - f_{xx}(x_0, y_0)f_{yy}(x_0, y_0) < 0$$
and $f_{xx}(x_0, y_0) > 0$ then $f(x_0, y_0)$ is a local minimum.
If
$$f_{xy}(x_0, y_0)^2 - f_{xx}(x_0, y_0)f_{yy}(x_0, y_0) > 0$$
then the quadratic below changes sign
$$(x - x_0)^2 \left[f_{yy}(x_0, y_0) \frac{(y - y_0)^2}{(x - x_0)^2} + 2f_{xy}(x_0, y_0) \frac{y - y_0}{x - x_0} + f_{xx}(x_0, y_0) \right]$$

Let us say it changes sign at $(y - y_0)/(x - x_0) = r$. Then there are directions along which (x, y) approaches (x_0, y_0) where the quadratic term is positive and other directions where the quadratic term is negative. Thus in any neighborhood of (x_0, y_0) there are points (x, y) with

$f(x, y) > f(x_0, y_0)$ and other points (x, y) with $f(x, y) < f(x_0, y_0)$. So $(x_0, y_0, f(x_0, y_0))$ is a saddle point.

Example 8.15 Find all local maxima and minima of the function
$$f(x, y) = 2x^2 - xy + 2y^2 - 20x$$

SOLUTION The equations
$$f_x(x, y) = 4x - y - 20 = 0 \qquad f_y(x, y) = -x + 4y = 0$$
imply the equations
$$x = 4y$$
$$16y - y - 20 = 0$$
and hence $\qquad y = \tfrac{4}{3} \quad$ and $\quad x = \tfrac{16}{3}$

Now
$$f_{xx} = 4 \qquad f_{xy} = -1 \qquad f_{yy} = 4$$
and so $\qquad f_{xy}^2 - f_{xx} f_{yy} = -15 < 0 \qquad$ with $f_{xx} > 0$

Hence $f(\tfrac{4}{3}, \tfrac{16}{3})$ is a local minimum.

Now suppose we wish to maximize or minimize $z = f(x, y)$, where (x, y) is constrained to satisfy $g(x, y) = 0$. Assuming that these functions have continuous partial derivatives, we can visualize $g(x, y) = 0$ as a curve in the (x, y) plane together with the level curves of $z = f(x, y)$ (see Fig. 8.7).

Intuitively, if we move the level curves in the direction of increasing z, the largest z occurs at a point where a level curve of $z = f(x, y)$ is tangent to the curve $g(x, y) = 0$. Since the gradient of f is normal to $f(x, y) = c$ and the

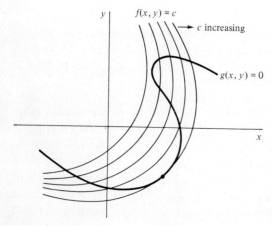

Figure 8.7

gradient of g is normal to $g(x, y) = 0$, the gradients should be in the same or opposite directions. Algebraically $\nabla f = -\lambda \nabla g$ for some constant λ. Thus, to find the maximum of $f(x, y)$ subject to $g(x, y) = 0$ find all solutions of the equations

$$\nabla f + \lambda \nabla g = 0$$
$$g(x, y) = 0$$

Local maxima and minima will be among these solutions. If the set

$$S = \{(x, y) \mid g(x, y) = 0\}$$

is a closed bounded subset of $\mathbf{R}^2$ then the absolute maximum and absolute minimum of $f(x, y)$ on S exist and are among these solutions.

Example 8.16 Find the absolute maximum and absolute minimum value of

$$f(x, y) = 9 - (x - 1)^2 - y^2 \qquad \text{on } x^2 + y^2 = 4$$

SOLUTION $\nabla f + \lambda \nabla g = 0$ becomes

$$-2(x - 1) + \lambda\, 2x = 0$$
$$-2y + \lambda\, 2y = 0$$

Hence
$$(\lambda - 1)x = -1 \qquad (1)$$
$$(\lambda - 1)y = 0 \qquad (2)$$

with the constraint

$$x^2 + y^2 = 4 \qquad (3)$$

From (1), $\lambda \neq 1$; from (2), $y = 0$. From (3), $x = \pm 2$. Solutions are

$$x = 2 \qquad y = 0 \qquad \lambda = \tfrac{1}{2}$$

and
$$x = -2 \qquad y = 0 \qquad \lambda = \tfrac{3}{2}$$

The values of $f(x, y)$ are $f(2, 0) = 8$, $f(-2, 0) = 0$. The absolute maximum is $f(2, 0)$ and the absolute minimum is $f(-2, 0)$. Figure 8.8 shows the curve $x^2 + y^2 - 4 = 0$ together with the level curves of $z = 9 - (x - 1)^2 - y^2$. Note the points of tangency.

Theorem 8.19 Let $f, g \in C^1(D)$, where $D \subseteq \mathbf{R}^3$ is an open region. If the function defined by $f(x, y, z)$, where x, y, z are constrained by $g(x, y, z) = 0$ for all $(x, y, z) \in D$, has a local maximum or local minimum at (x_0, y_0, z_0), where $\nabla g(x_0, y_0, z_0) \neq 0$, then there is a constant λ such that

$$\nabla f + \lambda \nabla g = 0 \qquad \text{at } (x_0, y_0, z_0)$$

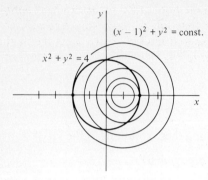

Figure 8.8

PROOF Since $\nabla g(x_0, y_0, z_0) \neq 0$, at least one of the first-order partial derivatives of g is not zero at (x_0, y_0, z_0). Assume that $g_z(x_0, y_0, z_0) \neq 0$. (The proof is similar for other cases.) By the implicit-function theorem there is a neighborhood $N_\delta(x_0, y_0)$ and a function $\varphi(x, y)$ such that

$$g(x, y, \varphi(x, y)) = 0$$

for all $x, y \in N_\delta(x_0, y_0)$ and $z_0 = \varphi(x_0, y_0)$. Also $\varphi \in \mathcal{C}^1(N_\delta(x_0, y_0))$, and

$$\varphi_x(x_0, y_0) = -\frac{g_x(x_0, y_0, z_0)}{g_z(x_0, y_0, z_0)} \quad (1)$$

$$\varphi_y(x_0, y_0) = -\frac{g_y(x_0, y_0, z_0)}{g_z(x_0, y_0, z_0)} \quad (2)$$

Also, $f(x, y, \varphi(x, y))$ has an extreme value at (x_0, y_0, z_0), and so by the chain rule

$$f_x(x_0, y_0, z_0) + f_z(x_0, y_0, z_0)\varphi_x(x_0, y_0) = 0 \quad (3)$$

$$f_y(x_0, y_0, z_0) + f_z(x_0, y_0, z_0)\varphi_y(x_0, y_0) = 0 \quad (4)$$

Substituting (1) and (2) into (3) and (4) and defining

$$\lambda = -\frac{f_z(x_0, y_0, z_0)}{g_z(x_0, y_0, z_0)}$$

we obtain

$$f_x(x_0, y_0, z_0) + \lambda g_x(x_0, y_0, z_0) = 0$$

and

$$f_y(x_0, y_0, z_0) + \lambda g_y(x_0, y_0, z_0) = 0$$

From the definition of λ

$$f_z(x_0, y_0, z_0) + \lambda g_z(x_0, y_0, z_0) = 0$$

Hence $\quad \nabla f + \lambda \, \nabla g = 0 \quad$ at (x_0, y_0, z_0)

Theorem 8.19 provides us with a method, called the *method of Lagrange multipliers*, for finding extrema of a function of three variables subject to one constraint. For higher dimensions and multiple constraints the method is as follows. To maximize or minimize $z = f(x_1, x_2, x_3, \ldots, x_n)$ subject to constraints

$$g_1(x_1, \ldots, x_n) = 0$$
$$g_2(x_1, \ldots, x_n) = 0$$
$$\cdots\cdots\cdots\cdots\cdots$$
$$g_k(x_1, \ldots, x_n) = 0$$

solve the following equations simultaneously

$$\nabla f - \sum_{i=1}^{k} \lambda_i \nabla g_i = 0$$

$$g_j(x_1, \ldots, x_n) = 0 \qquad j = 1, \ldots, k$$

The numbers $\lambda_1, \lambda_2, \ldots, \lambda_k$ are called the *Lagrange multipliers*. Note that they yield $n + k$ scalar equations in $n + k$ unknowns. The practicality of this method is limited by the fact that these equations, which are usually nonlinear, are difficult to solve.

We close this section with remarks on solving maximum and minimum problems on closed bounded regions. To find the absolute maximum or minimum of f on a closed, bounded region $D \subset \mathbf{R}^2$, first find all solutions of $\nabla f = 0$ in the interior of D; second find all maxima and minima of f constrained to the boundary (the method of Lagrange multipliers can be used here); then tabulate $f(x, y)$ at all these points. The largest of these values is the absolute maximum of f on D; the smallest is the absolute minimum of f on D.

EXERCISES

8.55 Give an example of each of the following cases: $\nabla f(x_0, y_0) = 0$ and $f_{xx}(x_0, y_0)f_{yy}(x_0, y_0) - f_{xy}(x_0, y_0)^2 = 0$ and
  (a) $f(x_0, y_0)$ is a local maximum.
  (b) $f(x_0, y_0)$ is a local minimum.
  (c) $(x_0, y_0, f(x_0, y_0))$ is a saddle point.

8.56 Find all local maxima and local minima of
  (a) $f(x, y) = x^2 - 2xy + 3y^2 - 4x$
  (b) $f(x, y) = -2x^2 + 2xy - 3y^2 + 3x - 4y$
  (c) $f(x, y) = xye^{-x}$

8.57 Given $f(x, y) = px^2 + 2qxy + ry^2$, where p, q, r are constants, find sufficient conditions on p, q, r for $(0, 0)$ to be the only point where $\nabla f = 0$ and
  (a) $f(0, 0)$ is a local maximum.
  (b) $f(0, 0)$ is a local minimum.
  (c) $(0, 0, f(0, 0))$ is a saddle point.

8.58 A long strip of sheet metal w centimeters wide is made into a trough by bending up a strip of width x along each edge so as to form equal angles θ with the base. Find x and θ so that the trough will have the largest volume.

8.59 The temperature T at any point (x, y, z) is given by $T = 200xyz^2$. Find the highest and lowest temperature on the surface of the sphere $x^2 + y^2 + z^2 = 1$ (a) without and (b) with Lagrange multipliers.

8.60 Find the shortest distance from the origin to the plane $ax + by + cz = d$, where $d \neq 0$.

8.61 State and prove the theorem for the method of Lagrange multipliers for three dimensions and two constraints.

8.62 The problem "minimize $\sum_{j=1}^{n} c_j x_j$ subject to $\sum_{j=1}^{n} a_{ij} x_j = b_i, i = 1, \ldots, m$, where c_j, a_{ij}, and b_i are constants for $i = 1, \ldots, m$ and $j = 1, \ldots, n$," is called a *linear programming problem*. Show that the method of Lagrange multipliers yields the constraints of the so-called *dual linear programming problem*: "maximize $\sum_{i=1}^{m} b_i w_i$ subject to $\sum_{i=1}^{m} a_{ij} w_i = c_j, j = 1, \ldots, n$."

8.63 Given n positive numbers $c_1, \ldots, c_n$, find the maximum value of $\sum_{i=1}^{n} c_i x_i$ if the variables $x_i, i = 1, 2, \ldots, n$ are restricted so that $\sum_{i=1}^{n} x_i^2 = 1$.

8.64 Given n points $(x_1, y_1), \ldots, (x_n, y_n)$, where x_i are distinct, find m and b such that the function

$$f(m, b) = \sum_{i=1}^{n} (mx_i + b - y_i)^2$$

has a minimum. Interpret graphically.

8.65 Let $f \in C^3(D)$, where $D \subseteq \mathbf{R}^3$ is an open region, and let $(x_0, y_0, z_0) \in D$. Assuming the following at (x_0, y_0, z_0), prove that f has a local minimum at (x_0, y_0, z_0):

(a) $\nabla f = 0$ (b) $f_{xx} > 0$ (c) $\begin{vmatrix} f_{xx} & f_{xy} \\ f_{yx} & f_{yy} \end{vmatrix} > 0$ (d) $\begin{vmatrix} f_{xx} & f_{xy} & f_{xz} \\ f_{yx} & f_{yy} & f_{yz} \\ f_{zx} & f_{zy} & f_{zz} \end{vmatrix} > 0$

CHAPTER
NINE
MULTIPLE INTEGRALS

9.1 THE DOUBLE INTEGRAL

In this section we first define the double integral of a bounded function f defined on a rectangle R. We then extend this notion to functions over more general regions in $\mathbf{R}^2$. The development here parallels the definition of the Riemann integral of a bounded function defined on a closed interval $[a, b]$ given in Chap. 6; we therefore omit proofs which are merely imitations of the single-variable case.

A rectangle in $\mathbf{R}^2$ is any set of the form

$$R = \{(x, y) \in \mathbf{R}^2 \mid a \leq x \leq b \text{ and } c \leq y \leq d\}$$

where $a < b$ and $c < d$ are real numbers. In the following we assume that the real-valued function f is defined and bounded on R. A *partition* of R is constructed by passing lines through R parallel to the y axis: $x = x_i$ for $i = 0, 1, 2, \ldots, n$, and parallel to the x axis: $y = y_j$ for $j = 0, 1, 2, \ldots, m$, with $a = x_0 < x_1 < x_2 < \cdots < x_{n-1} < x_n = b$ and $c = y_0 < y_1 < y_2 < \cdots < y_{m-1} < y_m = d$. We denote the subrectangles

$$R_{ij} = \{(x, y) \in R \mid x_{i-1} \leq x \leq x_i \text{ and } y_{j-1} \leq y \leq y_j\}$$

for each $i = 1, 2, \ldots, n$ and $j = 1, 2, \ldots, m$. It is easy to verify that the area of R, $A(R) = (b - a)(d - c)$, equals the sum of the areas of the subrectangles R_{ij}, denoted $\sum_{i=1}^{n} \sum_{j=1}^{m} A(R_{ij})$ or simply $\sum_{i,j} A(R_{ij})$ (see Exercise 9.1).

We let $M = \sup_{(x, y) \in R} f(x, y)$ and $m = \inf_{(x, y) \in R} f(x, y)$, and for the partition $P = \{R_{ij}\}$ we define

$$M_{ij} = \sup_{(x, y) \in R_{ij}} f(x, y) \quad \text{and} \quad m_{ij} = \inf_{(x, y) \in R_{ij}} f(x, y)$$

261

The *upper sum* and *lower sum* are then defined respectively by

$$U(P, f) = \sum_{i,j} M_{ij} A(R_{ij}) \quad \text{and} \quad L(P, f) = \sum_{i,j} m_{ij} A(R_{ij})$$

For each partition P of the rectangle R we have associated with P the two numbers $U(P, f)$ and $L(P, f)$. It is immediate from the above definitions that for each partition P of R we have

$$mA(R) \leq L(P, f) \leq U(P, f) \leq MA(R)$$

Thus the set of all upper and of all lower sums, as P varies, are bounded sets of real numbers. We define the *upper double integral* of f over R to be

$$\overline{\iint} f = \inf_P U(P, f)$$

similarly, we define the *lower double integral* of f over R by

$$\underline{\iint} f = \sup_P L(P, f)$$

These both exist as long as f is bounded on R.

A partition P^* is called a *refinement* of the partition P if P^* has in its construction at least all those lines which go into the construction of P. Then each of the subrectangles in the P^* partition is contained in one of the subrectangles of the P partition. As in the one-variable case, it is not difficult to show (see Exercise 9.2) that under refinement upper sums decrease and lower sums increase; that is, $U(P^*, f) \leq U(P, f)$ and $L(P^*, f) \geq L(P, f)$. Now, if P_1 and P_2 are any partitions of R and P is a common refinement of P_1 and P_2 then

$$L(P_1, f) \leq L(P, f) \leq U(P, f) \leq U(P_2, f)$$

By the arbitrariness of P_1 we have that

$$\underline{\iint} f \leq U(P_2, f)$$

but since P_2 was arbitrary it follows that

$$\underline{\iint} f \leq \overline{\iint} f$$

We are now ready to define an integrable function on the rectangle R and to assign a value to the double integral of such a function.

Definition 9.1 A bounded function $f: R \to \mathbf{R}$ is called *integrable* provided $\underline{\iint} f = \overline{\iint} f$. In this case we define the double integral of f over R to be

$$\iint_R f = \underline{\iint} f = \overline{\iint} f$$

Our first theorem is analogous to Theorem 6.1; we leave the proof to the reader (Exercise 9.3).

Theorem 9.1 Suppose f is bounded on the rectangle R. Then f is integrable on R if and only if given any $\varepsilon > 0$ there is a partition P of R such that $U(P, f) - L(P, f) < \varepsilon$.

Example 9.1 Let $R = \{(x, y) \mid 0 \leq x \leq 1, 0 \leq y \leq 1\}$; define $f(x, y) = 1$ if x and y are each rational numbers in $[0, 1]$, and let $f(x, y) = 0$ for all other points (x, y) in R. If P is any partition of R, it is clear that each subrectangle R_{ij} necessarily contains points where f is 1 and also points where f is 0. Therefore $M_{ij} = 1$ and $m_{ij} = 0$ for each $i = 1, 2, \ldots, n$; $j = 1, 2, \ldots, m$. Thus $U(P, f) = \sum_{i,j} 1 \cdot A(R_{ij}) = A(R) = 1$ and $L(P, f) = \sum_{i,j} 0 \cdot A(R_{ij}) = 0$. Since P was an arbitrary partition of R, it follows from Theorem 9.1 that f is not integrable on R, and so $\iint_R f$ does not exist.

Example 9.2 Let $R = \{(x, y) \mid 0 \leq x \leq 1, 0 \leq y \leq 1\}$ and define f by

$$f(x, y) = \begin{cases} 1 & \text{if } x = 1 \\ 0 & \text{if } 0 \leq x < 1 \end{cases}$$

Let $\varepsilon > 0$ be arbitrary and choose the partition P in such a way that $x_n - x_{n-1} < \varepsilon$. Then $M_{nj} = 1$ for each $j = 1, 2, \ldots, m$ and $M_{ij} = 0$ for $i = 1, 2, \ldots, n - 1$. Hence

$$U(P, f) = \sum_{i,j} M_{ij} A(R_{ij}) = \sum_{j=1}^{m} M_{nj} A(R_{nj})$$

$$= \sum_{j=1}^{m} A(R_{nj}) = \sum_{j=1}^{m} (x_n - x_{n-1})(y_j - y_{j-1})$$

$$= (x_n - x_{n-1}) \sum_{j=1}^{m} (y_j - y_{j-1}) = x_n - x_{n-1} < \varepsilon$$

Since $m_{ij} = 0$ for each $i = 1, 2, \ldots, n$; $j = 1, 2, \ldots, m$, $L(P, f) = 0$, and so by Theorem 9.1, f is integrable on R. We note that since $L(P, f) = 0$ for every partition P, $\iint f = 0$ and so $\iint_R f = 0$.

We define r_{ij} to be the length of a diagonal of R_{ij}, and we define the norm of a partition P of R to be

$$||P|| = \max_{\substack{1 \leq i \leq n \\ 1 \leq j \leq m}} r_{ij}$$

Let P be any partition of R and choose $\xi_{ij} \in R_{ij}$ for each $i = 1, 2, \ldots, n$; $j = 1, 2, \ldots, m$. We form the Riemann sum

$$S(P, f, \xi) = \sum_{i,j} f(\xi_{ij}) A(R_{ij})$$

Observe that for each selection of the points $\xi = \{\xi_{ij}\}$

$$L(P,f) \leq S(P,f,\xi) \leq U(P,f)$$

The following theorem is the analog of Theorem 6.5.

Theorem 9.2 The double integral of f on R exists and equals γ if and only if given any $\varepsilon > 0$ there is a $\delta > 0$ such that $|S(P,f,\xi) - \gamma| < \varepsilon$ for every Riemann sum $S(P,f,\xi)$ with $||P|| < \delta$.

In the previous theorem it is important that each subrectangle have a diameter sufficiently small. If, for example, we had required only that $A(R_{ij})$ be small for each $i = 1, 2, \ldots, n;\ j = 1, 2, \ldots, m$, we could not guarantee that $S(P,f,\xi)$ would be arbitrarily close to $\iint_R f$. Consider the function of Example 9.2. If we let $n = 1$ and let $y_j = j/m$ for an arbitrary fixed natural number m then $R_{ij} = [0,1] \times [(j-1)/m, j/m]$ and so $A(R_{ij}) = 1/m$ for $j = 1, 2, \ldots, m$. Thus each subrectangle in P will have arbitrarily small area (by choosing m sufficiently large), but if we choose $\xi_{ij} \in R_{ij}$ so that ξ_{ij} lies on the line $x = 1$ then

$$S(P,f,\xi) = \sum_{i,j} f(\xi_{ij}) A(R_{ij}) = \sum_{j=1}^{m} f(\xi_{1j}) A(R_{1j}) = \sum_{j=1}^{m} \frac{1}{m} = 1$$

On the other hand, the value $\gamma = \iint_R f = 0$, and so the inequality $|S(P,f,\xi) - \gamma| < \varepsilon$ fails when $0 < \varepsilon < 1$.

It is important to know that there is a healthy class of functions defined on R for which $\iint_R f$ exists. Toward this end let B be any bounded subset of $\mathbf{R}^2$. We can enclose B in a rectangle R. Let $P = \{R_{ij}\}$ be any partition of R; we classify the subrectangles R_{ij} as follows:

Class C1. Each point of R_{ij} is an interior point of B.
Class C2. Each point of R_{ij} is an exterior point of B.
Class C3. R_{ij} contains at least one boundary point of B (see Fig. 9.1).

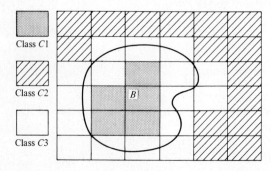

Figure 9.1

We define the outer content of B to be

$$\mathcal{O}(B) = \inf_{P} \sum_{C1 \cup C3} A(R_{ij})$$

where we have summed the areas of all these subrectangles belonging to class C1 or class C3. The inner content of B is defined by

$$\mathfrak{I}(B) = \sup_{P} \sum_{C1} A(R_{ij})$$

where we have summed the areas of all those subrectangles belonging to class C1. We ask the reader to verify that the definitions of $\mathcal{O}(B)$ and $\mathfrak{I}(B)$ do not depend on the choice of the rectangle R used to enclose the bounded set B (Exercise 9.8).

Definition 9.2 The bounded set $B \subset \mathbf{R}^2$ is said to have *area* provided $\mathfrak{I}(B) = \mathcal{O}(B)$. In this case we define the area of B to be $A(B) = \mathfrak{I}(B) = \mathcal{O}(B)$.

Let B be a bounded subset of $\mathbf{R}^2$ and let P_1 and P_2 be any partitions of the rectangle $R \supseteq B$. Then the sum of the areas of the subrectangles belonging to class C1 or class C3 for the P_1 partition is not less than the sum of the areas of the subrectangles belonging to class C1 for the P_2 partition. Thus $\mathcal{O}(B) \geq \mathfrak{I}(B)$. It follows that a set B has zero area if and only if given any $\varepsilon > 0$ there is a partition P of R such that the sum of the areas of the subrectangles of P belonging to class C1 or class C3 is less than ε. That is, B has zero area if and only if $\mathcal{O}(B) = 0$. If $A(B) = 0$, however, there will be no subrectangles of P belonging to class C1 (see Exercise 9.7). Therefore, if a set B has zero area then for any given $\varepsilon > 0$ there is a partition P of R such that the sum of the areas of those R_{ij} which contain at least one point of B is less than ε.

It is clear that a finite set has zero area; in the following we show that the graph of a continuous function defined on a closed, bounded interval is a set of zero area.

Theorem 9.3 If $f \in C[a, b]$ then the graph of f,

$$\mathrm{gr}\,(f) = \{(x, f(x)) \mid x \in [a, b]\}$$

is a set of zero area.

Proof Choose the rectangle $R = [a, b] \times [c, d]$ in such a way that both $M = \sup_x f(x)$ and $m = \inf_x f(x)$, where $x \in [a, b]$, are contained in the interval $[c, d]$. Let $\varepsilon > 0$ be given. By the uniform continuity of f on $[a, b]$ there is a $\delta > 0$ such that $|f(u) - f(v)| < \varepsilon/[3(b - a)]$ for any points $u, v \in [a, b]$ with $|u - v| < \delta$. Choose the partition P of R in such a way that $\Delta x_i = x_i - x_{i-1} < \delta$ for $i = 1, 2, \ldots, n$ and $\Delta y_j = y_j - y_{j-1}$

$< \varepsilon/[3(b-a)]$ for $j = 1, 2, \ldots, m$. Since gr (f) is a closed subset of $\mathbf{R}^2$ (continuity of f) with no interior points, there are only two types of subrectangles R_{ij}: those of class C2, which contain no points of gr (f), and those of class C3, which contain at least one point of gr (f). Since $\Delta x_i < \delta$ for each $i = 1, 2, \ldots, n$, the uniform continuity of f guarantees that for each i the sum of the areas of the subrectangles R_{ij} of class C3 is less than $\varepsilon/(b-a)\,\Delta x_i$. Therefore the sum of the areas of all C3 class subrectangles is less than

$$\sum_{i=1}^{n} \frac{\varepsilon}{b-a} \Delta x_i = \frac{\varepsilon}{b-a} \sum_{i=1}^{n} \Delta x_i = \frac{\varepsilon}{b-a} (b-a) = \varepsilon$$

Since $\varepsilon > 0$ was arbitrary, it follows that $\mathcal{O}(\text{gr }(f)) = 0$ and so gr (f) has zero area.

Suppose B is the set of all points (x, y) with $0 \leq x \leq 1$ and $0 \leq y \leq 1$ and both x and y rational. In this case every point in the rectangle $R = [0, 1] \times [0, 1]$ is a boundary point of B, and so for an arbitrary partition $P = \{R_{ij}\}$ of R each subrectangle R_{ij} is of class C3. It follows that $\mathcal{O}(B) = 1$ and $\mathfrak{I}(B) = 0$. Therefore B does not have area.

Now suppose B is any bounded subset of $\mathbf{R}^2$. If we let R be a rectangle containing B and if we let $P = \{R_{ij}\}$ be any partition of R then the subrectangles of class C3 cover the boundary of B, denoted ∂B. Also

$$\sum_{C3} A(R_{ij}) = \sum_{C1 \cup C3} A(R_{ij}) - \sum_{C1} A(R_{ij})$$

Now since ∂B is a closed set and so contains all its boundary points,

$$\mathcal{O}(\partial B) = \inf_{P} \sum_{C3} A(R_{ij})$$

Therefore, the outer content of ∂B is precisely the outer content of B minus the inner content of B; that is, $\mathcal{O}(\partial B) = \mathcal{O}(B) - \mathfrak{I}(B)$. This gives us the following lemma.

Lemma A bounded set $B \subset \mathbf{R}^2$ has area if and only if the boundary of B has zero area.

If D is a bounded region in $\mathbf{R}^2$ which is bounded by straight-line segments (polygonal lines) or graphs of continuous functions then D has area. The regular regions from geometry (rectangles, triangles, circles, ellipses, etc.) all have area, and these areas agree with their well-known formulas. Also, the region under the graph of a nonnegative continuous function $g \in \mathcal{C}(a, b]$ and bounded by the segment $[a, b]$ on the x axis and the vertical lines $x = a$ and $x = b$ has area given by the formula $A = \int_a^b g$ (see Exercise 9.6).

The following theorem gives us a large class of bounded functions f defined on the rectangle R for which $\iint_R f$ exists.

Theorem 9.4 If f is bounded on the rectangle R and continuous except on a set E of zero area then $\iint_R f$ exists.

PROOF Let $\varepsilon > 0$ be given. By Theorem 9.1 it suffices to find a partition P of R such that $U(P,f) - L(P,f) < \varepsilon$. Since E has zero area, there is a partition $\hat{P} = \{\hat{R}_{ij}\}$ such that $\sum_{K1} A(\hat{R}_{ij}) < \varepsilon/4M^*$, where K1 is the class of all those subrectangles $\hat{R}_{ij}$ in $\hat{P}$ for which $\hat{R}_{ij} \cap E \neq \emptyset$ and $M^* > |f(x,y)|$ for all $(x,y) \in R$. Let K2 be the class of all those $\hat{R}_{ij}$ in $\hat{P}$ satisfying $\hat{R}_{ij} \cap E = \emptyset$. We define $S_1 = \bigcup_{K1} \hat{R}_{ij}$ and $S_2 = \bigcup_{K2} \hat{R}_{ij}$. Clearly S_1 and S_2 are closed, bounded subsets of $\mathbf{R}^2$. Now f is uniformly continuous on S_2 by Exercise 8.18. Therefore there is a $\delta > 0$ such that $|f(x,y) - f(x',y')| < \varepsilon/2A(R)$ whenever $(x,y), (x',y') \in S_2$ with $|(x,y) - (x',y')| < \delta$. We refine the partition $\hat{P}$ to get a partition $P = \{R_{ij}\}$ such that $\|P\| < \delta$. Now

$$U(P,f) - L(P,f) = \sum_{i,j} M_{ij} A(R_{ij}) - \sum_{i,j} m_{ij} A(R_{ij})$$

$$= \sum_{i,j} (M_{ij} - m_{ij}) A(R_{ij})$$

$$= \sum_{R_{ij} \subseteq S_1} (M_{ij} - m_{ij}) A(R_{ij}) + \sum_{R_{ij} \subseteq S_2} (M_{ij} - m_{ij}) A(R_{ij})$$

$$< 2M^* \sum_{R_{ij} \subseteq S_1} A(R_{ij}) + \frac{\varepsilon}{2A(R)} \sum_{R_{ij} \subseteq S_2} A(R_{ij})$$

$$= 2M^* \sum_{\hat{R}_{ij} \in K1} A(\hat{R}_{ij}) + \frac{\varepsilon}{2A(R)} \sum_{\hat{R}_{ij} \in K2} A(\hat{R}_{ij})$$

$$= 2M^* \frac{\varepsilon}{4M^*} + \frac{\varepsilon}{2A(R)} A(R) = \frac{\varepsilon}{2} + \frac{\varepsilon}{2} = \varepsilon$$

Therefore, $\iint_R f$ exists.

If f is a continuous function on the rectangle R then certainly $\iint_R f$ exists. The reader has probably seen the interpretation of the double integral of f on R for a nonnegative continuous function as the volume of that portion of $\mathbf{R}^3$ under the surface $z = f(x,y)$ and above the rectangle R. If $f(x,y) = k > 0$ for every $(x,y) \in R$, for example, then $\iint_R f = k\, A(R)$, which is just the volume of the box in $\mathbf{R}^3$ with base dimensions those of R and height equal to k.

In order to define the double integral of f over more general subsets of $\mathbf{R}^2$ than rectangles we let B be a bounded subset of $\mathbf{R}^2$ which has area. Suppose f is defined and bounded on B. We further suppose that f is continuous at each interior point of B (if B has interior points). Now let R be any rectangle in $\mathbf{R}^2$ which contains B and define the auxiliary function f^* by

$$f^*(x, y) = \begin{cases} f(x, y) & \text{if } (x, y) \in B \\ 0 & \text{if } (x, y) \in R - B \end{cases}$$

Clearly f^* is bounded on R and continuous at each interior point of B. f^* is also continuous at each exterior point of B since it is identically zero throughout a neighborhood of such a point. Therefore, the set E of points of discontinuity of f^* is a subset of the boundary of B, and the boundary of B has zero area by the lemma. It follows that E has zero area (see Exercise 9.5) and so by Theorem 9.4 $\iint_R f^*$ exists. We now define the double integral of f on B by $\iint_B f = \iint_R f^*$. We ask the reader to verify (see Exercise 9.12) that the definition of $\iint_B f$ is independent of the choice of the rectangle R.

The above definition and Theorem 9.4 establish the following result.

Theorem 9.5 If B is a bounded subset of $\mathbf{R}^2$ having area and f is bounded on B and continuous except on a set E having zero area then $\iint_B f$ exists.

For most purposes it suffices to consider closed, bounded regions in $\mathbf{R}^2$ which have area for the subsets over which we shall be integrating. The following theorem is a consequence of the preceding definition of the double integral of a bounded function on a closed, bounded region D which has area. We assume that f and g are each bounded on D and continuous at every interior point of D so that $\iint_D f$ and $\iint_D g$ both exist.

Theorem 9.6 Let D be a closed, bounded region in $\mathbf{R}^2$ which has area and let f and g be bounded on D and continuous at every interior point of D. Then
(a) $\iint_D (f + g) = \iint_D f + \iint_D g$.
(b) $\iint_D kf = k \iint_D f$ for any real number k.
(c) $\iint_D f \leq \iint_D g$ provided $f(x, y) \leq g(x, y)$ for each $(x, y) \in D$ except for those in a set E of zero area.
(d) $\iint_D |f|$ exists, and $\left| \iint_D f \right| \leq \iint_D |f|$.

Suppose D is a closed, bounded region in $\mathbf{R}^2$ which has area. Let $f(x, y) = 1$ for each $(x, y) \in D$; then f is bounded on D, and f is continuous at each interior point of D. As above, we choose the rectangle R with $D \subseteq R$ and define the auxiliary function f^* on R by

$$f^*(x, y) = \begin{cases} 1 & \text{if } (x, y) \in D \\ 0 & \text{if } (x, y) \in R - D \end{cases}$$

Then the double integral of f on D is defined by

$$\iint_D f = \iint_R f^*$$

Now if $P = \{R_{ij}\}$ is any partition of R then

$$U(P, f^*) = \sum_{R_{ij} \cap D \neq \emptyset} A(R_{ij})$$

where the above summation is over all subrectangles R_{ij} with $R_{ij} \cap D \neq \emptyset$. Since D is closed (and so contains each of its boundary points), the subrectangles in the above sum are precisely those which either consist entirely of interior points of D (the class C1 subrectangles) or contain a boundary point of D (the class C3 subrectangles). Therefore

$$U(P, f^*) = \sum_{C1 \cup C3} A(R_{ij})$$

It follows that

$$A(D) = \mathcal{O}(D) = \inf_P \sum_{C1 \cup C3} A(R_{ij}) = \inf_P U(P^*, f)$$

$$= \overline{\iint} f^* = \iint_R f^* = \iint_D f$$

Therefore, the area of D can be written as

$$A(D) = \iint_D 1$$

If f is nonnegative and continuous on D, it is natural to define the volume of that portion of $\mathbf{R}^3$ under the surface $z = f(x, y)$ and above the region D to be $\iint_D f$.

We conclude this section with the following mean-value theorem for double integrals.

Theorem 9.7 Suppose that D is a closed, bounded region in $\mathbf{R}^2$ having area and that f is continuous on D. Then there is a point $(x_0, y_0) \in D$ such that $\iint_D f = f(x_0, y_0) A(D)$.

PROOF Since f is continuous on the closed, bounded region D, there are points $(x_1, y_1), (x_2, y_2) \in D$ with $f(x_1, y_1) = M = \sup_{(x, y) \in D} f(x, y)$ and $f(x_2, y_2) = m = \inf_{(x, y) \in D} f(x, y)$. Then

$$m\, A(D) = \iint_D m \leq \iint_D f \leq \iint_D M = M\, A(D)$$

Since regions cannot have zero area (see Exercise 9.7), $A(D) > 0$ and so

$$m \le \frac{1}{A(D)} \iint_D f \le M$$

It follows from the intermediate-value theorem (see Theorem 8.7) that there is a point $(x_0, y_0) \in D$ with

$$f(x_0, y_0) = \frac{1}{A(D)} \iint_D f$$

Clearly

$$\iint_D f = f(x_0, y_0) A(D)$$

EXERCISES

9.1 Prove that if $P = \{R_{ij}\}$ is a partition of the rectangle R then $\sum_{i,j} A(R_{ij}) = A(R)$.

9.2 Prove that if P^* is a refinement of the partition P then $U(P^*, f) \le U(P, f)$ and $L(P^*, f) \ge L(P, f)$.

9.3 Prove Theorem 9.1.

9.4 If $[a, b] \times [c, d]$ is a rectangle in $\mathbf{R}^2$, a partition P of this rectangle is called *regular* if

$$\Delta x_i = x_i - x_{i-1} = \frac{b-a}{n} \quad \text{for } i = 1, 2, \ldots, n$$

and

$$\Delta y_j = y_j - y_{j-1} = \frac{d-c}{m} \quad \text{for } j = 1, 2, \ldots, m$$

In this case

$$\|P\| = \sqrt{\left(\frac{b-a}{n}\right)^2 + \left(\frac{c-d}{m}\right)^2}$$

Using the fact that $\iint_R (2x + 3y)$ exists (Theorem 9.4), where $R = [0, 1] \times [0, 1]$, and the notion of regular partitions of R, evaluate the double integral.

9.5 (a) Prove that if B is a bounded subset of $\mathbf{R}^2$ with zero area and $B_1 \subseteq B$ then B_1 has zero area.

(b) Prove that if both B_1 and B_2 are bounded subsets of $\mathbf{R}^2$ with zero area then $B_1 \cup B_2$ has zero area.

9.6 Show that the concept of area defined in this section agrees with the notion of area for the region under the graph of a nonnegative continuous function g defined on $[a, b]$, as given by $\int_a^b g$ in Chap. 6. Illustrate with the example $g(x) = x^2$ on $[1, 2]$.

9.7 (a) Prove that a finite set has zero area.

(b) Give an example of a bounded infinite set with zero area.

(c) Prove that if B is a bounded subset of $\mathbf{R}^2$ with nonempty interior then B cannot have zero area.

9.8 Verify that the definition of area (Definition 9.2) of a bounded subset $B \subset \mathbf{R}^2$ does not depend on the rectangle R used initially to enclose B.

9.9 Give an example of a bounded subset of $\mathbf{R}^2$ which has nonempty interior but does not have area.

9.10 Let $f: [a, b] \to \mathbf{R}$ be a bounded function.
(a) Prove that if the graph of f has area then it has zero area.
(b) Give an example of a bounded function f whose graph does not have area.

9.11 Let R and R_1 be rectangles and let D be closed, bounded region having area with $D \subseteq R \subseteq R_1$. Suppose f is bounded on D and continuous at each interior point of D. Let $f^*(x, y) = f(x, y)$ for each $(x, y) \in D$ and let $f^*(x, y) = 0$ if $(x, y) \notin D$. Prove that

$$\iint_R f^* = \iint_{R_1} f^*$$

9.12 Use Exercise 9.11 to establish that the definition of $\iint_D f$ is independent of the rectangle R used in the definition.

9.13 Let f be bounded on the bounded set $B \subset \mathbf{R}^2$, where B has zero area. Prove that $\iint_B f = 0$.

9.14 Suppose D is a closed, bounded region in $\mathbf{R}^2$ and D has area. Prove that if f is bounded on D and continuous on D except for a subset E with zero zrea then $\iint_D f$ exists.

9.15 Determine which of the following functions are integrable on $R = [0, 1] \times [0, 1]$:

(a) $f(x, y) = x^2 + y^2$ (b) $f(x, y) = [\![x - y]\!]$

(c) $f(x, y) = \begin{cases} 0 & \text{if } (x, y) = \left(\dfrac{1}{n}, \dfrac{1}{m}\right) \text{ for some } n, m \in \mathbf{N} \\ 1 & \text{otherwise} \end{cases}$

(d) $f(x, y) = \begin{cases} 0 & \text{if } x, y \in \mathbf{Q} \\ 1 & \text{otherwise} \end{cases}$

(e) $f(x, y) = \begin{cases} \dfrac{1}{x + y} & (x, y) \neq (0, 0) \\ 0 & x = y = 0 \end{cases}$

(f) $f(x, y) = \begin{cases} \sin \dfrac{1}{x + y} & (x, y) \neq (0, 0) \\ 0 & x = y = 0 \end{cases}$

9.16 Suppose that f is nonnegative and continuous on the rectangle R. Prove that if $\iint_R f = 0$ then $f(x, y) = 0$ for every $(x, y) \in R$.

9.2 EVALUATION OF DOUBLE INTEGRALS

Evaluating double integrals by using the definition is for most functions a difficult if not impossible job. What is needed is a method similar to the fundamental theorem of calculus for functions of a single variable which will simplify this task. In the present section we develop a method called *iterated integration*, which will be quite useful in evaluating double integrals.

We begin by assuming that the function f is bounded on the rectangle R given by

$$R = \{(x, y) \in \mathbf{R}^2 \,|\, a \le x \le b \text{ and } c \le y \le d\}$$

If we fix $x \in [a, b]$ then $f(x, y)$ can be thought of as a function of the single real variable y defined on $[c, d]$. If this function (for each $x \in [a, b]$) is integrable on $[c, d]$ then this defines a function

$$g(x) = \int_c^d f(x, y)\, dy$$

on the interval $[a, b]$. If $g(x)$ is integrable on $[a, b]$ then the integral

$$\int_a^b g(x)\, dx = \int_a^b \left(\int_c^d f(x, y)\, dy \right) dx$$

is called an *iterated integral*, written more briefly as

$$\int_a^b \int_c^d f(x, y)\, dy\, dx$$

The following result allows us to evaluate double integrals by successively evaluating two integrals of functions of a single variable, that is, evaluating an iterated integral.

Theorem 9.8: Fubini's theorem Let f be a bounded function on the rectangle $R = [a, b] \times [c, d]$. If f is continuous on R except for the points in a set E of zero area and if for each $x \in [a, b]$, $f(x, y)$ is integrable on $[c, d]$ then

$$\iint_R f = \int_a^b \int_c^d f(x, y)\, dy\, dx$$

PROOF We first note that $\iint_R f$ exists according to Theorem 9.4. By the above hypothesis the function $g(x) = \int_c^d f(x, y)\, dy$ is well-defined on the interval $[a, b]$ and

$$|g(x)| = \left| \int_c^d f(x, y)\, dy \right| \le M(d - c) \qquad \text{where } M = \sup_{(x, y) \in R} f(x, y)$$

Therefore $g(x)$ is bounded on $[a, b]$. Let $\varepsilon > 0$ be given. In accordance with Theorem 6.5, the above result will be established if we show that whenever a partition of $[a, b]$ has sufficiently small norm any Riemann sum coming from this partition will be within ε of $\iint_R f$.

Now, since f is integrable on R, there is a $\delta > 0$ such that

$U(P, f) - L(P, f) < \varepsilon$ whenever P is a partition of R with $||P|| < \delta$. Let
$$a = x_0 < x_1 < x_2 < \cdots < x_{i-1} < x_i < x_{i+1} < \cdots < x_{n-1} < x_n = b$$
be any partition of $[a, b]$ with norm less than $\delta/2$ and let the intermediate points $\bar{x}_i \in [x_{i-1}, x_i]$ for $i = 1, 2, \ldots, n$ be arbitrary. Let
$$c = y_0 < y_1 < y_2 < \cdots < y_{j-1} < y_j < y_{j+1} < \cdots < y_{m-1} < y_m = d$$
be a partition of $[c, d]$ with norm less than $\delta/2$; then the partition $P = \{R_{ij}\}$ of R with $R_{ij} = [x_{i-1}, x_i] \times [y_{j-1}, y_j]$ satisfies $||P|| < \delta$. For $i = 1, 2, \ldots, n$ let ρ_{ij} be the mean value of $f(\bar{x}_i, y)$ on $[y_{j-1}, y_j]$, $j = 1, 2, \ldots, m$; that is,

$$\rho_{ij} = \frac{1}{\Delta y_j} \int_{y_{j-1}}^{y_j} f(\bar{x}_i, y) \, dy$$

Note that ρ_{ij} depends on the intermediate points $\bar{x}_i$ as well as i and j. Then
$$\inf_{(x,y)} f(x, y) = m_{ij} \leq \inf_y f(\bar{x}_i, y) \leq \rho_{ij} \leq \sup_y f(\bar{x}_i, y) \leq M_{ij} = \sup_{(x,y)} f(x, y)$$
where $(x, y) \in R_{ij} \quad y \in [y_{j-1}, y_j]$
and so
$$L(P, f) = \sum_{i,j} m_{ij} A(R_{ij}) \leq \sum_{i,j} \rho_{ij} A(R_{ij}) \leq \sum_{i,j} M_{ij} A(R_{ij}) = U(P, f)$$

Now, $g(\bar{x}_i) = \int_c^d f(\bar{x}_i, y) \, dy$ for $i = 1, 2, \ldots, n$ and so the Riemann sum

$$\sum_{i=1}^n g(\bar{x}_i) \Delta x_i = \sum_{i=1}^n \left(\int_c^d f(\bar{x}_i, y) \, dy \right) \Delta x_i$$
$$= \sum_{i=1}^n \left(\sum_{j=1}^m \int_{y_{j-1}}^{y_j} f(\bar{x}_i, y) \, dy \right) \Delta x_i$$
$$= \sum_{i=1}^n \left(\sum_{j=1}^m \rho_{ij} \Delta y_j \right) \Delta x_i$$
$$= \sum_{i=1}^n \sum_{j=1}^m \rho_{ij} \Delta y_j \Delta x_i = \sum_{i,j} \rho_{ij} A(R_{ij})$$

Consequently, $L(P, f) \leq \sum_{i=1}^n g(\bar{x}_i) \Delta x_i \leq U(P, f)$. But $L(P, f) \leq \iint_R f \leq U(P, f)$ always holds, and since $||P|| < \delta$, it follows that $U(P, f) - L(P, f) < \varepsilon$. Therefore

$$\left| \sum_{i=1}^n g(\bar{x}_i) \Delta x_i - \iint_R f \right| < \varepsilon$$

Since this holds for every Riemann sum coming from a partition of

$[a, b]$ with norm less than $\delta/2$, it follows that g is integrable on $[a, b]$ and
$$\int_a^b g(x)\, dx = \iint_R f$$
that is,
$$\iint_R f = \int_a^b \int_c^d f(x, y)\, dy\, dx$$

The following example shows how Fubini's theorem can be used to evaluate a double integral.

Example 9.3 Consider the double integral $\iint_R (x^2 + y^2)$, where R is the rectangle $[0, 1] \times [1, 2]$. By Fubini's theorem
$$\iint_R (x^2 + y^2) = \int_0^1 g(x)\, dx \quad \text{where } g(x) = \int_1^2 (x^2 + y^2)\, dy$$

Now
$$\int_1^2 (x^2 + y^2)\, dy = (x^2 y + \tfrac{1}{3} y^3)\Big|_1^2 = x^2 + \tfrac{7}{3}$$
and so
$$\iint_R (x^2 + y^2) = \int_0^1 (x^2 + \tfrac{7}{3})\, dx = (\tfrac{1}{3} x^3 + \tfrac{7}{3} x)\Big|_0^1 = \tfrac{8}{3}$$

Geometrically, this value represents the volume of the solid in $\mathbf{R}^3$ under the paraboloid $z = x^2 + y^2$ and over the rectangle $R = [0, 1] \times [1, 2]$.

There is an obvious counterpart to Fubini's theorem which says that if f is bounded on the rectangle $R = [a, b] \times [c, d]$ and f is continuous on R except at the points in a set E of zero area and if for each $y \in [c, d]$ the function $f(x, y)$ is an integrable function on $[a, b]$ then the function
$$h(y) = \int_a^b f(x, y)\, dx$$
defined on $[c, d]$ is integrable and $\int_c^d h(y)\, dy = \iint_R f$. As before, this is written in the form
$$\iint_R f = \int_c^d \int_a^b f(x, y)\, dx\, dy$$
We point out that if f is continuous on the rectangle R then $f(x, y)$ is

continuous on $[c, d]$ for each $x \in [a, b]$. Similarly, $f(x, y)$ is continuous on $[a, b]$ for each $y \in [c, d]$. Therefore these functions are certainly integrable on the appropriate intervals, and so

$$\iint_R f = \int_a^b \int_c^d f(x, y) \, dy \, dx = \int_c^d \int_a^b f(x, y) \, dx \, dy$$

If f is not continuous on the rectangle R then it is possible that $\int_c^d f(x, y) \, dy$ may fail to exist for certain values of $x \in [a, b]$.

Example 9.4 Let R be the rectangle $[0, 3] \times [0, 2]$ and let $f(x, y)$ be defined for each $(x, y) \in R$ by

$$f(x, y) = \begin{cases} 3 & \text{if } y \text{ is rational} \\ x^2 & \text{if } y \text{ is irrational} \end{cases}$$

For any $x \in [0, 3]$ with the exception of $x = \sqrt{3}$ the function $f(x, y)$ is 3 if y is rational and $x^2 \neq 3$ if y is irrational. Therefore $f(x, y)$ ($x \neq \sqrt{3}$) is discontinuous at each $y \in [0, 2]$. This implies of course that $\int_0^2 f(x, y) \, dy$ does not exist. On the other hand, for each $y \in [0, 2]$ the function $f(x, y)$ is either equal to 3 for every $x \in [0, 3]$ (if y is rational) or else is equal to x^2 for each $x \in [0, 3]$ (y irrational). In either case $f(x, y)$ is a continuous function of x on $[0, 3]$, and so $\int_0^3 f(x, y) \, dx$ exists. The function $h(y) = \int_0^3 f(x, y) \, dx$ is seen to be identically 9 on the interval $[0, 2]$, and so the iterated integral $\int_0^2 \int_0^3 f(x, y) \, dx \, dy$ exists (its value is 18). Therefore, it can certainly occur that one of the iterated integrals exists when the other does not. Of course, Fubini's theorem does not apply here since $f(x, y)$ is discontinuous at each point off the segment $x = \sqrt{3}, 0 \leq y \leq 2$. In this case the double integral $\iint_R f$ does not exist.

Suppose F and G are two continuous functions defined on the closed, bounded interval $[a, b]$ and $F(x) \leq G(x)$ for each $x \in [a, b]$. Let

$$D = \{(x, y) \mid a \leq x \leq b \text{ and } F(x) \leq y \leq G(x)\}$$

Then D is a closed, bounded region, and D has area according to Theorem 9.3 and the lemma of the preceding section. Now if f is continuous on D and we define $c = \inf_x F(x)$ and $d = \sup_x G(x)$ for $x \in [a, b]$ then we can define the auxiliary function f^* on the rectangle $R = [a, b] \times [c, d]$ by

$$f^*(x, y) = \begin{cases} f(x, y) & \text{if } (x, y) \in D \\ 0 & \text{if } (x, y) \in R - D \end{cases}$$

Since D is closed and bounded and f is continuous on D, it follows that f, and hence f^*, is bounded. By the definition of f^* the set of points where f^* is discontinuous is a subset of the set $\text{gr}(F) \cup \text{gr}(G)$. But $\text{gr}(F) \cup \text{gr}(G)$

has zero area (Theorem 9.3 and Exercise 9.7). Therefore the set E of points where f^* is discontinuous in R is a set with zero area. Moreover, for each fixed $x \in [a, b]$ the function $f^*(x, y)$ is continuous on $[c, d]$ with the possible exception of the points $y = F(x)$ and $y = G(x)$. Thus $\int_c^d f^*(x, y)\, dy$ exists for each $x \in [a, b]$. By Fubini's theorem

$$\iint_R f^* = \int_a^b \int_c^d f^*(x, y)\, dy\, dx$$

But $\iint_D f$ is defined to be $\iint_R f^*$ and

$$\int_c^d f^*(x, y)\, dy = \int_{F(x)}^{G(x)} f(x, y)\, dy$$

by the definition of f^*. This establishes the following theorem.

Theorem 9.9 If $F, G \in \mathcal{C}[a, b]$ with $F(x) \leq G(x)$ for all $x \in [a, b]$ and if $D = \{(x, y) \in \mathbf{R}^2 \,|\, a \leq x \leq b \text{ and } F(x) \leq y \leq G(x)\}$ and if f is continuous on D then

$$\iint_D f = \int_a^b \int_{F(x)}^{G(x)} f(x, y)\, dy\, dx$$

The following example illustrates how Theorem 9.9 can be used to evaluate a double integral.

Example 9.5 Evaluate the double integral $\iint_D (x^2 + y^2)$, where D is the closed region bounded by the graphs of $G(x) = x + 1$ and $F(x) = x^2 - 1$ (see Fig. 9.2).

SOLUTION By Theorem 9.9

$$\iint_D (x^2 + y^2) = \int_{-1}^{2} \int_{x^2-1}^{x+1} (x^2 + y^2)\, dy\, dx$$

$$= \int_{-1}^{2} \left(-\tfrac{1}{3}x^6 + \tfrac{4}{3}x^3 + 2x^2 + x + \tfrac{2}{3}\right) dx = \tfrac{117}{14}$$

This value represents the volume of the solid in $\mathbf{R}^3$ lying under the paraboloid $z = x^2 + y^2$ and above the region D.

Theorem 9.9 has a counterpart with the roles of x and y reversed. If $F, G \in \mathcal{C}[c, d]$ with $F(y) \leq G(y)$ for all $y \in [c, d]$ and if D is given by $D = \{(x, y) \in \mathbf{R}^2 \,|\, c \leq y \leq d \text{ and } F(y) \leq x \leq G(y)\}$ and if f is continuous on D then

$$\iint_D f = \int_c^d \int_{F(y)}^{G(y)} f(x, y)\, dx\, dy$$

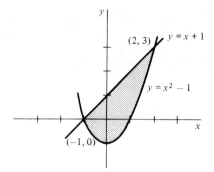

Figure 9.2

In some double integrals the region D allows one to set up the double integral as an iterated integral of either type. When this is possible, one of the iterated integrals may be easier to handle than the other.

Example 9.6 Evaluate the double integral $\iint_D xe^{x^2-y^2}$, where D is the closed, region bounded by the lines $y = x$, $y = x - 1$, $y = 0$ and $y = 1$ (see Fig. 9.3).

SOLUTION According to the above remarks, we can set the double integral up as follows:

$$\iint_D xe^{x^2-y^2} = \int_0^1 \int_y^{y+1} xe^{x^2-y^2}\,dx\,dy$$

$$= \int_0^1 \tfrac{1}{2}(e^{2y+1} - 1)\,dy = \tfrac{1}{4}(e^3 - e - 2)$$

The double integral could also have been written as

$$\int_0^1 \int_0^x xe^{x^2-y^2}\,dy\,dx + \int_1^2 \int_{x-1}^1 xe^{x^2-y^2}\,dy\,dx$$

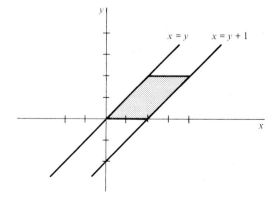

Figure 9.3

The reader will observe that the integration problem here is considerably more difficult.

Recall that in Chap. 8 we stated (but postponed the proof) that if a function f is continuous and has continuous second-order partial derivatives in an open region $D \subseteq \mathbf{R}^2$ then the mixed partials f_{xy} and f_{yx} are equal throughout D. This result can easily be proved using Fubini's theorem. We first need the following lemma.

Lemma If g is continuous throughout the open region $D \subseteq \mathbf{R}^2$ and $\iint_R g = 0$ for every rectangle R contained in D then $g(x, y) = 0$ for all $(x, y) \in D$.

PROOF Suppose the function g is continuous on the open region $D \subseteq \mathbf{R}^2$. Let $(x_0, y_0) \in D$ and suppose that $g(x_0, y_0) > 0$ (a similar argument applies to the case $g(x_0, y_0) < 0$). By the continuity of g at (x_0, y_0) and the fact that (x_0, y_0) is an interior point of D, there is a rectangle $R \subset D$ containing (x_0, y_0) such that $g(x, y) \geq \frac{1}{2}g(x_0, y_0)$ for every $(x, y) \in R$. Then

$$\iint_R g \geq \iint_R \tfrac{1}{2} g(x_0, y_0) = \tfrac{1}{2} g(x_0, y_0) \iint_R 1 = \tfrac{1}{2} g(x_0, y_0) A(R) > 0$$

Theorem 9.10 If D is an open region in $\mathbf{R}^2$ and if f_{xy} and f_{yx} are continuous in D then $f_{xy}(x, y) = f_{yx}(x, y)$ for every point $(x, y) \in D$.

PROOF By the above lemma it suffices to prove that $\iint_R (f_{xy} - f_{yx}) = 0$ for every rectangle R contained in D. Let $R = [a, b] \times [c, d]$; then, using Fubini's theorem, we have

$$\iint_R (f_{xy} - f_{yx}) = \iint_R f_{xy} - \iint_R f_{yx}$$

$$= \int_a^b \int_c^d f_{xy}(x, y)\, dy\, dx - \int_c^d \int_a^b f_{yx}(x, y)\, dx\, dy$$

$$= \int_a^b [f_x(x, d) - f_x(x, c)]\, dx - \int_c^d [f_y(b, y) - f_y(a, y)]\, dy$$

$$= [f(b, d) - f(b, c) - f(a, d) + f(a, c)]$$

$$\quad - [f(b, d) - f(a, d) - f(b, c) + f(a, c)] = 0$$

Again, we assume that f is defined on the rectangle $R = [a, b] \times [c, d]$ and that $f(x, y)$ is integrable on the interval $[c, d]$ for each $x \in [a, b]$. Then

$g(x) = \int_c^d f(x, y)\, dy$ is a well-defined function on $[a, b]$. If $f_x(x, y)$ exists and is continuous on R then the following theorem allows us to differentiate the function g by differentiating under the integral sign.

Theorem 9.11 Leibniz's rule If f is defined on the rectangle $R = [a, b] \times [c, d]$ and for each $x \in [a, b]$, $g(x) = \int_c^d f(x, y)\, dy$ exists and if $f_x(x, y)$ exists and is continuous on the rectangle R then $g(x)$ is differentiable on $[a, b]$ and $g'(x) = \int_c^d f_x(x, y)\, dy$.

PROOF Since f_x exists on R, for each $y \in [c, d]$ and each $h \ne 0$ it follows from the mean-value theorem that there is a t between x and $x + h$ with $f(x + h, y) - f(x, y) = h f_x(t, y)$. Consequently, for $h \ne 0$,

$$\frac{g(x+h) - g(x)}{h} = \frac{1}{h} \int_c^d [f(x+h, y) - f(x, y)]\, dy$$

$$= \frac{1}{h} \int_c^d h f_x(t, y)\, dy = \int_c^d f_x(t, y)\, dy$$

where t is between x and $x + h$. Now let $\varepsilon > 0$ be given. By the uniform continuity of f_x on R there is a $\delta > 0$ such that if $u, v \in R$ with $|u - v| < \delta$ then necessarily $|f_x(u) - f_x(v)| < \varepsilon/(d - c)$. Choose h satisfying $0 < |h| < \delta$. Then

$$\left| \frac{g(x+h) - g(x)}{h} - \int_c^d f_x(x, y)\, dy \right|$$

$$= \left| \int_c^d [f_x(t, y)\, dy - \int_c^d f_x(x, y)\, dy \right|$$

$$= \left| \int_c^d [f_x(t, y) - f_x(x, y)]\, dy \right| < \frac{\varepsilon}{d - c}(d - c) = \varepsilon$$

since t is between x and $x + h$ and so $|t - x| < |h| < \delta$. It follows that

$$g'(x) = \lim_{h \to 0} \frac{g(x+h) - g(x)}{h} = \int_c^d f_x(x, y)\, dy$$

We illustrate in the following example how Leibniz's rule for differentiating under the integral sign can be used to find a derivative.

Example 9.7 Let $g(x) = \int_1^2 (e^{xy}/y)\, dy$ be defined on any closed, bounded interval $[a, b]$ not containing 0. By Theorem 9.11

$$g'(x) = \int_1^2 e^{xy}\, dy = \frac{e^{2x} - e^x}{x} \quad \text{for all } x \in [a, b]$$

EXERCISES

9.17 If f is continuous on the rectangle given by $R = \{(x, y) \in \mathbf{R}^2 \,|\, a \leq x \leq b \text{ and } c \leq y \leq d\}$ then $f(x, y)$ is a function of the variable y for each x (fixed) in $[a, b]$. Prove that $f(x, y)$ is a continuous function on $[c, d]$.

9.18 Compute the double integrals:
  (a) $\iint_R (x^2 + y^2)$, where $R = [0, 1] \times [0, 1]$
  (b) $\iint_D (x^2 + y^2)$, where D is the triangle with vertices $(0, 0), (1, 0), (1, 1)$
  (c) $\iint_D (x^2 + y^2)$, where D is the triangle with vertices $(0, 0), (1, 0), (0, 1)$

9.19 Let f be defined on the rectangle $R = [0, 1] \times [0, 1]$ by

$$f(x, y) = \begin{cases} 1 & \text{if } x \text{ is rational} \\ 2y & \text{if } x \text{ is irrational} \end{cases}$$

Prove that one of the iterated integrals

$$\int_0^1 \int_0^1 f(x, y) \, dy \, dx \quad \text{and} \quad \int_0^1 \int_0^1 f(x, y) \, dx \, dy$$

exists but that the other does not. What can be said about $\iint_R f$?

9.20 Evaluate the double integral $\iint_D (2x^2 + y)$, where D is the closed region bounded by
  (a) The graphs of $F(x) = x^2$, $G(x) = x^3$ and the vertical lines $x = 2$ and $x = 3$.
  (b) The graphs of $F(y) = y$, $G(y) = y^2 + 1$ and the horizontal lines $y = 0$ and $y = 2$.

9.21 Suppose $f(x, y)$ is continuous on the rectangle $R = [0, 1] \times [0, 1]$. Find iterated integrals with the order of integration reversed but equal to the following iterated integrals:

  (a) $\int_0^1 \int_0^x f(x, y) \, dy \, dx$ (b) $\int_0^2 \int_0^{y^2} f(x, y) \, dx \, dy$ (c) $\int_{-1}^2 \int_{-x}^{2-x^2} f(x, y) \, dy \, dx$

9.22 Evaluate the double integral $\iint_D xy$, where D is the closed region bounded by the graphs of $\varphi(x) = x$ and $\psi(x) = x^3$.

9.23 Find the volume of the solid whose base is the closed region in $\mathbf{R}^2$ bounded by $y = 4 - x^2$ and $y = 3x$ while the top is bounded by the paraboloid $z = x^2 + y^2$.

9.24 Evaluate the iterated integrals:

  (a) $\int_0^1 \int_{2y}^2 e^{x^2} \, dx \, dy$ (b) $\int_0^1 \int_x^1 y \sin \pi y^3 \, dy \, dx$

9.25 Suppose f is continuous on the rectangle

$$R = \{(x, y) \in \mathbf{R}^2 \,|\, a \leq x \leq b \text{ and } c \leq y \leq d\}$$

Prove that the function $g(x) = \int_c^d f(x, y) \, dy$ is continuous on the closed interval $[a, b]$.

9.26 Find $g'(x)$ for:

  (a) $g(x) = \int_0^1 \ln(x^2 + y^2) \, dy$ (b) $g(x) = \int_0^1 \frac{\sin xy}{y} \, dy$

9.27 Suppose f is continuous on the rectangle $R = \{(x, y) \in \mathbf{R}^2 \,|\, a \leq x \leq b \text{ and } c \leq y \leq d\}$. Prove that $F_{xy} = f$ for all $(x, y) \in R$ where

$$F(x, y) = \int_c^y \int_a^x f(u, v) \, du \, dv$$

9.28 Differentiate the functions given by:

(a) $g(x) = \int_0^x e^{-x^2 y^2} \, dy$ (b) $g(x) = \int_{\sin x}^{e^x} \sqrt{1 + y^3} \, dy$

9.29 Assume that $u(x)$ and $v(x)$ have continuous derivatives for all x and that $f(x, y)$ has continuous first-order partial derivatives for all (x, y). Derive a formula for the derivative of

$$g(x) = \int_{u(x)}^{v(x)} f(x, y) \, dy$$

9.3 CHANGE OF VARIABLES IN AN INTEGRAL

In order to evaluate an integral it is sometimes necessary to make a change of variables. Just as we replaced $\int_a^b f$ by $\int_a^b f(x) \, dx$ in Chap. 6, here we use $\iint_D f(x, y) \, dx \, dy$ instead of $\iint_D f$ in order to indicate the variables of integration.

If we want to make the substitution $x = \varphi(u)$ in the integral $\int_a^b f(x) \, dx$, we first find an interval $[\alpha, \beta]$ of u values which is mapped onto $[a, b]$. Replace a, b by α, β, replace x by $\varphi(u)$, and replace dx with $\varphi'(u) \, du$ to get

$$\int_a^b f(x) \, dx = \int_\alpha^\beta f(\varphi(u)) \varphi'(u) \, du$$

Theorem 9.12 If φ is a continuously differentiable function on $[\alpha, \beta]$ and if f is continuous on the range of φ with $\varphi(\alpha) = a$ and $\varphi(\beta) = b$ then

$$\int_a^b f(x) \, dx = \int_\alpha^\beta f(\varphi(u)) \varphi'(u) \, du$$

PROOF Let $g(x) = \int_a^x f(t) \, dt$. Then $g(x)$ is differentiable on $[a, b]$ with derivative $f(x)$, and so

$$\frac{d}{du} g(\varphi(u)) = g'(\varphi(u)) \cdot \varphi'(u) = f(\varphi(u)) \cdot \varphi'(u)$$

By the fundamental theorem of calculus

$$\int_\alpha^\beta f(\varphi(u)) \cdot \varphi'(u) \, du = g(\varphi(u)) \Big|_\alpha^\beta = g(\varphi(\beta)) - g(\varphi(\alpha))$$

$$= g(b) - g(a) = \int_a^b f(x) \, dx$$

Example 9.8 Evaluate $\int_1^2 e^{x^2} x \, dx$.

SOLUTION Substitute $u = x^2$ or $x = \sqrt{u}$, $1 \leq x \leq 2$. This is a mapping of the interval $[1, 4]$ on the u axis to $[1, 2]$ on the x axis. The derivative $\varphi'(u) = 1/2\sqrt{u}$ is continuous on $[1, 4]$, and so

$$\int_1^2 e^{x^2} x \, dx = \int_1^4 e^u \sqrt{u} \, \frac{1}{2\sqrt{u}} \, du = \tfrac{1}{2}(e^4 - e)$$

The procedure for a change of variables in a double integral is similar and involves replacing the two variables x, y by the two variables u, v via

$$x = \varphi(u, v) \qquad y = \psi(u, v)$$

As we saw in Chap. 8, these functions define a transformation T from $\mathbf{R}^2$ (the (u, v) plane) into $\mathbf{R}^2$ (the (x, y) plane).

In order to write $\iint_D f(x, y) \, dx \, dy$ in terms of an integral with respect to the new variables u and v we first find the region D^* in the (u, v) plane which get mapped onto D in the (x, y) plane by the transformation T. The $dx \, dy$ part of the formula gets replaced by $|J(u, v)| \, du \, dv$, where $J(u, v)$ is the Jacobian of the transformation T. We get the formula

$$\iint_D f(x, y) \, dx \, dy = \iint_{D^*} f(\varphi(u, v), \psi(u, v)) |J(u, v)| \, du \, dv$$

Before stating a formal theorem on change of variables in a double integral, we give a geometric argument for the presence of the Jacobian in the formula.

Consider a large rectangle R containing D^* in the (u, v) plane. Partition R into subrectangles R_{ij} and consider the image of one such subrectangle under the transformation T. We let C_{i-1} be the image under the transformation T of the segment $u = u_{i-1}$, $v_{j-1} \leq v \leq v_j$. C_i is the image of the segment $u = u_i$, $v_{j-1} \leq v \leq v_j$; C'_{j-1} is the image of the segment $u_{i-1} \leq u \leq u_i$, $v = v_{j-1}$; and C'_j is the image of the segment $u_{i-1} \leq u \leq u_i$, $v = v_j$ (see Fig. 9.4). The two vectors (φ_u, ψ_u) and (φ_v, ψ_v) at $u = u_{i-1}$, $v = v_{j-1}$

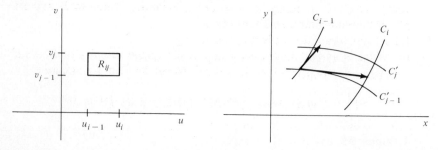

Figure 9.4

are tangent to C'_{j-1} and C_{i-1}, respectively, at the point $(x, y) = (\varphi(u_{i-1}, v_{j-1}), \psi(u_{i-1}, v_{j-1}))$ (see Exercise 9.30).

To construct a parallelogram with approximately the same area as the area of the image of R_{ij} under the transformation T we adjust the lengths of the tangent vectors (φ_u, ψ_u) and (φ_v, ψ_v). If u is held constant and v is incremented by Δv then Δx, the change in x, is approximately equal to $\varphi_v \, \Delta v$ and likewise Δy, the change in y, is approximately equal to $\psi_v \, \Delta v$. Similarly, if v is held constant and u is changed by a small amount Δu then Δx is approximately equal to $\varphi_u \, \Delta u$ and Δy is approximately equal to $\psi_u \, \Delta u$. Therefore we change the tangent vectors (φ_u, ψ_u) and (φ_v, ψ_v) to $(\varphi_u, \psi_u) \, \Delta u$ and $(\varphi_v, \psi_v) \, \Delta v$, respectively. Then the area of the image of the subrectangle R_{ij} is approximately equal to the area of the parallelogram with sides $(\varphi_u, \psi_u) \, \Delta u$ and $(\varphi_v, \psi_v) \, \Delta v$. The area of this parallelogram is given by $|(\varphi_u, \psi_u) \, \Delta u| |(\varphi_v, \psi_v) \, \Delta v| \sin \theta$, where $0 \le \theta \le \pi$ is the angle between the two vectors. But

$$|(\varphi_u, \psi_u) \, \Delta u| |(\varphi_v, \psi_v) \, \Delta v| \sin \theta$$
$$= |(\varphi_u, \psi_u, 0) \, \Delta u \times (\varphi_v, \psi_v, 0) \, \Delta v|$$
$$= |\varphi_u \psi_v - \varphi_v \psi_u| \, \Delta u \, \Delta v = |J(u, v)| \, \Delta u \, \Delta v$$

If $J(u, v) \ne 0$, we can interpret the absolute value of the Jacobian as a *magnification factor*. The area of R_{ij} in the (u, v) plane gets magnified by $|J(u, v)|$ under the transformation T. Therefore, if we want to evaluate a double integral over a region in the (x, y) plane in terms of the new variables u and v, the magnification factor $|J(u, v)|$ must be inserted in the double integral.

Theorem 9.13 Assume that T is a one-to-one transformation from the open set G in the (u, v) plane onto the open set H in the (x, y) plane, which is defined by the continuously differentiable functions

$$x = \varphi(u, v) \quad \text{and} \quad y = \psi(u, v)$$

Assume further that $J(u, v) \ne 0$ in G. Let D^* be a closed, bounded region in G whose image is D, a closed, bounded region in H and let $f(x, y)$ be bounded on D. Then
(a) D^* has area if and only if D has area.
(b) When D and D^* have area, $f(x, y)$ is integrable over D if and only if $f(\varphi(u, v), \psi(u, v))|J(u, v)|$ is integrable over D^*, and in this case

$$\iint_D f(x, y) \, dx \, dy = \iint_{D^*} f(\varphi(u, v), \psi(u, v)) |J(u, v)| \, du \, dv\dagger$$

$\dagger$ See J. M. H. Olmsted, *Real Variables*, Appleton-Century-Crofts, , New York, 1959.

The following example illustrates how the theorem can be used to make a change of variables in a double integral so that the integral can be evaluated.

Example 9.9 Evaluate $\iint_D \sin[(x-y)/(x+y)]\,dx\,dy$, where D is the region bounded by the coordinate axes and $x + y = 1$ in the first quadrant.

SOLUTION Consider the change of variables

$$u = x - y \qquad v = x + y \qquad \text{or} \qquad x = \tfrac{1}{2}(u+v) \qquad y = \tfrac{1}{2}(-u+v)$$

In the (u, v) plane the line $y = 0$ is $u = v$, the line $x = 0$ is $u = -v$, and $x + y = 1$ is $v = 1$. This gives us the boundary for D^* (see Fig. 9.5). The Jacobian is

$$\begin{vmatrix} \tfrac{1}{2} & \tfrac{1}{2} \\ -\tfrac{1}{2} & \tfrac{1}{2} \end{vmatrix} = \tfrac{1}{2}$$

Hence

$$\iint_D \sin\frac{x-y}{x+y}\,dx\,dy = \iint_{D^*} \left(\sin\frac{u}{v}\right)\frac{1}{2}\,du\,dv$$

It is desirable to write this as an interated integral where the first integration is with respect to u

$$\frac{1}{2}\int_0^1 \int_{-v}^{v} \sin\frac{u}{v}\,du\,dv = \frac{1}{2}\int_0^1 [-\cos 1 + \cos(-1)]v\,dv = 0$$

The hypothesis of Theorem 9.13 can be relaxed somewhat to allow the Jacobian to vanish on a set of zero area. In the next example the Jacobian vanishes on such a subset of the region, but the change of variables in the double integral is still valid.

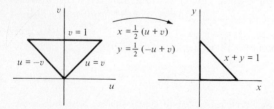

Figure 9.5

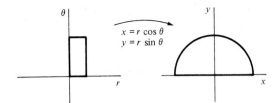

Figure 9.6

Example 9.10 Evaluate $\iint_D e^{x^2+y^2}\,dx\,dy$, where D is the closed region bounded by the semicircle $y = \sqrt{1-x^2}$ and the x axis.

SOLUTION It is natural to make the transformation

$$x = r\cos\theta \qquad y = r\sin\theta$$

Then

$$\iint_D e^{x^2+y^2}\,dx\,dy = \iint_{D^*} e^{r^2}r\,dr\,d\theta \qquad \text{where } r = |J(r,\theta)|$$

and

$$D^* = \{(r,\theta)\,|\,0 \le r \le 1,\, 0 \le \theta \le \pi\}$$

(see Fig. 9.6). Writing the double integral as an iterated integral, we obtain

$$\int_0^\pi \int_0^1 e^{r^2}r\,dr\,d\theta = \int_0^\pi \tfrac{1}{2}(e-1)\,d\theta = \tfrac{\pi}{2}(e-1)$$

The transformation in the preceding example is a useful transformation called a *polar coordinate transformation*. This transformation is one-to-one on the rectangle $\{(r,\theta)\,|\,0 \le r \le 1,\, 0 \le \theta \le \pi\}$ of Example 9.10, but this is not generally the case. If the closed region D^* is contained in the set $\{(r,\theta)\,|\,-\pi < \theta \le \pi\}$, for example, then the transformation is one-to-one and Theorem 9.13 applies.

EXERCISES

9.30 Prove that if lines $u = \text{const}$ and $v = \text{const}$ in the (u, v) plane are mapped into curves C and C' in the (x, y) plane, respectively, by the transformation

$$T: \begin{cases} x = \varphi(u,v) \\ y = \psi(u,v) \end{cases}$$

then the vector (φ_u, ψ_u) is tangent to C' and (φ_v, ψ_v) is tangent to C.

9.31 Express $\iint_D f(x, y)\, dx\, dy$ as a double integral in polar coordinates. Use Fubini's theorem to evaluate:

(a) $f(x, y) = x^2 - y^2$ and D is the closed region between the two semicircles $x^2 + y^2 = 1$, $y \geq 0$ and $x^2 + y^2 = 2$, $y \geq 0$

(b) $f(x, y) = \dfrac{xy}{x^2 + y^2}$, $D = \{(x,y) \mid x^2 + y^2 \leq 4,\ x \geq 0 \text{ and } y \geq 0\}$.

(c) $f(x, y) = e^{-(x^2 + y^2)}$, $D = \{(x, y) \mid x^2 + y^2 \leq 9\}$.

9.32 Find the volume under the paraboloid $z = 4 - x^2 - y^2$ and above the (x, y) plane using (a) rectangular coordinates and (b) polar coordinates.

9.33 Find a suitable substitution and evaluate

$$\int_0^1 \int_0^{1-x} e^{y/(x+y)}\, dy\, dx$$

Is there a problem with the integrand at $(0, 0)$? Convince yourself that your steps are valid.

9.34 Show that the Jacobian of a linear transformation from the (u, v) plane into the (x, y) plane is a constant. Interpret each of the following cases for that constant by examining the image of $[0, 1] \times [0, 1]$:

(a) $J > 0$ (b) $J < 0$ (c) $J = 0$

9.35 Evaluate

$$\iint_D \frac{y + 1}{x^2 + (y + 1)^2}\, dx\, dy \qquad \text{where } D = \{(x, y) \mid x^2 + y^2 \leq 1,\ y \geq 0\}$$

by first making a linear transformation and then a polar coordinate transformation.

9.4 IMPROPER INTEGRALS

In this section we extend the definition of the Riemann integral $\int_a^b f(x)\, dx$ to integrals over unbounded intervals and integrands which are unbounded at a point, that is, unbounded in every neighborhood of the point. If $b = \infty$ or $a = -\infty$ but f is bounded at each point in the interval then $\int_a^b f(x)\, dx$ is called an improper integral of the *first kind*. If a and b are real numbers but f is unbounded at some point in the interval $[a, b]$ then $\int_a^b f(x)\, dx$ is called an improper integral of the *second kind*. When a combination of these difficulties occurs, $\int_a^b f(x)\, dx$ is an improper integral of the *third kind*.

Definition 9.3 Let f be Riemann-integrable on $[a, R]$ for each $R > a$; then

$$\int_a^\infty f(x)\, dx = \lim_{R \to \infty} \int_a^R f(x)\, dx$$

Similarly, if f is Riemann-integrable on $[S, b]$ for all $S < b$ then

$$\int_{-\infty}^b f(x)\, dx = \lim_{S \to -\infty} \int_S^b f(x)\, dx$$

If the appropriate limit exists then the improper integral is said to converge; otherwise it diverges.

The integral test for infinite series from Chap. 7 suggests that much of the theory of convergence for improper integrals of the first kind is similar to that of infinite series. It is quite similar, but there are some differences also. One of the first results in Chap. 7 was that if $\sum_{n=1}^{\infty} a_n$ converges then $\lim_{n \to \infty} a_n = 0$. The analogous statement for improper integrals is false. The following example is a counterexample

Example 9.11 Let

$$f(x) = \begin{cases} 1 & \text{if } k - \dfrac{1}{k^2} \le x \le k \text{ where } k = 1, 2, \ldots \\ 0 & \text{otherwise} \end{cases}$$

Show that $\int_0^\infty f(x)\, dx$ converges but $\lim_{x \to \infty} f(x) \ne 0$.

SOLUTION As a function of R, $\int_0^R f(x)\, dx$ is monotone increasing

$$0 < \int_0^R f(x)\, dx < \sum_{k=1}^{\infty} \frac{1}{k^2}$$

and so $\int_0^R f(x)\, dx$ is bounded above. Therefore, $\lim_{R \to \infty} \int_0^R f(x)\, dx$ exists and $\int_0^\infty f(x)\, dx$ converges. The limit $\lim_{x \to \infty} f(x)$ does not exist because $\lim_{n \to \infty} f(n) = 1$ and $\lim_{n \to \infty} f(n + \tfrac{1}{2}) = 0$ for n a natural number.

Theorem 9.14: Comparison test If $0 \le f(x) \le g(x)$ for all x on $[a, \infty)$ and if $f(x)$ and $g(x)$ are continuous on $[a, \infty)$ then

(a) $\displaystyle\int_a^\infty g(x)\, dx$ converges implies that $\displaystyle\int_a^\infty f(x)\, dx$ converges.

(b) $\displaystyle\int_a^\infty f(x)\, dx$ diverges implies that $\displaystyle\int_a^\infty g(x)\, dx$ diverges.

PROOF For each $R > a$

$$0 \le \int_a^R f(x)\, dx \le \int_a^R g(x)\, dx$$

and each integral is a monotone increasing function of R. Hence if $\int_a^\infty g(x)\, dx$ converges, $\int_a^R f(x)\, dx$ is bounded above and so it converges. On the other hand, if $\int_a^\infty f(x)\, dx$ diverges then $\int_a^R g(x)\, dx$ is unbounded and hence diverges.

Example 9.12 Does $\int_0^\infty [1/(e^x + 1)]\, dx$ converge?

SOLUTION For x on $[0, \infty)$, $[1/(e^x + 1)] < 1/e^x$.

$$\int_0^\infty \frac{1}{e^x}\, dx = \lim_{R\to\infty} \int_0^R e^{-x}\, dx = \lim_{R\to\infty} \left. -e^{-x}\right|_0^R = \lim_{R\to\infty}(1 - e^{-R}) = 1$$

By the comparison test $\int_0^\infty [1/(e^x + 1)]\, dx$ converges.

An analogous theorem holds for improper integrals of the second kind. We define convergence of such an integral (see Exercise 9.36).

Definition 9.4 Assume that f is Riemann-integrable on any interval of the form $[a, c]$, where $a < c < b$ but f is unbounded at b. Then

$$\int_a^b f(x)\, dx = \lim_{c\to b-} \int_a^c f(x)\, dx$$

if this limit exists. Similarly, if f is Riemann-integrable on any interval of the form $[d, b]$ for $a < d < b$ but unbounded at a then

$$\int_a^b f(x)\, dx = \lim_{d\to a+} \int_d^b f(x)\, dx$$

if the limit exists. If the appropriate limit exists, the integral converges; otherwise it diverges. If the function f is unbounded at a point x_0, $a < x_0 < b$, then we say that the improper integral $\int_a^b f(x)\, dx$ converges if and only if $\int_a^{x_0} f(x)\, dx$ and $\int_{x_0}^b f(x)\, dx$ both converge.

Example 9.13 Does $\displaystyle\int_0^1 \frac{dx}{x^2 + x^{1/2}}$ converge?

SOLUTION We shall use a comparison test for integrals of the second kind which is similar to Theorem 9.14 and which is given in Exercise 9.36. Both

$$\frac{1}{x^2 + x^{1/2}} < \frac{1}{x^2} \quad \text{and} \quad \frac{1}{x^2 + x^{1/2}} < \frac{1}{x^{1/2}} \quad \text{on } (0, 1]$$

Therefore

$$\lim_{d\to 0+} \int_d^1 \frac{1}{x^2}\, dx = \lim_{d\to 0+} \left. -x^{-1}\right|_d^1 = \lim_{d\to 0+} \left(\frac{1}{d} - 1\right) = \infty$$

$$\lim_{d\to 0+} \int_d^1 \frac{1}{x^{1/2}}\, dx = \lim_{d\to 0+} \left. 2x^{1/2}\right|_d^1 = \lim_{d\to 0+} (2 - 2\sqrt{d}) = 2$$

By the comparison test (Exercise 9.36) with $1/(x^2 + x^{1/2}) < 1/x^{1/2}$ on $(0,1]$, $\int_0^1 [dx/(x^2 + x^{1/2})]$ converges.

Improper integrals of the third kind are first written as the sum of two integrals, one of the first kind plus one of the second kind. Consider the function from the previous example and extend the interval $\int_0^\infty [1/(x^2 + x^{1/2})]\, dx$. The interval is unbounded, and the integrand is unbounded at $x = 0$. We write

$$\int_0^1 \frac{1}{x^2 + x^{1/2}}\, dx + \int_1^\infty \frac{1}{x^2 + x^{1/2}}\, dx$$

We say that $\int_0^\infty [1/(x^2 + x^{1/2})]\, dx$ converges if both $\int_0^1 [1/(x^2 + x^{1/2})]\, dx$ and $\int_1^\infty [1/(x^2 + x^{1/2})]\, dx$ converge. In Example 9.13 we showed that $\int_0^1 [1/(x^2 + x^{1/2})]\, dx$ converges. The integral $\int_1^\infty [1/(x^2 + x^{1/2})]\, dx$ converges because $1/(x^2 + x^{1/2}) < 1/x^2$ for all x on $[1, \infty)$ and $\int_1^\infty (1/x^2)\, dx$ converges. Of course, the point $x = 1$ (for splitting up the interval) was chosen quite arbitrarily. The same conclusion would be reached for any other point chosen.

A similar technique is used for integrals over the entire real line. To evaluate $\int_{-\infty}^\infty f(x)\, dx$ consider separately $\int_{-\infty}^a f(x)\, dx$ and $\int_a^\infty f(x)\, dx$ for any conveniently chosen real number a. The original integral converges if and only if both these integrals converge. In certain applications a less stringent criterion is used, called the *Cauchy principal value*

$$\text{CPV} \int_{-\infty}^\infty f(x)\, dx = \lim_{R \to \infty} \int_{-R}^R f(x)\, dx$$

if the limit exists (see Exercise 9.42).

Just as for infinite series, there is a concept of uniform convergence for integrals involving a second independent variable or parameter. To ensure integrability we assume that $f(x, y)$ is continuous on $I \times [a, \infty)$ for some interval I.

Definition 9.5 The integral $\int_a^\infty f(x, y)\, dy$ converges uniformly to $g(x)$ on the interval I if for any $\varepsilon > 0$ there is $R > a$ such that

$$\left| \int_a^r f(x, y)\, dy - g(x) \right| < \varepsilon \qquad \text{when } r > R \text{ and } x \in I$$

Just as for uniform convergence of an infinite series of functions, there is a Cauchy criterion for integrals. It is particularly useful because it does not require knowledge of the limit function, and it leads to an M test which is similar to the Weierstrass M test for uniform convergence of infinite series.

Theorem 9.15 Cauchy criterion Let $f(x, y)$ be continuous on $I \times [a, \infty)$ for some interval I. $\int_a^\infty f(x, y)\, dy$ converges uniformly on I if and only if for any $\varepsilon > 0$ there is a real number T such that $\left|\int_r^s f(x, y) \, dy\right| < \varepsilon$ for all $s > r > T$ and $x \in I$.

PROOF First assume that $\int_a^\infty f(x, y) \, dy$ converges uniformly to $g(x)$ on I. Then given $\varepsilon > 0$ there is a real number R such that

$$\left|\int_a^r f(x, y) \, dy - g(x)\right| < \frac{\varepsilon}{2} \quad \text{when } r > R \text{ and } x \in I$$

Let $R < r < s$ and $x \in I$. Then

$$\left|\int_r^s f(x, y) \, dy\right| = \left|\int_a^s f(x, y) \, dy - \int_a^r f(x, y) \, dy\right|$$

$$= \left|\int_a^s f(x, y) \, dy - g(x) + g(x) - \int_a^r f(x, y) \, dy\right|$$

$$\leq \left|\int_a^s f(x, y) \, dy - g(x)\right| + \left|\int_a^r f(x, y) \, dy - g(x)\right|$$

$$< \frac{\varepsilon}{2} + \frac{\varepsilon}{2} = \varepsilon$$

Next assume that for any $\varepsilon > 0$ there is a real number T such that $\left|\int_r^s f(x, y) \, dy\right| < \varepsilon$ when $s > r > T$ and $x \in I$. Then the sequence

$$g_n(x) = \int_a^n f(x, y) \, dy$$

is a Cauchy sequence of functions on I. By Theorem 7.10 $\{g_n(x)\}$ converges uniformly on I to some function $g(x)$. It follows that $\int_a^\infty f(x, y) \, dy$ converges uniformly to $g(x)$ on I.

Theorem 9.16: M test If f is continuous on $I \times [a, \infty)$ and if $|f(x, y)| \leq M(y)$ for all $x \in I$ and $y \in [a, \infty)$ and if $\int_a^\infty M(y) \, dy$ converges then $\int_a^\infty f(x, y) \, dy$ converges uniformly on I.

PROOF Since $\int_a^\infty M(y) \, dy$ converges for any $\varepsilon > 0$ there is a T such that $0 \leq \int_r^s M(y) \, dy < \varepsilon$ for all $r, s > T$

$$\left|\int_r^s f(x, y) \, dy\right| \leq \int_r^s |f(x, y)| \, dy \leq \int_r^s M(y) \, dy < \varepsilon$$

By the Cauchy criterion $\int_a^\infty f(x, y) \, dy$ converges uniformly on I.

MULTIPLE INTEGRALS **291**

Uniform convergence of improper integrals of the second and third kinds can be defined in a similar fashion. There is a Cauchy criterion for each and an M test for each.

Definition 9.6 Let $f(x, y)$ be continuous on $I \times (a, b]$; then $\int_a^b f(x, y) \, dy$ converges uniformly to $g(x)$ on I if for any $\varepsilon > 0$ there is a $\delta > 0$ such that $|\int_d^b f(x, y) \, dy - g(x)| < \varepsilon$ whenever $a < d < a + \delta$ and $x \in I$.

Definition 9.7 An improper integral of the third kind converges uniformly on I if it can be written as the sum of an improper integral of the first kind and an improper integral of the second kind, each of which converges uniformly on I.

Example 9.14 Show that $\Gamma(x) = \int_0^\infty y^{x-1} e^{-y} \, dy$ (a) converges for each fixed $x > 0$ and (b) converges uniformly on $[\delta, L]$ for any δ, L satisfying $0 < \delta < L < \infty$. This function $\Gamma(x)$, called the *gamma function*, has applications to statistical theory.

SOLUTION (a) Assume $0 < x < 1$; then this is an improper integral of the third kind

$$\int_0^\infty y^{x-1} e^{-y} \, dy = \int_0^1 y^{x-1} e^{-y} \, dy + \int_1^\infty y^{x-1} e^{-y} \, dy$$

The first integral on the right-hand side converges because $0 \leq y^{x-1} e^{-y} \leq y^{x-1}$ for $0 \leq y \leq 1$, $0 < x < 1$ and $\int_0^1 y^{x-1} \, dy$ converges. The second integral converges because $0 < y^{x-1} e^{-y} \leq e^{-y}$ for $0 < x < 1$, $y \geq 1$ and $\int_1^\infty e^{-y} \, dy$ converges. Finally, if $x \geq 1$, $\int_0^\infty y^{x-1} e^{-y} \, dy$ is an integral of the first kind which converges (see Exercise 9.43).

(b) To establish uniform convergence on $[\delta, L]$ again we consider

$$\Gamma_1(x) = \int_0^1 y^{x-1} e^{-y} \, dy \quad \text{and} \quad \Gamma_2(x) = \int_1^\infty y^{x-1} e^{-y} \, dy$$

If $x \in [\delta, L]$ then

$$|y^{x-1} e^{-y}| = y^{x-1} e^{-y} \leq y^{\delta-1} \quad \text{for } 0 \leq y \leq 1$$

and

$$|y^{x-1} e^{-y}| < y^{L-1} e^{-y} \quad \text{for } y \geq 1$$

Since $\int_0^1 y^{\delta-1} \, dy$ and $\int_1^\infty y^{L-1} e^{-y} \, dy$ both converge, $\int_0^1 y^{x-1} e^{-y} \, dy$ and $\int_1^\infty y^{x-1} e^{-y} \, dy$ converge uniformly on $[\delta, L]$ and $\int_0^\infty y^{x-1} e^{-y} \, dy$ converges uniformly on $[\delta, L]$.

The next three theorems illustrate the importance of uniform convergence. The first says that the uniformly convergent integral of a continuous function is continuous.

Theorem 9.17 If $f(x, y)$ is continuous on $[a, b] \times [c, \infty)$ and $\int_c^\infty f(x, y)\,dy$ converges uniformly on $[a, b]$ to $g(x)$ then $g(x)$ is continuous on $[a, b]$.

PROOF We show that given any $\varepsilon > 0$ there is a $\delta > 0$ such that $|g(x_1) - g(x_0)| < \varepsilon$ when $|x_1 - x_0| < \varepsilon$ and $x_0, x_1 \in [a, b]$. First choose r so that

$$\left| \int_a^r f(x, y)\,dy - g(x) \right| < \frac{\varepsilon}{3} \quad \text{for any } x \in [a, b]$$

Then

$$|g(x_1) - g(x_0)| < \frac{2\varepsilon}{3} + \left| \int_a^r f(x_1, y) - f(x_0, y)\,dy \right|$$

$$< \frac{2\varepsilon}{3} + \int_a^r |f(x_1, y) - f(x_0, y)|\,dy$$

Next pick $\delta > 0$ such that

$$|f(x_1, y) - f(x_0, y)| < \frac{\varepsilon}{3(r - a)}$$

for all $x_0, x_1 \in [a, b]$ and $y \in [a, r]$ (see Exercise 8.18). Finally it follows that $|g(x_1) - g(x_0)| < \varepsilon$ for any $x_0, x_1 \in [a, b]$.

The next theorem uses the continuity established in the previous theorem to show that the order of integration of a Riemann-integral and an improper integral can be interchanged.

Theorem 9.18 If $f(x, y)$ is continuous on $[a, b] \times [c, \infty)$ and if $\int_c^\infty f(x, y)\,dy$ converges uniformly on $[a, b]$ then $\int_c^\infty \int_a^b f(x, y)\,dx\,dy$ converges and

$$\int_a^b \int_c^\infty f(x, y)\,dy\,dx = \int_c^\infty \int_a^b f(x, y)\,dx\,dy$$

PROOF By Theorem 9.17 for each $r > c$

$$\int_a^b \int_c^r f(x, y)\,dy\,dx = \int_c^r \int_a^b f(x, y)\,dx\,dy$$

If $g(x) = \int_c^\infty f(x, y)\, dy$ then by the previous theorem $g(x)$ is continuous and $\int_a^b g(x)\, dx$ exists. Consider

$$\left| \int_a^b g(x)\, dx - \int_c^r \int_a^b f(x, y)\, dx\, dy \right| = \left| \int_a^b g(x)\, dx - \int_a^b \int_c^r f(x, y)\, dy\, dx \right|$$

$$\leq \int_a^b \left| g(x) - \int_c^r f(x, y)\, dy \right| dx$$

Choose r so large that

$$\left| g(x) - \int_c^r f(x, y)\, dy \right| < \frac{\varepsilon}{b - a} \quad \text{for all } x \in [a, b]$$

Then

$$\int_a^b \left| g(x) - \int_c^r f(x, y)\, dy \right| dx < \varepsilon$$

Hence

$$\left| \int_a^b g(x)\, dx - \int_c^r \int_a^b f(x, y)\, dx\, dy \right| < \varepsilon$$

and

$$\lim_{r \to \infty} \int_c^r \int_a^b f(x, y)\, dx\, dy = \int_a^b g(x)\, dx$$

Finally

$$\int_a^b \int_c^\infty f(x, y)\, dy\, dx = \int_c^\infty \int_a^b f(x, y)\, dx\, dy$$

The next theorem allows us to switch the order of the operations differentiation and integration. Notice that the proof uses the preceding theorems together with the fundamental theorem of calculus.

Theorem 9.19 Let $f(x, y)$ and $f_x(x, y)$ be continuous on $[a, b] \times [c, \infty)$. If $\int_c^\infty f(x, y)\, dy$ converges to $g(x)$ for each $x \in [a, b]$ and if $\int_c^\infty f_x(x, y)\, dy$ converges uniformly on $[a, b]$ then for each $x \in (a, b)$

$$g'(x) = \int_c^\infty f_x(x, y)\, dy$$

PROOF Since $\int_c^\infty f_x(x, y)\, dy$ converges uniformly on $[a, b]$ and $f_x(x, y)$ is continuous on $[a, b]$, it converges to a continuous function, say $F(x)$.

Let $x_0 \in (a, b)$; then

$$\int_a^{x_0} F(x)\, dx = \int_a^{x_0} \int_c^\infty f_x(x, y)\, dy\, dx$$

$$= \int_c^\infty \int_a^{x_0} f_x(x, y)\, dx\, dy \qquad \text{by Theorem 9.18}$$

$$= \int_c^\infty [f(x_0, y) - f(a, y)]\, dy \qquad \text{by the fundamental theorem of calculus}$$

$$= g(x_0) - g(a)$$

Since $F(x)$ is continuous on $[a, b]$ and $\int_a^{x_0} F(x)\, dx = g(x_0) - g(a)$ for each $x_0 \in (a, b)$, it follows that $g'(x_0) = F(x_0)$, that is,

$$g'(x) = \int_c^\infty f_x(x, y)\, dy$$

As an application of uniform convergence, consider the *Laplace transform* of a function $f(x)$:

$$\mathcal{L}(f(x)) = \hat{f}(s) = \int_0^\infty e^{-sx} f(x)\, dx$$

which is defined at each s for which the integral converges. We say that $f(x)$ is of *exponential order* if $f(x)$ is continuous on $[0, \infty)$ and if there are real numbers α, c, R such that $|f(x)| \le ce^{\alpha x}$ for all $x > R$.

We shall use the theorems of this section to show that if $f(x)$ is of exponential order then $\hat{f}(s)$ exists and has a derivative on (α, ∞) and $\hat{f}'(s) = \mathcal{L}(-xf(x))$.

First we use the comparison test: $|e^{-sx} f(x)| < ce^{-(s-\alpha)x}$, and $\int_0^\infty ce^{-(s-\alpha)x}\, dx$ converges for each $s > \alpha$. Therefore, $\int_0^\infty |e^{-sx} f(x)|\, dx$, $s > \alpha$, converges, and by Exercise 9.39, $\int_0^\infty e^{-sx} f(x)\, dx$, $s > \alpha$, converges.

For $\delta > \alpha$, $\int_0^\infty e^{-\delta x} f(x)\, dx$ converges, and

$$|e^{-sx} f(x)| = e^{-sx} |f(x)| \le e^{-\delta x} |f(x)| = |e^{-\delta x} f(x)|$$

when $s \ge \delta$. Hence by the M test $\int_0^\infty e^{-sx} f(x)\, dx$ converges uniformly on any interval $[\delta, \infty)$, where $\delta > \alpha$. It follows from Theorem 9.17 that $\hat{f}(s)$ is continuous on any interval $[a, b]$, where $\alpha < a < b$.

Finally to show that $\hat{f}(s)$ is differentiable at any s_0, with $s_0 > \alpha$, first choose an interval $[a, b]$ with $\alpha < a < s_0 < b$. With $g(s, x) = e^{-sx} f(x)$,

$$\int_0^\infty g(s, x)\, dx \qquad \text{converges for } s \text{ on } [a, b]$$

$$\int_0^\infty g_s(s, x)\, dx = \int_0^\infty e^{-sx}(-x) f(x)\, dx$$

converges uniformly on $[a, b]$ because $-xf(x)$ is also of exponential order and both $g(s, x)$ and $g_s(s, x)$ are continuous on $[a, b]$. By Theorem 9.19, $\hat{f}(s_0)$ exists and

$$\hat{f}'(s_0) = \int_0^\infty g_s(s_0, x)\, dx$$

Thus $\qquad \hat{f}'(s) = \mathcal{L}(-xf(x)) \qquad$ for each $s > \alpha$

EXERCISES

9.36 Prove that if $0 \le f(x) \le g(x)$ on $(a, b]$ and if f and g are Riemann-integrable on $[d, b]$ for all $a < d < b$ then
(a) $\int_a^b g(x)\, dx$ converges implies that $\int_a^b f(x)\, dx$ converges.
(b) $\int_a^b f(x)\, dx$ diverges implies that $\int_a^b g(x)\, dx$ diverges.

9.37 (a) For what values of p does $\int_1^\infty x^p\, dx$ converge?
(b) For what values of p does $\int_0^1 x^p\, dx$ converge?

9.38 Which of the following improper integrals converge? Identify each as first, second, or third kind.

(a) $\displaystyle\int_0^\infty \sin x\, dx \qquad$ (b) $\displaystyle\int_1^\infty \frac{1}{x + \ln x}\, dx \qquad$ (c) $\displaystyle\int_{-\infty}^0 e^{-x^2}\, dx \qquad$ (d) $\displaystyle\int_0^\infty \frac{1}{x^3 + x^{1/3}}\, dx$

9.39 Define $\int_a^\infty f(x)\, dx$ converges absolutely if $\int_a^\infty |f(x)|\, dx$ converges. Prove that if $\int_a^\infty f(x)\, dx$ converges absolutely then $\int_a^\infty f(x)\, dx$ converges.

9.40 Define $\int_a^\infty f(x)\, dx$ converges conditionally if $\int_a^\infty f(x)\, dx$ converges but $\int_a^\infty |f(x)|\, dx$ diverges. Prove that $\int_0^\infty (\sin x/x)\, dx$ converges conditionally.

9.41 Which of the following improper integrals converge absolutely? Conditionally?

(a) $\displaystyle\int_{-\infty}^\infty \frac{x^3}{x^8 + 1}\, dx \qquad$ (b) $\displaystyle\int_1^\infty \frac{\sin x}{x^2}\, dx \qquad$ (c) $\displaystyle\int_0^1 \frac{\sin x}{x^2}\, dx \qquad$ (d) $\displaystyle\int_{-\infty}^{-2} \frac{1}{\sqrt{x - x^3}}\, dx$

9.42 Which of the following integrals converge? Which Cauchy principal values converge?

(a) $\displaystyle\int_{-\infty}^\infty e^{-x^2}\, dx \qquad$ (b) $\displaystyle\int_{-\infty}^\infty x^{1/3}\, dx$

9.43 Prove that for each fixed $x \ge 1$, $\int_0^\infty y^{x-1} e^{-y}\, dy$ converges. *Hint:* Show that for y sufficiently large $y^{x-1} \le e^{(1/2)y}$.

9.44 Prove that $\int_0^\infty y^{x-1} e^{-y}\, dy$ diverges if $x < 0$.

9.45 Prove that if $\int_a^\infty f(x)\, dx$ converges then for any $\varepsilon > 0$ there is a real number X such that $\left|\int_r^\infty f(x)\, dx\right| < \varepsilon$ for all $r > X$.

9.46 Prove the following properties of the gamma function:
(a) $\Gamma(x + 1) = x \cdot \Gamma(x)$, $x > 0$.
(b) $\Gamma(n + 1) = n!$ if n is a nonnegative integer.

9.47 Find the Laplace transforms of each of the following functions:
(a) $f(x) = e^{ax}$, $a = \text{const}$ (b) $f(x) = x$ (c) $f(x) = \sin x$

9.48 Prove that if $f(x)$ is of exponential order then so is $x^n f(x)$, where n is any positive integer.

9.49 Prove that if $\hat{f}(s)$, $s > \alpha$, is the Laplace transform of $f(x)$, a function of exponential order, then $\hat{f}(s)$ has derivatives of all orders for $s > \alpha$. Find formulas for these derivatives.

9.50 Let A be the set of all those functions which are continuous on $[0, \infty)$ and for which $\int_0^\infty |f(x)|\, dx$ converges.

(a) Prove that if $f \in A$ then $\int_0^\infty f(x) \sin sx\, dx$ converges uniformly on $(-\infty, \infty)$ to some function $\varphi(s)$ and $\int_0^\infty f(x) \cos sx\, dx$ converges uniformly on $(-\infty, \infty)$ to some function $\psi(s)$.

(b) Are the functions $\varphi(s)$ and $\psi(s)$ necessarily continuous on $(-\infty, \infty)$?

(c) Prove or disprove:

$$\varphi'(s) = \int_0^\infty x f(x) \cos sx\, dx \quad \text{and} \quad \psi'(s) = -\int_0^\infty x f(x) \sin sx\, dx$$

for all s on $(-\infty, \infty)$.

CHAPTER
TEN

METRIC SPACES

10.1 DEFINITION AND EXAMPLES

The real number system **R** is richly endowed, having both an algebraic structure and a topological structure. We can summarize the algebraic properties by saying that **R** together with the binary operations of addition and multiplication is a field. But the algebra in **R** is entwined with an order relation which leads to the notion of distance between any two real numbers, and it is this notion of distance that generates a topological structure in **R**. In this chapter we develop some of these topological properties in a somewhat more general setting.

For many areas of mathematics the concept of *neighboring* or *near* is important. It allows us the possibility of considering such things as limit operations and continuity. The most natural and concrete way of introducing this concept is by defining a distance function or metric in a set. The metric should be defined so that it behaves in a manner consistent with our ideas of how distance should behave; in particular, it should agree with the important properties of the distance function defined on the real line **R**.

We begin with a nonempty set S and define the notion of distance between any two elements in S.

Definition 10.1 A *metric space* is a nonempty set S together with a real-valued function $d: S \times S \to \mathbf{R}$ which satisfies the following four conditions:
(a) $d(x, y) \geq 0$ for all $x, y \in S$.
(b) $d(x, y) = 0$ if and only if $x = y$.
(c) $d(x, y) = d(y, x)$ for all $x, y \in S$.
(d) $d(x, y) \leq d(x, w) + d(w, y)$ for all $x, y, w \in S$.

The function d is called a *metric* on S, and $d(x, y)$ is to be thought of as the distance from x to y. A nonempty set S equipped with a metric d is referred to as the metric space (S, d). Naturally, we want the distance between two elements of S, or two points, to be a nonnegative real number and the distance should be positive when, and only when, the points are distinct. Also, the distance from x to y should always be the same as the distance from y to x. Property (d) called the *triangle inequality*, says that the distance from x to y must never exceed the distance from x to a third point w plus the distance from w to y. In the following we look at some examples of metric spaces, a number of which should be familiar to the reader.

Example 10.1 Our first example is the usual metric on $\mathbf{R}$,

$$d: \mathbf{R} \times \mathbf{R} \to \mathbf{R}$$

defined by

$$d(x, y) = |x - y| \quad \text{for all } x, y \in \mathbf{R}$$

Properties (a) to (d) are known to hold, and so we have the metric space $(\mathbf{R}, d)$.

The following example shows that any nonempty set S can be made into a metric space.

Example 10.2 Let S be a nonempty set and define

$$d: S \times S \to \mathbf{R}$$

by

$$d(x, y) = \begin{cases} 1 & \text{if } x \neq y \\ 0 & \text{if } x = y \end{cases}$$

We leave it to the reader to verify that (S, d) is a metric space. d is called the *discrete metric* on S, and (S, d) is referred to as *discrete space*.

Example 10.3 Let $S = \mathbf{R} \times \mathbf{R}$ and let d be the usual distance function defined in the plane

$$d(x, y) = \sqrt{(x_1 - y_1)^2 + (x_2 - y_2)^2}$$

for any two points $x = (x_1, x_2)$ and $y = (y_1, y_2)$ in $\mathbf{R} \times \mathbf{R}$. Then

($\mathbf{R} \times \mathbf{R}$, d) is a metric space. We refer to d as the *usual metric* on $\mathbf{R} \times \mathbf{R}$. We can also define a function d' by

$$d'(x, y) = |x_1 - y_1| + |x_2 - y_2|$$

for any two points $x = (x_1, x_2)$ and $y = (y_1, y_2)$ in $\mathbf{R} \times \mathbf{R}$. Properties (a) to (d) of Definition 10.1 are easily seen to hold, and so ($\mathbf{R} \times \mathbf{R}, d'$) is a metric space. d' is sometimes referred to as the *rectangular metric*. Since d and d' are different functions, we get two distinct metric spaces ($\mathbf{R} \times \mathbf{R}$, d) and ($\mathbf{R} \times \mathbf{R}$, d'). In the next section we shall see that topologically these spaces are really the same, but strictly speaking they are different metric spaces because d and d' are different metrics. We note that these metrics satisfy the inequalities

$$d(x, y) \le d'(x, y) \le \sqrt{2}\, d(x, y)$$

for all points $x = (x_1, x_2)$ and $y = (y_1, x_2)$ in $\mathbf{R} \times \mathbf{R}$.

The next example generalizes the usual metric d to the set $\mathbf{R}^n$ ($n \in N$) of all n-tuples $x = (x_1, x_2, \ldots, x_n)$ of real numbers.

Example 10.4 Let $S = \mathbf{R}^n$ for any natural number n and define

$$d: \mathbf{R}^n \times \mathbf{R}^n \to \mathbf{R}$$

by

$$d(x, y) = \sqrt{\sum_{k=1}^{n} (x_k - y_k)^2}$$

where $x = (x_1, x_2, \ldots, x_n)$ and $y = (y_1, y_2, \ldots, y_n)$ are elements in $\mathbf{R}^n$. ($\mathbf{R}^n$, d) is called *euclidean n-space*, and d is referred to as the *euclidean metric* on $\mathbf{R}^n$. Properties (a) to (c) are obvious and property (d), the triangle inequality, follows from the Cauchy-Schwarz inequality (Sec. 8.1).

Let (S, d) be a metric space and $A \subseteq S$. We say the set A is *bounded* if there is a positive real number M such that $d(x, y) \le M$ for all points x, $y \in A$. The empty set $\emptyset$ is always bounded, and (S, d) is called a *bounded metric space* provided S is bounded. For example, discrete space is a bounded metric space, but euclidean n-space is not a bounded space. Every metric space can be made into a bounded space while preserving a number of topological characteristics of the space (see Exercise 10.8).

Example 10.5 Let $\mathcal{B}$ be the set of all bounded sequences of real numbers and define the function

$$d^*: \mathcal{B} \times \mathcal{B} \to \mathbf{R}$$

by

$$d^*(x, y) = \sup_n |x_n - y_n|$$

where $x = \{x_n\}$ and $y = \{y_n\}$ are elements in $\mathcal{B}$. Clearly $d^*(x, y) \geq 0$ for all $x, y \in \mathcal{B}$. If $d^*(x, y) = 0$ then $\sup_n |x_n - y_n| = 0$, and so $|x_n - y_n| = 0$ for every natural number n. Hence $x = y$. Conversely, if $x = y$ then $|x_n - y_n| = 0$ for all n, and so $d^*(x, y) = \sup_n |x_n - y_n| = 0$. Clearly $d^*(x, y) = \sup_n |x_n - y_n| = \sup_n |y_n - x_n| = d^*(y, x)$. Let $w = \{w_n\}$; then for each n, $|x_n - y_n| \leq |x_n - w_n| + |w_n - y_n|$, and so

$$d^*(x, y) = \sup_n |x_n - y_n| \leq \sup_n (|x_n - w_n| + |w_n - y_n|)$$

$$\leq \sup_n |x_n - w_n| + \sup_n |w_n - y_n| = d^*(x, w) + d^*(w, y)$$

A metric space (S_1, d_1) is called a *subspace* of the metric space (S, d) provided $\emptyset \neq S_1 \subseteq S$ and $d_1(x, y) = d(x, y)$ for all points $x, y \in S_1$. In other words, d_1 is the restriction of d to the set $S_1 \times S_1$. In Example 10.1, (Q, d), where Q is the set of all rationals and d is the usual metric, is a subspace of $(\mathbf{R}, d)$. In Example 10.5 if we let S be the subset of $\mathcal{B}$ consisting of all square-summable sequences, that is, sequences $\{x_n\}$ with $\sum_{n=1}^{\infty} x_n^2 < \infty$, then (S, d^*) is a subspace of $(\mathcal{B}, d^*)$. In our next example we show that there is another quite natural way of defining a metric on S, but in this case we do not get a subspace of $(\mathcal{B}, d^*)$.

Example 10.6 Let S be the set of all sequences of real numbers $\{x_n\}$ for which $\sum_{n=1}^{\infty} x_n^2 < \infty$, that is, the series $\sum_{n=1}^{\infty} x_n^2$ converges. Define the function

$$d: S \times S \to \mathbf{R}$$

by

$$d(x, y) = \sqrt{\sum_{n=1}^{\infty} (x_n - y_n)^2}$$

where $x = \{x_n\}$ and $y = \{y_n\}$ are elements in S. (S, d) is a metric space (see Exercise 10.9) and an example of a special kind of metric space known as *Hilbert space*.

Since a sequence of real numbers is just a real-valued function defined on the set $\mathbf{N}$ of natural numbers, Example 10.5 can be generalized as follows.

Example 10.7 Let $\mathcal{B}(X)$ be the set of all bounded real-valued functions defined on the nonempty set X; that is,

$$\mathcal{B}(X) = \{f : X \to \mathbf{R} \mid f \text{ is bounded on } X\}$$

Let

$$d^* : \mathcal{B}(X) \times \mathcal{B}(X) \to \mathbf{R}$$

be defined by

$$d^*(f, g) = \sup_{x \in X} |f(x) - g(x)|$$

We ask the reader to verify that $(\mathcal{B}(X), d^*)$ is a metric space. Notice that if $X = N$ then $(\mathcal{B}(X), d^*)$ is the space of Example 10.5.

We conclude this section with a final example which will recur in later sections.

Example 10.8 Let $\mathcal{C}[0, 1]$ be the set of all real-valued continuous functions defined on $[0, 1]$. Define

$$\tilde{d}(f, g) = \int_0^1 |f(x) - g(x)| \, dx$$

for all points f and g contained in $\mathcal{C}[0, 1]$. Then $(\mathcal{C}[0, 1], \tilde{d})$ is a metric space. Clearly $\tilde{d}(f, g) \geq 0$ for all $f, g \in \mathcal{C}[0, 1]$ and $\tilde{d}(f, g) = 0$ if $f(x) = g(x)$ for every $x \in [0, 1]$. Suppose $\tilde{d}(f, g) = 0$; then we have $\int_0^1 |f(x) - g(x)| \, dx = 0$, and this implies that $|f(x) - g(x)| = 0$ for every $x \in [0, 1]$ since $|f(x) - g(x)|$ is a nonnegative continuous function. Therefore, $f = g$. Clearly,

$$\tilde{d}(f, g) = \int_0^1 |f(x) - g(x)| \, dx = \int_0^1 |g(x) - f(x)| \, dx = \tilde{d}(g, f)$$

Suppose $h \in \mathcal{C}[0, 1]$; then for each $x \in [0, 1]$

$$|f(x) - g(x)| \leq |f(x) - h(x)| + |h(x) - g(x)|$$

and so

$$\tilde{d}(f, g) = \int_0^1 |f(x) - g(x)| \, dx \leq \int_0^1 |f(x) - h(x)| \, dx$$
$$+ \int_0^1 |h(x) - g(x)| \, dx = \tilde{d}(f, h) + \tilde{d}(h, g)$$

EXERCISES

10.1 Let S be any nonempty set and let $d: S \to \mathbf{R}$ be defined by

$$d(x, y) = \begin{cases} 1 & \text{if } x \neq y \\ 0 & \text{if } x = y \end{cases}$$

Verify that (S, d) is a metric space.

10.2 Prove that $d(x, y) \leq d'(x, y) \leq \sqrt{2} \, d(x, y)$ for all points $x, y \in \mathbf{R} \times \mathbf{R}$, where d is the usual metric and d' is the rectangular metric.

10.3 Let $S = \mathbf{R}^n$ for any natural number n and let d be defined as in Example 10.4. Prove that $(\mathbf{R}^n, d)$ is a metric space.

10.4 Verify that each of the following functions is a metric on $\mathbf{R}^n$:
(a) $d'(x, y) = \sum_{k=1}^n |x_k - y_k|$ for all points $x = (x_1, x_2, \ldots, x_n)$ and $y = (y_1, y_2, \ldots, y_n)$ in $\mathbf{R}^n$. d' is called the *rectangular metric* on $\mathbf{R}^n$.
(b) $d^*(x, y) = \max_{1 \le k \le n} |x_k - y_k|$ for all points $x = (x_1, x_2, \ldots, x_n)$ and $y = (y_1, y_2, \ldots, y_n)$ in $\mathbf{R}^n$.

10.5 Prove that the following inequalities hold for all points $x, y \in \mathbf{R}^n$:
$$d^*(x, y) \le d(x, y) \le \sqrt{n}\, d^*(x, y)$$
where d^* is the metric from part (b) Exercise 10.4 and d is the euclidean metric on $\mathbf{R}^n$.

10.6 Suppose (S, d_1) and (S, d_2) are metric spaces and k is a positive real number. Which of the following are metric spaces?
(a) (S, d_1^2) (b) (S, kd_1) (c) $(S, d_1 + d_2)$
(d) $(S, d_1 d_2)$ (e) $(S, \max(d_1, d_2))$ (f) $(S, \min(d_1, d_2))$

10.7 Let S be any nonempty set and suppose that $d: S \times S \to \mathbf{R}$ satisfies
$$d(x, y) = 0 \text{ if and only if } x = y$$
and
$$d(x, y) \le d(x, z) + d(y, z) \quad \text{for all } x, y, z \in S$$
Prove that (S, d) is a metric space.

10.8 Let (S, d) be a metric space and suppose that $\rho: S \times S \to \mathbf{R}$ is defined by
$$\rho(x, y) = \frac{d(x, y)}{1 + d(x, y)}$$
for all points $x, y \in S$. Prove that (S, ρ) is a metric space, that it is bounded, and that $\rho(x, y) \le d(x, y)$ for all $x, y \in S$.

10.9 Prove that the example of a Hilbert space (Example 10.6) is a metric space.

10.10 Let $\mathcal{B}(X) = \{f : X \to \mathbf{R} \mid f \text{ is bounded on } X\}$ and let $d^*(f, g) = \sup_{x \in X} |f(x) - g(x)|$ for all points $f, g \in \mathcal{B}(X)$. Prove that $(\mathcal{B}(X), d^*)$ is a metric space.

10.11 A set $\mathcal{F}$ of real-valued functions defined on X is said to be *uniformly bounded* if there is a positive real number M such that $|f(x)| \le M$ for every $x \in X$ and for every $f \in \mathcal{F}$. Show that a set $\mathcal{F}$ in the metric space $(\mathcal{B}(X), d^*)$ of Example 10.7 is bounded if and only if $\mathcal{F}$ is a uniformly bounded set of functions.

10.12 Let $\mathcal{R}[0, 1]$ be the set of all Riemann-integrable functions defined on $[0, 1]$ and let $\tilde{d}$ be defined by
$$\tilde{d}(f, g) = \int_0^1 |f(x) - g(x)|\, dx$$
for all points $f, g \in \mathcal{R}[0, 1]$. Explain why $(\mathcal{R}[0, 1], \tilde{d})$ is not a metric space.

10.2 OPEN AND CLOSED SETS; TOPOLOGY

In this section we develop some of the topological properties of a metric space (S, d). We begin by defining the basic open set, the open sphere.

Definition 10.2 For any $x_0 \in S$ and any $r > 0$ the *open sphere* of radius r centered at x_0 is defined by
$$S_r(x_0) = \{x \in S \mid d(x, x_0) < r\}$$

The open sphere of radius r centered at x_0 consists of all points in the space S which are at a distance less than r from the point x_0. Clearly $S_\rho(x_0) \subseteq S_r(x_0)$ whenever $0 < \rho \leq r$. The *closed sphere* of radius r centered at x_0 is defined by

$$S_r[x_0] = \{x \in S \mid d(x, x_0) \leq r\}$$

It is clear that $S_r(x_0) \subseteq S_r[x_0]$ for every $x_0 \in S$ and for every $r > 0$. In the space $(\mathbf{R}, d)$, where $d(x, y) = |x - y|$ for all $x, y \in \mathbf{R}$, referred to as the *real line with the usual metric*, the open sphere of radius $r > 0$ centered at $x_0 \in \mathbf{R}$ is the neighborhood $N_r(x_0) = (x_0 - r, x_0 + r)$, while the closed sphere is the closed interval $[x_0 - r, x_0 + r]$.

Definition 10.3 A set $G \subseteq S$ is called *open* if for each $x_0 \in G$ there is an $r > 0$ with $S_r(x_0) \subseteq G$.

The empty set $\emptyset$ and the entire space S both satisfy the condition of Definition 10.3, and so both are open sets. We verify that every open sphere is an open set. Let $a \in S_r(x_0)$ and define $\eta = r - d(a, x_0) > 0$. Now, if $y \in S_\eta(a)$ then $d(y, a) < \eta$, and so

$$d(y, x_0) \leq d(y, a) + d(a, x_0) < \eta + d(a, x_0) = r$$

Thus $y \in S_r(x_0)$, and hence $S_\eta(a) \subseteq S_r(x_0)$. This establishes that $S_r(x_0)$ is open.

In what follows we assume that Γ is a nonempty set.

Theorem 10.1 If G_α is an open set for each $\alpha \in \Gamma$ then $\bigcup_{\alpha \in \Gamma} G_\alpha$ is open.

PROOF Let $y \in \bigcup_{\alpha \in \Gamma} G_\alpha$; then there is an $\alpha_0 \in \Gamma$ such that $y \in G_{\alpha_0}$. Since G_{α_0} is open, there is an $r > 0$ such that $S_r(y) \subseteq G_{\alpha_0}$. Then $S_r(y) \subseteq \bigcup_{\alpha \in \Gamma} G_\alpha$, and so $\bigcup_{\alpha \in \Gamma} G_\alpha$ is open.

As we observed back in Chap. 2, the intersection of open sets on the real line may fail to be open. For example, the intersection of the open intervals $(-1/n, 1/n)$ for $n = 1, 2, 3, \ldots$ is the singleton set $\{0\}$, which is clearly not an open subset of $\mathbf{R}$.

The following theorem states that a finite intersection of open sets is always open.

Theorem 10.2 If G_k is an open set for $k = 1, 2, \ldots, n$ then $\bigcap_{k=1}^{n} G_k$ is open.

PROOF Let $y \in \bigcap_{k=1}^{n} G_k$; then $y \in G_k$ for each $k = 1, 2, \ldots, n$. Since each G_k is open, there is an $r_k > 0$ such that $S_{r_k}(y) \subseteq G_k$ for $k = 1, 2, \ldots, n$. Let $r = \min_{1 \leq k \leq n} r_k$; then $S_r(y) \subseteq S_{r_k}(y) \subseteq G_k$ for each

$k = 1, 2, \ldots, n$. It follows that $S_r(y) \subseteq \bigcap_{k=1}^n G_k$, and so $\bigcap_{k=1}^n G_k$ is open.

We note that just as every open set on the real line is a union of open intervals, every open set in the metric space (S, d) can be written as a union of open spheres. If $G \subseteq S$ is open then for each $x \in G$ there is an $r_x > 0$ such that $S_{r_x}(x) \subseteq G$. Hence

$$G = \bigcup_{x \in G} S_{r_x}(x)$$

Conversely, the union of open spheres is always an open set in (S, d); this follows from Theorem 10.1 and the fact that an open sphere is an open set.

A point $x_0 \in S$ is called a *limit point* of the set $A \subseteq S$ if every open sphere centered at x_0 contains at least one point of A different from x_0. We leave it to the reader to verify that x_0 is a limit point of A if and only if each open sphere centered at x_0 contains infinitely many points of A (see Exercise 10.14). If we consider the euclidean plane with the usual metric and if we let $A = \{(1 + 1/n, 1 - 1/n) \mid n \in \mathbf{N}\}$ then $(1, 1)$ is a limit point of A and is, in fact, the only limit point of A. If $B = \{(1, r) \mid r \in \mathbf{Q}\}$ then $(1, y)$ is a limit point of B for every $y \in \mathbf{R}$. This example illustrates the fact that, in general, a set will contain some of its limit points but there may be limit points of the set which fail to be in the set.

Definition 10.4 A set $F \subseteq S$ is called *closed* if F contains all its limit points.

We note that the empty set $\emptyset$ and the space S are closed sets in every metric space (S, d). On the real line the set $\mathbf{N}$ of natural numbers is a closed set, as is every closed interval; the set $\mathbf{Q}$ of rational numbers is not closed. In the euclidean plane with the usual metric the closed unit square $U = \{(x, y) \mid 0 \leq x \leq 1, \ 0 \leq y \leq 1\}$ is a closed set; the sets $A = \{(1 + 1/n, 1 - 1/n) \mid n \in \mathbf{N}\}$ and $B = \{(1, r) \mid r \in \mathbf{Q}\}$ are not closed. Our next theorem characterizes closed sets in terms of open sets.

Theorem 10.3 A set $F \subseteq S$ is closed if and only if $S - F$ is open.

PROOF Assume first that F is closed. If $S - F = \emptyset$ then it is open, and so we may assume that $S - F \neq \emptyset$. Let $y \in S - F$; since $y \notin F$, y is not a limit point of F. Thus there is an open sphere centered at y which is disjoint from F. Clearly this open sphere is contained in $S - F$, and so $S - F$ is open. Conversely, suppose that $S - F$ is open. If $y \in S - F$ then there is an open sphere $S_r(y) \subseteq S - F$. Clearly then y is not a limit point of F. It follows that if x is a limit point of F then necessarily $x \in F$ and so F is closed.

The following result is a consequence of the previous three theorems. We leave the proof to the reader (see Exercise 10.17).

Theorem 10.4
(a) If F_α is a closed set for each $\alpha \in \Gamma$ then $\bigcap_{\alpha \in \Gamma} F_\alpha$ is closed.
(b) If F_k is a closed set for $k = 1, 2, \ldots, n$ then $\bigcup_{k=1}^{n} F_k$ is closed.

Consider the discrete space (S, d) where S is any nonempty set and d is defined by

$$d(x, y) = \begin{cases} 1 & \text{if } x \neq y \\ 0 & \text{if } x = y \end{cases}$$

For any $x \in S$ and $0 < r < 1$, $S_r(x) = \{x\}$. Therefore every singleton set is an open set. By Theorem 10.1 every subset of S is open, and so by Theorem 10.3 every subset of S is also closed. We point out that in discrete space there are no limit points and so every set satisfies the condition of Definition 10.4.

If A is a set in some metric space (S, d) then we denote the set of all limit points of A by A_0. If $A = \{1, \frac{1}{2}, \frac{1}{3}, \ldots\}$ on the real line with the usual metric then $A_0 = \{0\}$. Similarly, if $B = [0, 1)$ then $B_0 = [0, 1]$. If A is any set in discrete space then $A_0 = \emptyset$. We define the *closure* of a set A to be the union of A and its set of limit points A_0, written $\bar{A} = A \cup A_0$. By Definition 10.4 a set A is closed provided $A_0 \subseteq A$. It follows that a set A is closed if and only if $\bar{A} = A$.

Theorem 10.5 For any set A in the metric space (S, d), $\bar{A}$ is a closed set.

PROOF We show that $S - \bar{A}$ is open. Since $\emptyset$ is an open set, we may assume that $S - \bar{A} \neq \emptyset$. Let $y \in S - \bar{A}$ then $y \notin A$, and y is not a limit point of A. Hence there is an $r > 0$ such that $S_r(y)$ contains no point of A. Since $S_r(y)$ is an open set, for each $x \in S_r(y)$ there is an open sphere centered at x and contained in $S_r(y)$. Therefore every $x \in S_r(y)$ has an open sphere centered at x which is disjoint from A. It follows that no limit point of A lies in $S_r(y)$. Consequently $S_r(y) \cap \bar{A} = \emptyset$, and so $S_r(y) \subseteq S - \bar{A}$. This proves that $S - \bar{A}$ is an open set; $\bar{A}$ is closed by Theorem 10.3.

The closure of the set A is the smallest closed set containing A in the sense that $\bar{A}$ is a closed set containing A, and if F is any closed set containing A then F necessarily contains $\bar{A}$. That $\bar{A}$ is a closed set containing A follows from the definition of $\bar{A}$ and Theorem 10.5. Now if F is any closed set containing A then a point $y \in S - F$ is neither in A (since $A \subseteq F$) nor in A_0 (since $S - F$ is an open set by Theorem 10.3). Thus $\bar{A} \cap (S - F) = \emptyset$, and so $\bar{A} \subseteq F$. This result allows us to represent $\bar{A}$ as the intersection of all closed sets in (S, d) which contain the set A.

We define the *diameter* of a nonempty set A to be

$$d(A) = \sup_{x, y \in A} d(x, y)$$

The empty set has no defined diameter and a nonempty set A is easily seen to be bounded if and only if $d(A) < \infty$; that is, its diameter is a nonnegative real number. Clearly a set has diameter 0 if and only if the set is a singleton set. The following theorem shows that the closure of A contains no points which fail to be "close to A." We leave the proof to the reader (see Exercise 10.21).

Theorem 10.6 For any nonempty set A in the metric space (S, d), $d(\bar{A}) = d(A)$.

If A and B are any two nonempty sets then we define the distance from A to B to be

$$d(A, B) = \inf_{x \in A, y \in B} d(x, y)$$

In particular, if $A = \{a\}$ then $d(a, B) = \inf_{y \in B} d(a, y)$ is the distance from the point $a \in S$ to the set B. We note that if x_0 is a limit point of $A \subseteq S$ then given any $\varepsilon > 0$ there is a point y of $A - \{x_0\}$ in the open sphere of radius ε centered at x_0. Since $d(x_0, y) < \varepsilon$, $d(x_0, A - \{x_0\}) < \varepsilon$. But $\varepsilon > 0$ was arbitrary and so $d(x_0, A - \{x_0\}) = 0$. Conversely, if $d(x_0, A - \{x_0\}) = 0$ then every open sphere $S_r(x_0)$ centered at x_0 contains a point of $A - \{x_0\}$, for otherwise we would have $d(x_0, y) \geq r$ for every $y \in A - \{x_0\}$ and so $d(x_0, A - \{x_0\}) \geq r$. It follows that x_0 is a limit point of A if and only if $d(x_0, A - \{x_0\}) = 0$. Similar to the preceding theorem, the next result shows that $\bar{A}$ contains all those points close to A.

Theorem 10.7 For any nonempty set A in the metric space (S, d)

$$\bar{A} = \{x \in S \mid d(x, A) = 0\}$$

PROOF Let $x_0 \in \bar{A}$. If $x_0 \in A$ then surely $d(x_0, A) = 0$. Suppose $x_0 \in A_0 - A$ and let $\varepsilon > 0$ be given. Then there is a point $y \in A \cap S_\varepsilon(x_0)$. Since $d(x_0, y) < \varepsilon$, we must have $d(x_0, A) < \varepsilon$. The arbitrariness of ε establishes that $d(x_0, A) = 0$. Conversely, suppose $d(x_0, A) = 0$. If $x_0 \in A$ then $x_0 \in \bar{A}$ and we are finished. Suppose $x_0 \notin A$ and consider any open sphere $S_r(x_0)$ centered at x_0. Since $d(x_0, A) = 0$, there is a point $y \in A$ with $d(x_0, y) < r$. Thus $y \in S_r(x_0)$ and $y \neq x_0$ since $x_0 \notin A$. It follows that x_0 is a limit point of A and consequently $x_0 \in \bar{A}$.

Two nonempty sets A and B are called *close* provided $d(A, B) = 0$.

Clearly if A and B intersect, that is, have a point in common, then they are close; but disjoint sets can be close as well. On the real line a set A and its complement $\mathbf{R} - A$ (provided neither is empty) are always close. On the other hand, in discrete space two sets are close if and only if they intersect. Later we shall show that if A is closed and B is closed and bounded and we are in $(\mathbf{R}, d)$, where d is the usual metric, then A and B are close if and only if they intersect. The following example shows that this does not hold in every metric space.

Example 10.9 Let $S = \{1, \frac{1}{2}, \frac{1}{3}, \frac{1}{4}, \frac{1}{5}, \ldots\}$ and $d(x, y) = |x - y|$ for all $x, y \in S$. (S, d) is a subspace of the real line. Let $A = \{1, \frac{1}{3}, \frac{1}{5}, \ldots\}$ and $B = \{\frac{1}{2}, \frac{1}{4}, \frac{1}{6}, \ldots\}$. Then each of the sets A and B is closed (they have no limit points) and bounded. They are clearly disjoint, but $d(A, B) = 0$.

A set D in the metric space (S, d) is said to be *dense* in S provided D has at least one point in common with every open sphere. We saw earlier that the rationals $\mathbf{Q}$ are dense in the space $(\mathbf{R}, d)$, where d is the usual metric. If (S, d) is discrete space then the only dense subset of S is S itself.

Theorem 10.8 D is dense in S if and only if $\bar{D} = S$.

PROOF Suppose D is dense in S and let $y \in S - D$. Then every open sphere centered at y contains a point of D different from y. By the definition of limit point, $y \in D_0$, and so $D \cup D_0 = S$; that is, $\bar{D} = S$. Conversely, suppose D is not dense in S. Then there exists an open sphere $S_r(x)$ disjoint from D. Hence $D \subseteq S - S_r(x)$, and since $S - S_r(x)$ is closed, it follows that $\bar{D} \subseteq S - S_r(x)$. Clearly $\bar{D}$ is then a proper subset of S.

A metric space (S, d) is called *separable* provided there is a countable set which is dense in S. The rationals are a countable dense subset of $\mathbf{R}$, and so the real line is a separable metric space. If we let $S = [0, 1]$ (or any uncountable set) and if d is the discrete metric on S then (S, d) is not separable since the only dense subset of discrete space (S, d) is the space S itself.

EXERCISES

10.13 Let x_1 and x_2 be distinct points in the metric space (S, d). Verify that there are open spheres $S_{r_1}(x_1)$ and $S_{r_2}(x_2)$ which are disjoint.

10.14 Prove that x_0 is a limit point of $A \subseteq S$ if and only if each open sphere centered at x_0 contains infinitely many points of A.

10.15 Prove that in any metric space every finite set is closed.

10.16 Prove that in any metric space every closed sphere $S_r[x_0]$ is a closed set.

10.17 Prove Theorem 10.4 using Theorems 10.1 to 10.3 and the De Morgan laws of set theory:

(a) $S - \bigcup_{\alpha \in \Gamma} B_\alpha = \bigcap_{\alpha \in \Gamma} (S - B_\alpha)$ (b) $S - \bigcap_{\alpha \in \Gamma} B_\alpha = \bigcup_{\alpha \in \Gamma} (S - B_\alpha)$

where $B_\alpha \subseteq S$ for every $\alpha \in \Gamma$.

10.18 Show by example that a set which fails to be closed need not be open.

10.19 Show by example that a nonempty proper subset of S in the metric space (S, d) can be both open and closed.

10.20 Give an example to show that the closure of the open sphere of radius r centered at x_0 is not necessarily equal to the closed sphere of radius r centered at x_0. Prove that we always have $\overline{S_r(x_0)} \subseteq S_r[x_0]$.

10.21 Prove Theorem 10.6.

10.22 Prove that a set D is dense in the metric space (S, d) if and only if S is the only closed set containing D.

10.23 Prove that for any natural number n, $\mathbf{R}^n$ with the euclidean metric is a separable metric space.

10.24 Verify that $(\mathbf{R} - \mathbf{Q}, d)$, the irrationals with the usual metric, is a separable metric space.

10.25 Prove that every subspace of a separable metric space is again a separable metric space.

10.26 Prove that the metric space $(\mathscr{B}, d^*)$ of Example 10.5 is a nonseparable metric space.

10.27 Prove that the example of a Hilbert space (Example 10.6) is a separable metric space.

10.28 Prove that if (S, d) is a separable metric space then there is a countable collection $\mathscr{G}$ of open sets such that every open set $G \subseteq S$ can be written as a union of sets in $\mathscr{G}$.

10.29 Prove that if (S, d) is a separable metric space and G_α ($\alpha \in \Gamma$) is a collection of open sets with $\bigcup_{\alpha \in \Gamma} G_\alpha = S$ then there is a countable number of these open sets, say $G_{\alpha_1}, G_{\alpha_2}, G_{\alpha_3}, \ldots$, such that $\bigcup_{n=1}^{\infty} G_{\alpha_n} = S$.

10.30 Suppose (S, d) and (S, ρ) are metric spaces, k is a positive real number, and $d(x, y) \leq k \cdot \rho(x, y)$ for every pair $x, y \in S$. Prove that the space (S, ρ) is richer in open sets in the sense that every set G which is open in the metric space (S, d) is also open in the metric space (S, ρ).

10.31 In a metric space (S, d) the class of all open sets is referred to as the *topology* of the space. Verify that $(\mathbf{R}^2, d)$ and $(\mathbf{R}^2, d')$, where d is the usual metric and d' the rectangular metric, have the same topology.

10.32 Suppose (S_1, d) is a subspace of the metric space (S, d). Prove that a set $H \subseteq S_1$ is open if and only if $H = G \cap S_1$ for some open set $G \subseteq S$.

10.3 CONVERGENCE AND COMPLETENESS

In this section we begin by discussing sequences $\{x_n\}$ of points in a metric space (S, d).

Definition 10.5 A sequence $\{x_n\}$ is said to *converge* to the point $x_0 \in S$ if given any $\varepsilon > 0$ there is a natural number N such that $x_n \in S_\varepsilon(x_0)$ whenever $n > N$.

We denote convergence of the sequence $\{x_n\}$ to x_0 by $\lim_{n \to \infty} x_n = x_0$

or more briefly $x_n \to x_0$. x_0 is called the *limit* of the sequence $\{x_n\}$. A sequence $\{x_n\}$ converges to x_0 if given any open sphere centered at x_0, x_n is in this open sphere for all but finitely many natural numbers n. Exercise 10.13 guarantees that if a sequence converges then its limit is unique.

Definition 10.6 A sequence $\{x_n\}$ is called a *Cauchy sequence* if given any $\varepsilon > 0$ there is a natural number N such that $d(x_m, x_n) < \varepsilon$ whenever $m, n > N$.

The following theorem tells us that every convergent sequence in a metric space (S, d) satisfies the condition of Definition 10.6.

Theorem 10.9 If $\{x_n\}$ is convergent then $\{x_n\}$ is a Cauchy sequence.

PROOF Suppose $\lim_{n \to \infty} x_n = x_0$ for some $x_0 \in S$. Let $\varepsilon > 0$ be given. There is a natural number N such that $x_n \in S_{\varepsilon/2}(x_0)$ whenever $n > N$. Hence if $m, n > N$, we have $x_m, x_n \in S_{\varepsilon/2}(x_0)$, and so

$$d(x_m, x_n) \le d(x_m, x_0) + d(x_0, x_n) < \frac{\varepsilon}{2} + \frac{\varepsilon}{2} = \varepsilon$$

It follows that $\{x_n\}$ is a Cauchy sequence.

If we consider the metric space (S, d), where $S = \{1, \frac{1}{2}, \frac{1}{3}, \frac{1}{4}, \frac{1}{5}, \ldots\}$ and $d(x, y) = |x - y|$ for all $x, y \in S$, then the sequence $\{1/n\}$ is a Cauchy sequence but $\{1/n\}$ does not converge. Therefore the converse to Theorem 10.9 does not hold. As the above example illustrates, a convergent sequence depends upon there being a point in the space to which the sequence converges. Cauchy sequences do not share this dependence.

Definition 10.7 A metric space (S, d) is called *complete* if every Cauchy sequence in the space converges.

We saw back in Chap. 1 that the real line **R** with the usual metric is complete but that the subspace of rationals is not complete. Suppose (S, d) is discrete space. Then $\{x_n\}$ is Cauchy if and only if there is a natural number N such that $x_m = x_n$ whenever $m, n > N$. Now $\lim_{n \to \infty} x_n = x_{N+1}$, and so discrete space is complete. In particular, if $S = \{1, \frac{1}{2}, \frac{1}{3}, \frac{1}{4}, \ldots\}$ and d is the discrete metric then (S, d) is complete. On the other hand, S with the usual metric is not a complete metric space since here the sequence $\{1/n\}$ is a Cauchy sequence and, as noted above, it does not converge. It is interesting to observe that although these two metric spaces have exactly the same class of open sets only one of the spaces is complete.

Theorem 10.10 If (S, d) is a complete metric space and $\emptyset \neq A \subseteq S$ then the subspace (A, d) is complete if and only if A is closed (in the space (S, d)).

PROOF Suppose (A, d) is complete and let x_0 be a limit point of A. For each $n \in \mathbf{N}$ there is a point $y_n \in A \cap S_{1/n}(x_0)$. Clearly $\{y_n\}$ is a Cauchy sequence and $y_n \to x_0$. Since (A, d) is complete, it must follow that $x_0 \in A$. Therefore A is closed. Conversely, suppose A is a closed set in the complete metric space (S, d) and suppose $\{x_n\}$ is a Cauchy sequence in (A, d). Then $\{x_n\}$ is a Cauchy sequence in (S, d) and necessarily converges (in (S, d)). Thus there is an $x_0 \in S$ such that $x_n \to x_0$. But then $x_0 \in \bar{A}$ (since $x_n \in A$ for all n); and since A is closed, $x_0 \in A$. Consequently, the Cauchy sequence $\{x_n\}$ converges in (A, d), and so (A, d) is complete.

We revisit two metric spaces discussed in Sec. 10.1.

Example 10.10 Let $\mathcal{B}[0, 1]$ be the set of all bounded real-valued functions defined on $[0, 1]$ and let

$$d^*(f, g) = \sup_{x \in [0, 1]} |f(x) - g(x)| \qquad \text{for all } f, g \in \mathcal{B}[0, 1].$$

Then $(\mathcal{B}[0, 1], d^*)$ is a metric space (see Example 10.7). We show that $(\mathcal{B}[0, 1], d^*)$ is complete.

SOLUTION Let $\{f_n\}$ be a Cauchy sequence in $(\mathcal{B}[0, 1], d^*)$. Then for each $x \in [0, 1]$ we have

$$|f_n(x) - f_m(x)| \leq d^*(f_n, f_m).$$

It follows that $\{f_n(x)\}$ is a Cauchy sequence of real numbers. Since $\mathbf{R}$ is complete, $\{f_n(x)\}$ converges. We define

$$f(x) = \lim_{n \to \infty} f_n(x)$$

for each $x \in [0, 1]$. To show that $f \in \mathcal{B}[0, 1]$ note that there is a natural number N such that $d^*(f_m, f_n) < 1$ whenever $m, n > N$. Thus, $|f_m(x) - f_n(x)| < 1$ for all $x \in [0, 1]$ whenever $m, n > N$. Therefore if $n > N$, we have $|f_n(x) - f_{N+1}(x)| < 1$ for all $x \in [0, 1]$. Since f_{N+1} is bounded on $[0, 1]$ there is an $M > 0$ such that $|f_{N+1}(x)| \leq M$ for all $x \in [0, 1]$. Then if $n > N$, we have $|f_n(x)| < 1 + |f_{N+1}(x)| \leq 1 + M$ for all $x \in [0, 1]$. Since $f(x) = \lim_{n \to \infty} f_n(x)$, for each $x \in [0, 1]$ there is an $n > N$ with $|f_n(x) - f(x)| < 1$. Thus

$$|f(x)| \leq |f(x) - f_n(x)| + |f_n(x)| < 1 + (1 + M) = 2 + M.$$

Therefore $f(x)$ is bounded on $[0, 1]$. We show that $\{f_n\}$ converges to f in $(\mathcal{B}[0, 1], d^*)$. Let $\varepsilon > 0$ be given. Since $\{f_n\}$ is Cauchy, there is a natural number N with $d^*(f_n, f_m) < \varepsilon/2$ whenever $m, n > N$. Now for each $x \in [0, 1]$ choose $k > N$ such that $|f(x) - f_k(x)| < \varepsilon/2$. Then if $n > N$, we have

$$|f(x) - f_n(x)| \leq |f(x) - f_k(x)| + |f_k(x) - f_n(x)| < \varepsilon/2 + \varepsilon/2 = \varepsilon$$

Therefore $d^*(f, f_n) < \varepsilon$ whenever $n > N$, and so $(\mathcal{B}[0, 1], d^*)$ is complete.

Consider the subset $\mathcal{C}[0, 1]$ of $\mathcal{B}[0, 1]$ consisting of all continuous real-valued functions defined on $[0, 1]$. We show that this is a closed set. Let g be a limit point of $\mathcal{C}[0, 1]$. For each natural number n choose $f_n \in \mathcal{C}[0, 1] \cap S_{1/n}(g)$. Clearly $\{f_n\}$ converges to g. We show g is continuous on $[0, 1]$. Let $x_0 \in [0, 1]$ be arbitrary and let $\varepsilon > 0$ be given. Choose $n \in \mathbf{N}$ so large that $1/n < \varepsilon/3$. Then $d^*(f_n, g) < 1/n < \varepsilon/3$, and so $|f_n(x) - g(x)| < \varepsilon/3$ for every $x \in [0, 1]$. Since $f_n \in \mathcal{C}[0, 1]$, there is a $\delta > 0$ such that $|f_n(x) - f_n(x_0)| < \varepsilon/3$ whenever $x \in [0, 1]$ and $|x - x_0| < \delta$. Thus if $x \in [0, 1]$ and $|x - x_0| < \delta$, we have

$$|g(x) - g(x_0)| \leq |g(x) - f_n(x)| + |f_n(x) - f_n(x_0)| + |f_n(x_0) - g(x_0)|$$
$$< \frac{\varepsilon}{3} + \frac{\varepsilon}{3} + \frac{\varepsilon}{3} = \varepsilon$$

This establishes that g is continuous at x_0, and the arbitrariness of x_0 guarantees that $g \in \mathcal{C}[0, 1]$. Since g was an arbitrary limit point of $\mathcal{C}[0, 1]$, we have that $\mathcal{C}[0, 1]$ is a closed set in the complete metric space $(\mathcal{B}[0, 1], d^*)$. By Theorem 10.10, the subspace $(\mathcal{C}[0, 1], d^*)$ is a complete metric space.

It is interesting to note that the set of all polynomial functions

$$p(x) = a_0 x^n + a_1 x^{n-1} + \cdots + a_{n-1} x + a_n$$

where $a_j \in \mathbf{R}$ for $j = 0, 1, 2, \ldots, n$, is dense in the metric space $(\mathcal{C}[0, 1], d^*)$. This means that given any function $f(x)$ which is continuous on $[0, 1]$ and given any $\varepsilon > 0$ there is a polynomial $p(x)$ for which $d^*(p, f) < \varepsilon$. Thus $|p(x) - f(x)| < \varepsilon$ for every $x \in [0, 1]$. This remarkable result is known as the *Weierstrass approximation theorem*.†

Example 10.11 Here we consider the set $\mathcal{C}[0, 1]$ of all continuous real-valued functions defined on $[0, 1]$, and we define

$$\tilde{d}(f, g) = \int_0^1 |f(x) - g(x)|\, dx$$

for all $f, g \in \mathcal{C}[0, 1]$. Then $(\mathcal{C}[0, 1], \tilde{d})$ is a metric space (see Example

---

† We refer the interested reader to G. F. Simmons, *Introduction to Topology and Modern Analysis*, McGraw-Hill, New York, 1963 for its proof.

10.8). We show that $(\mathcal{C}[0, 1], \tilde{d})$ is not complete by noting the following sequence. Let $f_1(x) = 1$ for all $x \in [0, 1]$, and for $n \geq 2$ define

$$f_n(x) = \begin{cases} 0 & \text{if } 0 \leq x \leq \frac{1}{2} - \frac{1}{n} \\ nx + 1 - \frac{n}{2} & \text{if } \frac{1}{2} - \frac{1}{n} < x < \frac{1}{2} \\ 1 & \text{if } \frac{1}{2} \leq x \leq 1 \end{cases}$$

We leave it to the reader (see Exercise 10.41) to verify that $\{f_n\}$ is a Cauchy sequence in $(\mathcal{C}[0, 1], \tilde{d})$ but $\{f_n\}$ does not converge (see Fig. 10.1).

Theorem 10.11 If (S, d) is a complete metric space and $\emptyset \neq A_n \subseteq S$ is closed for $n = 1, 2, 3, \ldots$ and $A_1 \supseteq A_2 \supseteq A_3 \supseteq A_4 \supseteq \ldots$ with $\lim_{n \to \infty} d(A_n) = 0$ then $\bigcap_{n=1}^{\infty} A_n$ is a singleton set.

PROOF If $x, y \in \bigcap_{n=1}^{\infty} A_n$ then $d(x, y) \leq d(A_n)$ for every natural number n implies that $d(x, y) = 0$ and so $x = y$. Therefore $\bigcap_{n=1}^{\infty} A_n$ cannot contain two distinct points. It suffices to show that $\bigcap_{n=1}^{\infty} A_n$ is non-empty. Choose $x_n \in A_n$ for every $n \in \mathbf{N}$. Since $\lim_{n \to \infty} d(A_n) = 0$ and $x_k \in A_n$ for $k \geq n$, $\{x_n\}$ is a Cauchy sequence. But (S, d) is complete, and so there is an $x_0 \in S$ such that $\lim_{n \to \infty} x_n = x_0$. We show that $x_0 \in \bigcap_{n=1}^{\infty} A_n$. Let $n^* \in \mathbf{N}$ be arbitrary. Then $x_k \in A_{n^*}$ for $k \geq n^*$. It follows that $x_0 \in \bar{A}_{n^*}$. But A_{n^*} is closed, and so $x_0 \in A_{n^*}$. Since n^* was arbitrary, $x_0 \in \bigcap_{n=1}^{\infty} A_n$.

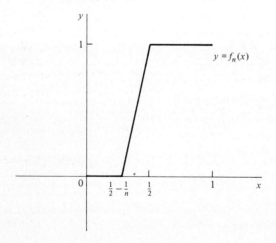

Figure 10.1

A set A in the metric space (S, d) is called *nowhere dense* in S if for every nonempty open set $G \subseteq S$ there is a nonempty open set $G^* \subseteq G$ such that $G^* \cap A = \emptyset$. It is clear that if A is nowhere dense in S and $B \subseteq A$ then B is nowhere dense in S. Equivalently, if A fails to be nowhere dense in S and $A \subseteq B$ then B fails to be nowhere dense in S.

Theorem 10.12 A is nowhere dense in S if and only if $S - \bar{A}$ is dense in S.

PROOF Assume that A is nowhere dense in S and let G be any nonempty open set. There is a nonempty open set $G^* \subseteq G$ such that $G^* \cap A = \emptyset$. It follows that $G^* \cap \bar{A} = \emptyset$. Why? Thus $G^* \subseteq S - \bar{A}$, and so $G \cap (S - \bar{A}) \neq \emptyset$. Since G was arbitrary, $S - \bar{A}$ intersects every nonempty open set and so is dense in (S, d).

Conversely, suppose $S - \bar{A}$ is dense in S. Then for every nonempty open set G, $G \cap (S - \bar{A}) \neq \emptyset$. Let $G^* = G \cap (S - \bar{A})$; clearly G^* is open, nonempty, and contained in G. Moreover $G^* \subseteq S - \bar{A} \subseteq S - A$, and so $G^* \cap A = \emptyset$. Hence A is nowhere dense in S.

Theorem 10.13 Let (S, d) be a complete metric space. If $\{A_n\}$ is a sequence of nowhere dense sets in S then $S - \bigcup_{n=1}^{\infty} A_n$ is dense in S.

PROOF We prove that $S - \bigcup_{n=1}^{\infty} A_n$ is dense in S by showing that this set has a point in common with every nonempty open set G. Since A_1 is nowhere dense in (S, d), there is a nonempty open set $G_1^* \subseteq G$ such that $G_1^* \cap A_1 = \emptyset$. Let $x_1 \in G_1^*$ and choose $r_1 > 0$ such that $S_{r_1}(x_1) \subseteq G_1^*$. Let $a_1 = \min(r_1/2, 1/2)$. Then $F_1 = S_{a_1}[x_1] \subseteq S_{r_1}(x_1) \subseteq G_1^* \subseteq G$, F_1 is a nonempty closed set, $F_1 \cap A_1 = \emptyset$, and $d(F_1) \leq 2a_1 \leq 1$. Since A_2 is nowhere dense in (S, d), there is a nonempty open set $G_2^* \subseteq S_{a_1}(x_1)$ such that $G_2^* \cap A_2 = \emptyset$. Let $x_2 \in G_2^*$ and choose $r_2 > 0$ such that $S_{r_2}(x_2) \subset G_2^*$. Let $a_2 = \min(r_2/2, 1/4)$. Then $F_2 = S_{a_2}[x_2] \subseteq S_{r_2}(x_2) \subseteq G_2^* \subseteq S_{a_1}(x_1) \subseteq F_1$. F_2 is a nonempty closed set, $F_2 \cap A_2 = \emptyset$, and $d(F_2) \leq 2a_2 \leq \tfrac{1}{2}$. Having chosen closed sets $F_1, F_2, \ldots, F_k$, we note that A_{k+1} is nowhere dense in (S, d) and so there is a nonempty open set $G_{k+1}^* \subseteq S_{a_k}(x_k)$ such that $G_{k+1}^* \cap A_{k+1} = \emptyset$. Let $x_{k+1} \in G_{k+1}^*$ and choose $r_{k+1} > 0$ such that $S_{r_{k+1}}(x_{k+1}) \subseteq G_{k+1}^*$. Let $a_{k+1} = \min(r_{k+1}/2, 1/2k)$. Then

$$F_{k+1} = S_{a_{k+1}}[x_{k+1}] \subseteq S_{r_{k+1}}(x_{k+1}) \subseteq G_{k+1}^* \subseteq S_{a_k}(x_k) \subseteq F_k$$

F_{k+1} is a nonempty closed set, $F_{k+1} \cap A_{k+1} = \emptyset$, and $d(F_{k+1}) \leq 2a_{k+1} \leq 1/k$. This gives a sequence $F_1 \supseteq F_2 \supseteq F_3 \supseteq \ldots$ of nonempty closed sets with $d(F_n) \leq 1/n$ and $F_n \cap A_n = \emptyset$ for $n = 1, 2, \ldots$. Since (S, d) is complete, it follows from Theorem 10.11 that $\bigcap_{n=1}^{\infty} F_n = \{x_0\}$ for some $x_0 \in S$. Clearly $x_0 \in G$ and $x_0 \notin A_n$ for $n = 1, 2, \ldots$. Therefore

$x_0 \in S - \bigcup_{n=1}^{\infty} A_n$, and so $G \cap (S - \bigcup_{n=1}^{\infty} A_n) \neq \emptyset$. It follows that $S - \bigcup_{n=1}^{\infty} A_n$ is dense in (S, d).

A set B in the metric space (S, d) is said to be of the *first category* if B can be written as the union of a sequence of sets, $B = \bigcup_{n=1}^{\infty} A_n$, where each A_n is nowhere dense in S. A set is said to be of the *second category* if it is not of the first category in (S, d). Note that in discrete space the only nowhere dense set is the empty set (see Exercise 10.46), and so every non-empty set is of the second category. In particular, the set **J** of integers is of the second category in the space **R** of real numbers imposed with the discrete metric. On the other hand, if we consider $(\mathbf{R}, d)$, where d is the usual euclidean metric, then the set **J** is nowhere dense and hence a set of the first category in this space. This illustrates that a set is not of the first or second category in and of itself; rather its category classification also depends on the metric space to which the set belongs. It is surprising to note that as a consequence of our next result, a useful tool in many existence proofs in mathematics, the set of integers **J** is a set of the second category in the complete metric space $(\mathbf{J}, d)$.

Theorem 10.14: Baire's theorem If (S, d) is a complete metric space then S is of the second category.

PROOF Suppose A_n $(n = 1, 2, 3, \ldots)$ is any sequence of nowhere dense sets. Then by Theorem 10.13, $S - \bigcup_{n=1}^{\infty} A_n$ is dense in S and so cannot be empty. Therefore $\bigcup_{n=1}^{\infty} A_n$ must be a proper subset of S, showing that it is impossible to write S as a union of a sequence of nowhere dense sets. Therefore, S must be of the second category.

EXERCISES

10.33 Prove that the limit of a convergent sequence $\{x_n\}$ in a metric space (S, d) is unique.

10.34 Suppose $\lim_{n \to \infty} x_n = x_0$ and $\lim_{n \to \infty} y_n = y_0$, where $\{x_n\}$ and $\{y_n\}$ are sequences in the metric space (S, d) and $x_0, y_0 \in S$. Prove that the sequence of real numbers $\{d(x_n, y_n)\}$ converges to the real number $d(x_0, y_0)$.

10.35 Prove that every Cauchy sequence $\{x_n\}$ in any metric space (S, d) is bounded.

10.36 Suppose that $\{x_n\}$ is a Cauchy sequence and $\{x_{n_k}\}$ is a convergence subsequence of $\{x_n\}$. Prove that $\{x_n\}$ converges.

10.37 Let A be a nonempty set in the metric space (S, d). Prove that $x_0 \in \bar{A}$ if and only if there is a sequence $\{x_n\}$ of points in A with $x_n \to x_0$.

10.38 Using the Weierstrass approximation theorem as stated in this section, prove that the set of all polynomial functions is dense in the space $(\mathcal{C}[a, b], d^*)$, where $[a, b]$ is any closed, bounded interval of real numbers, $\mathcal{C}[a, b]$ is the set of all continuous real-valued functions defined on $[a, b]$, and

$$d^*(f, g) = \sup_{x \in [a, b]} |f(x) - g(x)| \quad \text{for all } f, g \in \mathcal{C}[a, b]$$

10.39 Show that the set of all polynomials is not dense in the space $(\mathcal{B}[a, b], d^*)$ of all bounded real-valued functions defined on $[a, b]$.

10.40 Can an arbitrary function f continuous on $\mathbf{R}$ be approximated in the sense of the Weierstrass approximation theorem by a polynomial? Justify your answer.

10.41 Prove that the sequence $\{f_n\}$ from Example 10.11 is a Cauchy sequence in $(\mathcal{C}[0, 1], \tilde{d})$ but does not converge.

10.42 Give an example of a complete metric space (S, d) and a sequence of nonempty closed sets $\{A_n\}$ in S with $A_1 \supseteq A_2 \supseteq A_3 \supseteq \ldots$ such that $\bigcap_{n=1}^{\infty} A_n = \emptyset$.

10.43 Let x be an element in the metric space (S, d). Prove that $\{x\}$ is nowhere dense if and only if x is a limit point of S.

10.44 Prove that if x is not a limit point of S then $\{x\}$ is open. Such a point x is called an *isolated point* in the metric space (S, d).

10.45 Prove that if (S, d) is a complete metric space and each $x \in S$ is a limit point of S then S is uncountable.

10.46 Prove that if A is nowhere dense in discrete space then $A = \emptyset$.

10.47 Prove that a closed set in the metric space (S, d) either is nowhere dense in S or else contains some nonempty open set.

10.48 Prove that if G is an open set dense in the metric space (S, d) then $S - G$ is nowhere dense in S.

10.49 Prove that if G_n is an open set which is dense in the complete metric space (S, d) for $n = 1, 2, 3, \ldots$ then $\bigcap_{n=1}^{\infty} G_n$ is dense in S.

10.50 Use Exercise 10.47 and Baire's theorem to prove the following *uniform-boundedness principle*. If (S, d) is a complete metric space and if $\mathcal{F}$ is a family of real-valued continuous functions defined on S such that the set $\{f(x) | f \in \mathcal{F}\}$ is bounded for every $x \in S$ then there is a nonempty open set $G \subseteq S$ and an $M > 0$ such that $|f(x)| \le M$ for every $x \in G$ and for every $f \in \mathcal{F}$.

10.4 CONTINUITY AND COMPACTNESS

In this section we generalize the notion of a continuous real-valued function defined on $\mathbf{R}$ to functions from one metric space into another. Suppose (S, d) and (T, ρ) are metric spaces and $f: S \to T$.

Definition 10.8 f is said to be *continuous* at the point $x_0 \in S$ provided for every given $\varepsilon > 0$ there is a $\delta > 0$ such that $\rho(f(x), f(x_0)) < \varepsilon$ whenever $d(x, x_0) < \delta$.

Equivalently, f is continuous at x_0 if given any $\varepsilon > 0$ there is a $\delta > 0$ such that $f(S_\delta(x_0)) \subseteq S_\varepsilon(f(x_0))$; that is, every point lying in the open sphere of radius δ centered at x_0 has its image under f in the open sphere of radius ε centered at $f(x_0)$. A function $f: S \to T$ is said to be continuous on S if it is continuous at every point in S.

Example 10.12 If (S, d) is discrete space then every function $f: S \to T$ is continuous on S. For any $x_0 \in S$ if we choose $\delta < 1$ then $S_\delta(x_0) = \{x_0\}$, and so the condition $f(S_\delta(x_0)) \subseteq S_\varepsilon(f(x_0))$ holds for each positive ε.

Intuitively, f is continuous at x_0 if points close to x_0 in (S, d) are mapped to points close to $f(x_0)$ in (T, ρ). In discrete space the only point close to x_0 is x_0 itself. More generally, a point x_0 in a metric space (S, d) is called *isolated* if there is some $r > 0$ such that $S_r(x_0) = \{x_0\}$. It follows then that any function defined on S will be continuous at each isolated point x_0.

Theorem 10.15 f is continuous at x_0 if and only if for each sequence $\{x_n\}$ in S with $x_n \to x_0$ we have $f(x_n) \to f(x_0)$ in T.

PROOF Suppose f is continuous at x_0 and $\{x_n\}$ is a sequence in S with $x_n \to x_0$. Let $\varepsilon > 0$ be given. Then there is a $\delta > 0$ such that $f(S_\delta(x_0)) \subseteq S_\varepsilon(f(x_0))$. Since $x_n \to x_0$, there is a natural number N such that $x_n \in S_\delta(x_0)$ whenever $n > N$. But then $f(x_n) \in S_\varepsilon(f(x_0))$, and so $\rho(f(x_n), f(x_0)) < \varepsilon$ whenever $n > N$. This proves $f(x_n) \to f(x_0)$.

Conversely suppose f is not continuous at x_0. Then there exists an $\varepsilon_0 > 0$ for which $f(S_\delta(x_0)) \not\subseteq S_{\varepsilon_0}(f(x_0))$ for every $\delta > 0$. For every natural number n choose $x_n \in S_{1/n}(x_0)$ for which $f(x_n) \notin S_{\varepsilon_0}(f(x_0))$; that is, $\rho(f(x_n), f(x_0)) \geq \varepsilon_0$. Clearly $x_n \to x_0$ (since $d(x_n, x_0) < 1/n$ for $n = 1, 2, 3, \ldots$), but $\{f(x_n)\}$ does not converge to $f(x_0)$.

If x_0 is an isolated point in (S, d) and $x_n \to x_0$ then there is an $N \in \mathbf{N}$ such that $x_n = x_0$ whenever $n > N$. It follows that $f(x_n) = f(x_0)$ for $n > N$ and so $f(x_n) \to f(x_0)$.

Theorem 10.16 $f: S \to T$ is continuous on S if and only if for each open set $H \subseteq T$, $f^{-1}(H)$ is open in S.

PROOF Suppose that $f: S \to T$ is continuous on S and let H be an open set in T. If $f^{-1}(H) = \varnothing$, it is open; so assume $x_0 \in f^{-1}(H)$. Then $f(x_0) \in H$, and so there is an $\varepsilon > 0$ such that $S_\varepsilon(f(x_0)) \subseteq H$. Since f is continuous at x_0, there is a $\delta > 0$ with $f(S_\delta(x_0)) \subseteq S_\varepsilon(f(x_0))$. Then $f(S_\delta(x_0)) \subseteq H$, and so $S_\delta(x_0) \subseteq f^{-1}(H)$. Therefore $f^{-1}(H)$ is an open subset of S.

Conversely, suppose that $f^{-1}(H)$ is open in S for every open set H in T. Let $x_0 \in S$ and consider $S_\varepsilon(f(x_0))$. Then $f^{-1}(S_\varepsilon(f(x_0)))$ is open and contains x_0. Hence there is a $\delta > 0$ such that $S_\delta(x_0) \subseteq f^{-1}(S_\varepsilon(f(x_0)))$. Clearly $f(S_\delta(x_0)) \subseteq S_\varepsilon(f(x_0))$, and so f is continuous at x_0. Since $x_0 \in S$ was arbitrary, f is continuous on S.

A function $f: S \to T$ which is one-to-one and onto T is called a *homeomorphism* provided both f and f^{-1} are continuous on S and T, respectively. By Theorem 10.16, if $f: S \to T$ is a homeomorphism then the preimage of every open set H in T is open in S and the image of every open set G in S is open in T. Briefly, a homeomorphism induces a one-to-one correspondence between the open sets in S and the open sets in T. If there exists a homeomorphism between two metric spaces (S, d) and (T, ρ) then the spaces are topologically equivalent. Such spaces are called *homeomorphic*. But not all metric properties are shared by homeomorphic spaces, as the following example shows.

Example 10.13 Let $S = \{1, \frac{1}{2}, \frac{1}{3}, \frac{1}{4}, \ldots\}$ with d the usual metric on subsets of $\mathbf{R}$ and let $T = \mathbf{N}$ with ρ the usual metric. Then the function $f: S \to T$ defined by $f(1/n) = n$ is a homeomorphism from (S, d) to (T, ρ); in fact, every subset of S and every subset of T is open in the respective spaces. But (S, d) is a bounded metric space and (T, ρ) is not. Also, (T, ρ) is a complete metric space but (S, d) is not.

A function $f: S \to T$ which satisfies $d(x_1, x_2) = \rho(f(x_1), f(x_2))$ for every pair of points $x_1, x_2 \in S$ is called an *isometry*. It is clear that if f is an isometry then necessarily f is one-to-one. If, in addition, f is onto T then the spaces (S, d) and (T, ρ) are called *isometric*. We ask the reader to verify that an isometry f from S onto T is also a homeomorphism (see Exercise 10.57). Therefore, if two spaces are isometric then necessarily they are homeomorphic. Example 10.13 illustrates that spaces (S, d) and (T, ρ) can be homeomorphic without being isometric. Basically, isometric spaces possess all the same metric properties. Such spaces are metrically identical and really differ only in the names of their elements.

Definition 10.9 A function $f: S \to T$ is called *uniformly continuous* if given any $\varepsilon > 0$ there is a $\delta > 0$ such that $\rho(f(x_1), f(x_2)) < \varepsilon$ whenever $d(x_1, x_2) < \delta$.

Every function $f: S \to T$ which is uniformly continuous on S is necessarily continuous on S. We note that the function from Example 10.13 is not uniformly continuous and so homeomorphisms are not in general uniformly continuous functions. We ask the reader to verify that isometries are always uniformly continuous functions (see Exercise 10.63). It is also clear from Example 10.13 that the continuous image of a Cauchy sequence may fail to be a Cauchy sequence. The following theorem shows that this will not occur if the function is uniformly continuous.

Theorem 10.17 If $f: S \to T$ is uniformly continuous and $\{x_n\}$ is a Cauchy sequence in S then $\{f(x_n)\}$ is a Cauchy sequence in T.

PROOF Let $\varepsilon > 0$ be given. Then there is a $\delta > 0$ such that $\rho(f(u), f(v)) < \varepsilon$ whenever $d(u, v) < \delta$. Since $\{x_n\}$ is a Cauchy sequence in S, there is a natural number N such that $d(x_m, x_n) < \delta$ whenever

$m, n > N$. Hence $\rho(f(x_m), f(x_n)) < \varepsilon$ for all $m, n > N$. It follows that $\{f(x_n)\}$ is a Cauchy sequence in T.

We say that a family of sets $\mathcal{G}$ in the metric space (S, d) (each element of $\mathcal{G}$ is a subset of S) is a *cover* of the set $A \subseteq S$ provided each element in A is a member of at least one of the sets in $\mathcal{G}$. If in addition each member of $\mathcal{G}$ is an open set in (S, d) then $\mathcal{G}$ is called an *open cover* of the set A. If $\mathcal{G}_1 \subseteq \mathcal{G}$, where $\mathcal{G}$ is a cover of A, and if $\mathcal{G}_1$ is also a cover of A then we call $\mathcal{G}_1$ a *subcover*.

Definition 10.10 A set K in the metric space (S, d) is called *compact* if for each open cover $\mathcal{G}$ of K there is a finite set $\mathcal{G}_1 \subseteq \mathcal{G}$ which also is a cover of the set K.

A set is then compact if each open cover of the set admits a finite subcover. We call the metric space (S, d) compact provided the set S is compact. If $\emptyset \neq A \subseteq S$ then the open sets in the subspace (A, d) are precisely the sets of the form $G \cap A$, where G is open in (S, d) (see Exercise 10.32). Therefore A is compact if and only if (A, d) is a compact metric space.

Suppose A is a nonclosed set in the metric space (S, d) and let x_0 be a limit point of A which fails to be in A. Then the family of sets $\mathcal{G} = \{S - S_{1/n}[x_0] \mid n = 1, 2, \ldots\}$ is an open cover of A for which there is no finite subcover. Hence A cannot be compact. Suppose B is an unbounded set in the metric space (S, d). For any fixed $x_0 \in S$ the family of open spheres $\{S_n(x_0) \mid n = 1, 2, \ldots\}$ is an open cover of B for which there is no finite subcover. Thus B is not compact. We record these two results in the following theorem.

Theorem 10.18 If K is a compact set in the metric space (S, d) then K is both closed and bounded.

The Heine-Borel theorem tells us that in any euclidean space the converse of this theorem is also true; that is, if K is a closed, bounded subset of $\mathbf{R}^n$ then K is compact.

Example 10.14 Let $S = [0, 1]$ and suppose that (S, d) is discrete space. Then the set $A = \{1, \frac{1}{2}, \frac{1}{3}, \ldots\}$ is closed (why?) and bounded, but the open cover $\mathcal{G} = \{S_1(1/n) = \{1/n\} \mid n = 1, 2, \ldots\}$ shows that A is not compact.

Theorem 10.19 If (S, d) is a compact metric space and $F \subseteq S$ is closed then F is compact.

PROOF Suppose that $\mathcal{G}$ is an open cover of F. Then $\mathcal{G} \cup \{S - F\}$ is an open cover of S and by compactness of S there is a finite set $\mathcal{G}_1 \subseteq \mathcal{G} \cup \{S - F\}$ which covers S. Clearly $\mathcal{G}_1 - \{S - F\}$ is a finite subset of $\mathcal{G}$ which covers F, and so F is compact.

The above theorem shows that each closed subset of a compact metric space is compact. On the real line the reader should verify that the closed set $\mathbf{N}$ is not compact in $\mathbf{R}$. The closed set $F = (0, 1]$ is not compact in the metric space (S, d), where $S = (0, 2)$ and d is the usual metric. These fail of course because in each case the space is not compact. The next theorem shows that the continuous image of a compact metric space is again compact.

Theorem 10.20 If (S, d) is a compact metric space and f is a continuous function from S into the metric space (T, ρ) then $f(S)$, the range of f, is a compact set in T.

PROOF Let $\mathcal{G}$ be an open cover of $f(S)$. Then $\tilde{\mathcal{G}} = \{f^{-1}(G) \mid G \in \mathcal{G}\}$ is an open cover of S. By compactness of S there exists a finite number of sets $f^{-1}(G_i) \in \tilde{\mathcal{G}}$ ($i = 1, 2, \ldots, n$) which covers S. But then the sets G_i ($i = 1, 2, \ldots, n$) cover $f(S)$, and so $f(S)$ is compact.

The next property of a metric space will be shown to be equivalent to compactness (though it is not equivalent in more general types of spaces), but it is often more useful in establishing some of the properties that metric spaces may possess. It translates the cumbersome open-cover definition of compactness into a statement about sequences in the space.

Definition 10.11 A metric space (S, d) is called *sequentially compact* if each sequence $\{x_n\}$ in S has a convergent subsequence.

Lemma 1 If every infinite set $A \subseteq S$ has a limit point, then (S, d) is sequentially compact.

PROOF Suppose $\{x_n\}$ is an arbitrary sequence in (S, d). If the set $A = \{x_n \mid n \in \mathbf{N}\}$ is finite then $\{x_n\}$ has a constant (and hence convergent) subsequence. Consequently we may assume that A is infinite. Then by hypothesis A has a limit point x_0. Therefore, for each $r > 0$ the set $A \cap S_r(x_0)$ is infinite. Choose $x_{n_1} \in A \cap S_1(x_0)$; having chosen $x_{n_1}, x_{n_2}, \ldots, x_{n_k}$, choose $x_{n_{k+1}} \in A \cap S_{1/(k+1)}(x_0)$ with $n_{k+1} > n_k$. Clearly the subsequence $\{x_{n_k}\}$ of $\{x_n\}$ converges to x_0, and so (S, d) is sequentially compact.

The converse of Lemma 1 (which we shall not need) is also true, and so in metric spaces sequential compactness is equivalent to the condition that every infinite subset of the space has a limit point in the space.

Theorem 10.21 Every compact metric space (S, d) is sequentially compact.

PROOF Suppose A is an infinite subset of S which has no limit point in S. Then for each $x \in A$ there is an $r_x > 0$ such that $S_{r_x}(x) \cap A = \{x\}$; otherwise x would be a limit point of A. Clearly the family of sets $\{S_{r_x}(x) | x \in A\} \cup \{S - A\}$ is an open cover of S which admits no finite subcover, and this contradicts the compactness of S. Hence A must have a limit point in S, and so by the previous lemma (S, d) is sequentially compact.

It is now immediate that every compact metric space is complete, since if $\{y_n\}$ is a Cauchy sequence in the compact metric space (S, d) then $\{y_n\}$ necessarily has a convergent subsequence and so must itself converge. Therefore compactness implies completeness. On the other hand, infinite discrete space furnishes us with an example of a complete metric space which is not compact. Hence compactness is a stronger property than completeness. In order to determine which among the complete metric spaces are compact we note that simple boundedness is not enough. Again, infinite discrete space is bounded. We consider the following stronger condition.

For each positive real number ε an ε *net* in the metric space (S, d) is a nonempty finite set $H \subseteq S$ such that for any $x \in S$ there is a point $a \in H$ with $d(x, a) < \varepsilon$. In other words, each point in the space comes within ε distance of one of the points in H. It is easy to show that a space is bounded if and only if it has some ε net.

Definition 10.12 A metric space (S, d) is called *totally bounded* if it has an ε net for every positive ε.

Clearly every totally bounded metric space is bounded. Infinite discrete space is bounded but not totally bounded; it has no $\frac{1}{2}$ net. We shall see shortly that it is this stronger form of boundedness, called total boundedness, which is the property that complements completeness to guarantee compactness. The next theorem characterizes total boundedness in terms of sequences in the space, but first we need the following lemma.

Lemma 2 If A is an infinite subset in the totally bounded metric space (S, d) then for any $\varepsilon > 0$ there is an infinite set $B \subseteq A$ with $d(B) < \varepsilon$.

PROOF Suppose A is infinite and $\varepsilon > 0$ is given. Assume that

$H = \{x_1, x_2, \ldots, x_n\}$ is an $\varepsilon/3$ net in (S, d). Then $S = \bigcup_{j=1}^n S_{\varepsilon/3}(x_j)$, and so $A = \bigcup_{j=1}^n (A \cap S_{\varepsilon/3}(x_j))$. Thus at least one of the sets $A \cap S_{\varepsilon/3}(x_j)$, call it B, must be infinite. Clearly $B \subseteq A$ and $d(B) < \varepsilon$.

Theorem 10.22 A metric space (S, d) is totally bounded if and only if each sequence $\{x_n\}$ in S has a Cauchy subsequence.

PROOF First suppose that (S, d) is totally bounded and let $\{x_n\}$ be any sequence in S. If the set $A = \{x_n \mid n \in \mathbf{N}\}$ is finite then $\{x_n\}$ has a constant (and hence Cauchy) subsequence. We assume that A is infinite. By Lemma 2 there is an infinite subset $B_1 \subseteq A$ with $d(B_1) < 1$. Choose n_1 such that $x_{n_1} \in B_1$. Again by Lemma 2 there is an infinite set $B_2 \subseteq B_1$ with $d(B_2) < \frac{1}{2}$. Choose $n_2 > n_1$ such that $x_{n_2} \in B_2$. Having chosen infinite sets $B_k \subseteq B_{k-1} \subseteq \cdots \subseteq B_2 \subseteq B_1 \subseteq A$ with $d(B_j) < 1/j$ for $j = 1, 2, \ldots, k$ and positive integers $n_k > n_{k-1} > \cdots > n_2 > n_1$ with $x_{n_j} \in B_j$ $(j = 1, 2, \ldots, k)$, let B_{k+1} be an infinite subset of B_k with $d(B_{k+1}) < 1/(k+1)$ and choose $n_{k+1} > n_k$ such that $x_{n_{k+1}} \in B_{k+1}$. In this way we obtain a subsequence $\{x_{n_k}\}$ of the sequence $\{x_n\}$. To show that $\{x_{n_k}\}$ is Cauchy let $\varepsilon > 0$ be given. Choose k_0 so large that $1/k_0 < \varepsilon$. By our construction $x_{n_m} \in B_{k_0}$ for $m \geq k_0$, and so if $j, m > k_0$ then $d(x_{n_j}, x_{n_m}) < 1/k_0 < \varepsilon$. This proves that each sequence in (S, d) has a Cauchy subsequence.

Conversely, suppose that (S, d) is not totally bounded. Then there is an $\varepsilon_0 > 0$ for which there is no ε_0 net in S. Let $x_1 \in S$ be arbitrary; choose $x_2 \in S$ such that $d(x_2, x_1) \geq \varepsilon_0$. This is possible since the set $\{x_1\}$ cannot be an ε_0 net in S. Choose $x_3 \in S$ such that $d(x_3, x_1) \geq \varepsilon_0$ and $d(x_3, x_2) \geq \varepsilon_0$; again this is possible since $\{x_1, x_2\}$ is not an ε_0 net. Having obtained the points $\{x_1, x_2, \ldots, x_k\}$ with the property that $d(x_i, x_j) \geq \varepsilon_0$ for all $i \neq j$, where $i, j = 1, 2, \ldots, k$, we note that $\{x_1, x_2, \ldots, x_k\}$ cannot be an ε_0 net in S. Thus there is an $x_{k+1} \in S$ with $d(x_{k+1}, x_j) \geq \varepsilon_0$ for $j = 1, 2, \ldots, k$. Clearly the sequence $\{x_n\}$ has no Cauchy subsequence.

The subspace $S = (0, 1)$ of the real line is totally bounded but certainly not sequentially compact, for consider the sequence $\{1/n\}$. By Theorem 10.10, the complete subspaces (S, d) of the real line are precisely those for which S is a closed subset of $\mathbf{R}$. Thus $(0, 1)$ is not complete.

Theorem 10.23 The metric space (S, d) is sequentially compact if and only if it is complete and totally bounded.

PROOF If (S, d) is sequentially compact then each sequence $\{x_n\}$ in S has a convergent subsequence. Since this subsequence is evidently Cauchy, every sequence in (S, d) has a Cauchy subsequence and so by Theorem

10.22 (S, d) is totally bounded. If $\{y_n\}$ is a Cauchy sequence in (S, d) then since it has a convergent subsequence it must itself converge. Hence (S, d) is complete.

Conversely, suppose (S, d) is complete and totally bounded and let $\{x_n\}$ be any sequence in (S, d). By total boundedness $\{x_n\}$ has a Cauchy subsequence $\{x_{n_k}\}$ and by completeness $\{x_{n_k}\}$ converges. Therefore (S, d) is sequentially compact.

Corollary Every compact metric space is complete and totally bounded.

PROOF Follows from Theorems 10.21 and 10.23.

In order to show that each sequentially compact metric space is compact we need the notion of a Lebesgue number for an open covering.

Definition 10.13 Let $\mathcal{G}$ be an open cover of the metric space (S, d). A *Lebesgue number* for $\mathcal{G}$ is a number $\eta > 0$ such that if $A \subseteq S$ with $d(A) < \eta$ then there is a $G \in \mathcal{G}$ with $A \subseteq G$.

Not every open cover of a metric space has a Lebesgue number. For example, let $S = (0, 1)$ be a subspace of the real line and let $\mathcal{G} = \{(1/n, 1) | n = 2, 3, 4, \ldots\}$. For arbitrary $\eta > 0$ it is clear that the set $A = (0, \eta/2)$ satisfies $d(A) < \eta$ but A is not contained in any one member of $\mathcal{G}$.

Lemma 3 If (S, d) is a sequentially compact metric space then every open cover $\mathcal{G}$ of S has a Lebesgue number.

PROOF Suppose (S, d) is sequentially compact and let $\mathcal{G}$ be any open cover of S. If there is no Lebesgue number for $\mathcal{G}$ then for each natural number n there is a nonempty set $A_n \subseteq S$ with $d(A_n) < 1/n$ such that A_n is contained in no member of S. Choose $x_n \in A_n$ for $n = 1, 2, \ldots$; since (S, d) is sequentially compact, $\{x_n\}$ has a convergent subsequence $\{x_{n_k}\}$. Let $x_{n_k} \to x_0$ as $k \to \infty$ and choose $G_0 \in \mathcal{G}$ with $x_0 \in G_0$. Since G_0 is an open set, there is an $r > 0$ with $S_r(x_0) \subseteq G_0$, and since $\{x_{n_k}\}$ converges to x_0, x_{n_k} is contained in $S_{r/2}(x_0)$ for infinitely many natural numbers n_k. Choose $n_{k'}$ so large that $x_{n_{k'}} \in S_{r/2}(x_0)$ and $1/n_{k'} < r/2$. Now if $y \in A_{n_{k'}}$ then $d(y, x_0) < 1/n_{k'}$ (recall that $d(A_{n_{k'}}) < 1/n_{k'}$), and so $d(y, x_0) < r/2$. But this says that $y \in S_{r/2}(x_0)$. Thus $A_{n_{k'}} \subseteq S_{r/2}(x_0)$. But then $A_{n_{k'}} \subseteq S_r(x_0) \subseteq G_0$, and this contradicts the fact that for each natural number n, A_n is contained in no member of $\mathcal{G}$. It follows that $\mathcal{G}$ must have a Lebesgue number.

We are finally ready to prove the converse to Theorem 10.21 and establish that in metric spaces the notions of compactness and sequential compactness are equivalent.

Theorem 10.24 If (S, d) is a sequentially compact metric space then (S, d) is compact.

PROOF Let $\mathcal{G}$ be any open cover of S. By Lemma 3, $\mathcal{G}$ has a Lebesgue number $\eta > 0$, and by Theorem 10.23, (S, d) is totally bounded, and so there is an $\eta/3$ net $\{x_1, x_2, \ldots, x_n\}$. Now for each $k = 1, 2, \ldots, n$, $d(S_{\eta/3}(x_k)) \leq \frac{2}{3}\eta < \eta$, and so there is a $G_k \in \mathcal{G}$ with $S_{\eta/3}(x_k) \subseteq G_k$. Since $\bigcup_{k=1}^{\infty} S_{\eta/3}(x_k) = S$, it follows that $\{G_1, G_2, \ldots, G_n\}$ covers S, and so (S, d) is compact.

Corollary A metric space is compact if and only if it is sequentially compact.

Theorem 10.25 A metric space (S, d) is compact if and only if (S, d) is complete and totally bounded.

PROOF Follows from Theorem 10.23 and the above corollary.

EXERCISES

10.51 Prove that $f: S \to T$ is a continuous mapping if and only if for each closed set F in T, $f^{-1}(F)$ is closed in S.

10.52 Prove that if $f, g:(S, d) \to (T, \rho)$ are each continuous on S and $f(x) = g(x)$ for every $x \in A \subseteq S$ then $f(x) = g(x)$ for all $x \in \bar{A}$.

10.53 Let $S = [0, 1]$, d be the usual metric, and ρ be the discrete metric. Is the identity function i_s on S continuous? *Hint*: There are two answers.

10.54 Give an example of a function $f: S \to T$ which is one-to-one, onto T, and continuous on S but not a homeomorphism.

10.55 Give an example of a homeomorphism $f: S \to T$ and a Cauchy sequence $\{x_n\}$ in S for which $\{f(x_n)\}$ is not Cauchy in T.

10.56 Prove that every metric space is homeomorphic to a bounded metric space.

10.57 Prove that if the spaces (S, d) and (T, ρ) are isometric then they are homeomorphic.

10.58 Prove that if $f: S \to T$ is an isometry from S onto T then f^{-1} is an isometry from T onto S.

10.59 Prove that if $f: S \to T$ is an isometry from S onto T then for every Cauchy sequence $\{x_n\}$ in S, $\{f(x_n)\}$ is a Cauchy sequence in T.

10.60 Prove that if the spaces (S, d) and (T, ρ) are isometric then either they are both complete or neither is complete.

10.61 Let (S, d) be a metric space with $x_0 \in S$. Define $f: S \to \mathbf{R}$ by $f(x) = d(x, x_0)$. Prove that f is uniformly continuous on S.

10.62 Prove that the function in Example 10.13 is not uniformly continuous. Is its inverse uniformly continuous?

10.63 Prove that if $f: S \to T$ is an isometry from S onto T then f is uniformly continuous.

10.64 Suppose that every open set in the metric space (S, d_1) is also open in the space (S, d_2). Prove that the identity function $i_s : (S, d_2) \to (S, d_1)$ is continuous but need not be uniformly continuous.

10.65 Suppose that (S, d_1) and (S, d_2) are metric spaces and there is a number $k > 0$ such that $d_1(x, y) \leq k \cdot d_2(x, y)$ for every pair $x, y \in S$. Prove that the identity $i_s : (S, d_2) \to (S, d_1)$ is uniformly continuous.

10.66 Prove that if (S, d) and (T, ρ) are metric spaces with (T, ρ) complete and if $f: A \to T$ is uniformly continuous on $A \subset S$ then f has a unique extension to a uniformly continuous function $\bar{f}: \bar{A} \to T$. *Hint*: Use Theorem 10.17.

10.67 Let (S, d) be any metric space and let A and B be subsets of S. Prove that if A is closed, B is compact, and $d(A, B) = 0$ then $A \cap B \neq \emptyset$.

10.68 Let $S = \{1, \frac{1}{2}, \frac{1}{3}, \ldots\}$ and let d be the euclidean metric. Show that the set $A = \{1, \frac{1}{3}, \frac{1}{5}, \ldots\}$ is a closed, bounded set in (S, d) but that A is not compact. Explain why this does not contradict the Heine-Borel theorem.

10.69 A class of sets in a metric space (S, d) is said to have the *finite-intersection property* if every finite subclass has nonempty intersection. Prove that the space (S, d) is compact if and only if every class of closed sets in (S, d) with the finite-intersection property has a nonempty intersection.

10.70 Prove the converse of Lemma 1: If (S, d) is sequentially compact then every infinite set $A \subseteq S$ has a limit point.

10.71 Prove that a metric space (S, d) is bounded if and only if it has an ε net for some $\varepsilon > 0$.

10.72 Prove that boundedness and total boundedness are equivalent in euclidean space.

10.73 Prove that every compact metric space is separable.

10.74 Let (S, d) be a compact metric space and let (T, ρ) be any metric space. Prove that if $f: S \to T$ is one-to-one, continuous, and onto T then f is a homeomorphism.

10.75 The family $\mathcal{F}$ of functions from the metric space (S, d) to the metric space (T, ρ) is called *equicontinuous* on S if given any $\varepsilon > 0$ there is a $\delta > 0$ such that for every $f \in \mathcal{F}$, $\rho(f(x_1), f(x_2)) < \varepsilon$ whenever $d(x_1, x_2) < \delta$. Prove that if (S, d) is a compact metric space and the sequence $f_n: S \to T$ is equicontinuous on S and $f(x) = \lim_{n \to \infty} f_n(x)$ for each $x \in S$ then the sequence $\{f_n\}$ converges uniformly to f on S.

10.76 Prove Ascoli's theorem: A closed subspace $\mathcal{F}$ of $\mathcal{C}[0, 1]$ is compact if and only if $\mathcal{F}$ is equicontinuous and uniformly bounded.

10.5 APPLICATIONS

We begin with an application of Baire's theorem and show that there is an element in $\mathcal{C}[0, 1]$ which is nowhere differentiable. Recall that $\mathcal{C}[0, 1]$ is the set of all real-valued functions defined and continuous on the closed unit interval $[0, 1]$. We define a metric on $\mathcal{C}[0, 1]$ by

$$d(f, g) = \sup_{x \in [0, 1]} |f(x) - g(x)| \quad \text{for all } f, g \in \mathcal{C}[0, 1]$$

It was proved in Sec. 10.3 that $(\mathcal{C}[0, 1], d)$ is a complete metric space. By Baire's theorem, $\mathcal{C}[0, 1]$ is a set of the second category. To demonstrate the existence of a function $\psi \in \mathcal{C}[0, 1]$ which at no point in $[0, 1]$ has a derivative we show that the set of all functions $f \in \mathcal{C}[0, 1]$ which are differentiable at some point in $[0, 1]$ is a set of the first category in $\mathcal{C}[0, 1]$.

Now $f(x)$ is differentiable at $x_0 \in [0, 1]$ if

$$\lim_{h \to 0} \frac{f(x_0 + h) - f(x_0)}{h}$$

exists, where we consider only values of h for which $x_0 + h \in [0, 1]$. Since $f(x)$ is bounded on $[0, 1]$, it is clear that if the above limit exists then the difference quotient $[f(x_0 + h) - f(x_0)]/h$ must be bounded for all values of h with $x_0 + h \in [0, 1]$. Hence there is a natural number n for which

$$\left| \frac{f(x_0 + h) - f(x_0)}{h} \right| \leq n$$

whenever $h > 0$ and $x_0 + h \in [0, 1]$.

We define C_n to be the subset of $\mathcal{C}[0, 1]$ consisting of all functions f for which there is an $x_0 \in [0, 1]$ such that

$$\left| \frac{f(x_0 + h) - f(x_0)}{h} \right| \leq n$$

for every $h > 0$ with $x_0 + h \in [0, 1]$. Our remarks above then indicate that the set D of all functions $f \in \mathcal{C}[0, 1]$ which are differentiable at some $x_0 \in [0, 1]$ satisfies $D \subseteq \bigcup_{n=1}^{\infty} C_n$. We show that each set C_n is nowhere dense in $\mathcal{C}[0, 1]$. Let $\varepsilon > 0$ be given and let $g \in \mathcal{C}[0, 1]$ be arbitrary. Since $g(x)$ is uniformly continuous on $[0, 1]$, we can construct a function $k(x)$ whose graph is a broken line each of whose segments has slope greater in absolute value than n and which satisfies $|k(x) - g(x)| < \varepsilon$ for all $x \in [0, 1]$. Then $(\mathcal{C}[0, 1] - C_n) \cap S_\varepsilon(g) \neq \emptyset$. Therefore $\mathcal{C}[0, 1] - C_n$ is dense in $\mathcal{C}[0, 1]$. We next show that C_n is a closed set in $\mathcal{C}[0, 1]$. Let f be a limit point of C_n. Then there is a sequence $f_k \in C_n$ $(k = 1, 2, 3, \ldots)$ such that $d(f_k, f) < 1/k$. For each natural number k there is a point $x_k \in [0, 1]$ such that

$$\left| \frac{f_k(x_k + h) - f_k(x_k)}{h} \right| \leq n$$

for every $h > 0$ with $x_k + h \in [0, 1]$. We can assume without loss of generality that $x_k \to x^* \in [0, 1]$. (By the Bolzano-Weierstrass theorem the bounded sequence $\{x_k\}$ has a convergent subsequence and the limit x^* necessarily lies in $[0, 1]$.) We show that $f \in C_n$ by verifying that

$$\left| \frac{f(x^* + h) - f(x^*)}{h} \right| \leq n$$

for every $h > 0$ with $x^* + h \in [0, 1]$. Suppose $h > 0$ and $x^* + h \in [0, 1]$, and $\varepsilon > 0$ is arbitrary. Since $d(f_k, f) < 1/k$, there is a natural number N such that $|f(x) - f_k(x)| < \varepsilon h/4$ for every $x \in [0, 1]$ whenever $k \geq N$ (choose N so that $1/N < \varepsilon h/4$). Choose $k_0 \geq N$ such that $|f(x^*) - f(x_{k_0})| < \varepsilon h/4$ and $|f_{k_0}(x^* + h) - f_{k_0}(x_{k_0} + h)| < \varepsilon h/4$. This is possible by the continuity of the functions and the fact that $d(f_k, f) \to 0$ (see Exercise 10.77). Then we have

$$\left| \frac{f(x^* + h) - f(x^*)}{h} \right|$$

$$\leq \left| \frac{f(x^* + h) - f_{k_0}(x^* + h)}{h} \right| + \left| \frac{f_{k_0}(x^* + h) - f_{k_0}(x_{k_0} + h)}{h} \right|$$

$$+ \left| \frac{f_{k_0}(x_{k_0} + h) - f_{k_0}(x_{k_0})}{h} \right| + \left| \frac{f_{k_0}(x_{k_0}) - f(x_{k_0})}{h} \right| + \left| \frac{f(x_{k_0}) - f(x^*)}{h} \right|$$

$$< \frac{\varepsilon}{4} + \frac{\varepsilon}{4} + n + \frac{\varepsilon}{4} + \frac{\varepsilon}{4} = \varepsilon + n$$

Since $\varepsilon > 0$ was arbitrary,

$$\left| \frac{f(x^* + h) - f(x^*)}{h} \right| \leq n$$

Therefore $f \in C_n$, and so C_n is closed. Hence $\mathcal{C}[0, 1] - C_n$ is an open set which is dense in $\mathcal{C}[0, 1]$. It follows from Theorem 10.12 that C_n is nowhere dense in $\mathcal{C}[0, 1]$. Thus $\bigcup_{n=1}^{\infty} C_n$ is of the first category in $\mathcal{C}[0, 1]$, and so by Baire's theorem there exists a function $\psi \in \mathcal{C}[0, 1]$ such that $\psi \notin \bigcup_{n=1}^{\infty} C_n$. Since the set D of all functions $f \in \mathcal{C}[0, 1]$ which have a derivative at some point in $[0, 1]$ satisfies $D \subseteq \bigcup_{n=1}^{\infty} C_n$, it follows that $\psi \notin D$. Therefore ψ is a continuous function on $[0, 1]$ and ψ is nowhere differentiable. Actually there is an infinite set of such functions. If there were only a finite set E of such functions, $E \cup (\bigcup_{n=1}^{\infty} C_n)$ would be of the first category and still unequal to $\mathcal{C}[0, 1]$. Hence there is an infinite variety of continuous functions which have a derivative nowhere and which we therefore cannot visualize or graph.

As a second application, we prove an existence-uniqueness theorem of differential equations. First we need the notion of a contraction mapping.

Definition 10.14 A function $\varphi: (Y, d) \to (Y, d)$ is called a *contraction mapping* if there is a real number θ ($0 < \theta < 1$) such that

$$d(\varphi(y_1), \varphi(y_2)) \leq \theta \, d(y_1, y_2) \qquad \text{for all } y_1, y_2 \in Y$$

Theorem 10.26 If $\varphi: (Y, d) \to (Y, d)$ is a contraction mapping and if (Y, d) is a complete metric space then φ has a unique fixed point; that is, there is a unique $y^* \in Y$ such that $\varphi(y^*) = y^*$.

PROOF Define the sequence $y_{i+1} = \varphi(y_i)$, 1, 2, ..., for any $y_1 \in Y$.

$$d(y_2, y_3) = d(\varphi(y_1), \varphi(y_2)) \leq \theta d(y_1, y_2)$$

and

$$d(y_n, y_{n+1}) = d(\varphi(y_{n-1}), \varphi(y_n)) \leq \theta d(y_{n-1}, y_n)$$
$$\leq \theta^2 d(y_{n-2}, y_{n-1})$$
$$\cdots\cdots\cdots\cdots$$
$$\leq \theta^{n-1} d(y_1, y_2)$$

for any natural number $n > 1$. To show that $\{y_n\}$ is a Cauchy sequence let m, n be fixed natural numbers with $m < n$; then

$$d(y_m, y_n) \leq d(y_m, y_{m+1}) + d(y_{m+1}, y_{m+2}) + \cdots + d(y_{n-1}, y_n)$$
$$\leq (\theta^{m-1} + \theta^m + \cdots + \theta^{n-2}) d(y_1, y_2)$$
$$\leq \theta^{m-1} \frac{1}{1-\theta} d(y_1, y_2)$$

For any $\varepsilon > 0$ there is a natural number N such that

$$\theta^{m-1} \frac{1}{1-\theta} d(y_1, y_2) < \varepsilon \quad \text{when } m > N$$

Hence $\{y_n\}$ is a Cauchy sequence. Since (Y, d) is complete, $\{y_n\}$ converges, say to y^*.

Next we show that $y^* = \varphi(y^*)$. Consider

$$d(y^*, \varphi(y^*)) \leq d(y^*, y_n) + d(y_n, \varphi(y_n)) + d(\varphi(y_n), \varphi(y^*))$$
$$\leq (1 + \theta) d(y^*, y_n) + d(y_n, y_{n+1})$$

Given any $\varepsilon > 0$ we choose n so large that

$$d(y^*, y_n) < \frac{\varepsilon}{2(1+\theta)} \quad \text{and} \quad d(y_n, y_{n+1}) < \frac{\varepsilon}{2}$$

Then
$$d(y^*, \varphi(y^*)) < \varepsilon$$

By the arbitrariness of ε $\quad y^* = \varphi(y^*)$

To show that y^* is unique assume that $y' = \varphi(y')$ and consider

$$d(y^*, y') = d(\varphi(y^*), \varphi(y')) \leq \theta d(y^*, y')$$
$$(1 - \theta) d(y^*, y') \leq 0$$

Hence $\quad d(y^*, y') = 0 \quad$ and so $\quad y^* = y'$

This theorem can be applied to the following initial-value problem of differential equations. Find $y(x)$ which satisfies

$$\frac{dy}{dx} = f(x, y) \qquad y(x_0) = y_0$$

where f is some given function.

Theorem 10.27 Assume that f is continuous in the region $R = \{(x, y) \mid a \leq x \leq b, -\infty < y < \infty\}$ and let (x_0, y_0) be an interior point of R. Assume that f satisfies the following Lipschitz condition in R: There is a constant M such that $|f(x, y_1) - f(x, y_2)| \leq M|y_1 - y_2|$ for all $(x, y_1), (x, y_2) \in R$. Then the initial-value problem

$$\frac{dy}{dx} = f(x, y) \qquad y(x_0) = y_0$$

has a unique solution $y(x)$ in the interval $[x_0, x_0 + \alpha]$ for some $\alpha > 0$.

PROOF We first use Theorem 10.26 to show that the equation

$$y(x) = y_0 + \int_{x_0}^{x} f(t, y(t))\, dt$$

has a unique solution. Pick $\alpha > 0$ such that $\alpha M < 1$ and $x_0 + \alpha \leq b$. Define

$$\varphi(y) = y_0 + \int_{x_0}^{x} f(t, y(t))\, dt \qquad \text{for all } x \in [x_0, x_0 + \alpha]$$

Then, $\varphi(y)$ is a continuous function on $[x_0, x_0 + \alpha]$, and so φ is a mapping from $\mathcal{C}[x_0, x_0 + \alpha]$ into $\mathcal{C}[x_0, x_0 + \alpha]$.

$$|\varphi(y_1) - \varphi(y_2)| \leq \int_{x_0}^{x} |f(t, y_1(t)) - f(t, y_2(t))|\, dt$$

$$\leq \int_{x_0}^{x} M|y_1(t) - y_2(t)|\, dt \leq M|x - x_0| \sup_{x \in [x_0, x_0 + \alpha]} |y_1(x) - y_2(x)|$$

Then

$$|\varphi(y_1) - \varphi(y_2)| \leq \theta \cdot \sup_{x \in [x_0, x_0 + \alpha]} |y_1(x) - y_2(x)| \qquad \text{where } \theta = \alpha M < 1$$

and so

$$\sup_{x \in [x_0, x_0 + \alpha]} |\varphi(y_1) - \varphi(y_2)| \leq \theta \sup_{x \in [x_0, x_0 + \alpha]} |y_1 - y_2|$$

Hence φ is a contraction mapping from $\mathcal{C}[x_0, x_0 + \alpha]$ into $\mathcal{C}[x_0, x_0 + \alpha]$. By Theorem 10.26 there is a unique fixed point $y(x)$

which satisfies

$$y = y_0 + \int_{x_0}^{x} f(t, y(t))\, dt$$

By Exercise 10.78, y is the unique solution of the initial-value problem.

EXERCISES

10.77 Let $\{f_n\}$ be a Cauchy sequence in $\mathcal{C}[0, 1]$ and suppose that $\lim_{n \to \infty} x_n = x_0$, where $x_n \in [0, 1]$ for $n = 1, 2, 3, \ldots$. Prove that given any $\varepsilon > 0$ there is a natural number N such that $|f_n(x_n) - f_n(x_0)| < \varepsilon$ whenever $n > N$.

10.78 Assume that $f(x, y)$ is continuous in the region $R = \{(x, y) \mid a < x < b,\ -\infty < y < \infty\}$. Prove that any function $y(x)$, continuous on $[x_0, x_0 + \alpha]$, where $a < x_0 < x_0 + \alpha < b$, is a solution of

$$y(x) = y_0 + \int_{x_0}^{x} f(t, y(t))\, dt \qquad x \in [x_0, x_0 + \alpha]$$

if and only if $y(x)$ is a solution of

$$\frac{dy}{dx} = f(x, y) \qquad y(x_0) = y_0 \qquad \text{on} \qquad [x_0, x_0 + \alpha]$$

10.79 Prove that if $f_y(x, y)$ exists and is bounded in a rectangle R then $f(x, y)$ satisfies a Lipschitz condition in R.

10.80 Which of the following initial-value problems has a unique solution?

(a) $\dfrac{dy}{dx} = \dfrac{x}{y} \qquad y(1) = 0$ \qquad (b) $\dfrac{dy}{dx} = x|y| \qquad y(0) = 0$

(c) $\dfrac{dy}{dx} = \dfrac{x^2 - y}{x^2 + y^2} \qquad y(1) = 0$

10.81 Prove that a contraction mapping is a continuous function.

10.82 Prove that the set of all functions which are continuous and nowhere differentiable on $[0, 1]$ is a set of the second category in the space $\mathcal{C}[0, 1]$.

10.83 Let $\varphi: \mathcal{C}[0, 1] \to \mathcal{C}[0, 1]$ be defined by $\varphi(f) = \alpha \int_0^x t f(t)\, dt$, where α is a constant.

(a) Find α such that φ is a contraction mapping.

(b) For each of the values of α given by part (a) prove that φ has a unique fixed point.

(c) Find the fixed point of φ.

Solutions and Hints to Selected Exercises

Section 1.1

1.1 Verify that for each $x_0 \in X$, $(f^{-1})^{-1}(x_0) = f(x_0)$.
1.3 Fix $x^* \in X$ and define $h(y) = x^*$ for each $y \in Y$ which is not in the range of f; $h(y_0) = x_0$ if $f(x_0) = y_0$. Verify $h \circ f = i_x$.
1.6 For $z_0 \in Z$ there is $y_0 \in Y$ with $g(y_0) = z_0$; next there is $x_0 \in X$ with $f(x_0) = y_0$. Then $(g \circ f)(x_0) = z_0$.
1.7 (a) Equivalence relation   (b) Not transitive
   (c) Equivalence relation   (d) Not symmetric
1.8 Only i_x. The equivalence classes are the singleton sets $\{x\}$, $x \in X$.
1.10 Define $\varphi(x, y) = (y, x)$ and show that φ is a one-to-one correspondence.
1.11 Let $f: X \to \mathscr{P}(X)$ be any function. To show that f is not onto consider the set $A = \{x \in X \mid x \notin f(x)\}$.

Section 1.2

1.12 If $m \neq n$ then $m < n$ or $m > n$. For $m < n$ there is $q \in \mathbb{N}$ with $m + q = n$. Then $pm + pq = pn$ and $pm \neq pn$.
1.13 Use mathematical induction.
1.15 Use the result of Exercise 1.13.

Section 1.3

1.19 If $j \neq 0$, multiply both sides of the equation $j \cdot k = 0$ by j^{-1}.
1.21 Verify that an integer is odd if and only if it can be written as an even integer plus 1.
1.25 (a) $u < v$ implies $v - u > 0$ implies $(v + w) - (u + w) > 0$ implies $u + w < v + w$.
1.26 If $r \leq 0$, take $n = 1$; if $r = p/q > 0$ assume that $p > 0$ and $q > 0$ and then $p + 1 > p \geq p/q = r$.
1.27 If $r < s$ and they have the same sign then $rs > 0$ and so $1/rs > 0$. Then by Exercise 1.25, $r(1/rs) < s(1/rs)$.

Section 1.4

1.31 To show $U + (-U) = 0$ show that
$$U + -U \subseteq \{x \in Q \mid x > 0\} \quad \text{and} \quad U + -U \supseteq \{x \in Q \mid x > 0\}$$
For the latter use Exercise 1.26 to show there are rationals $b \in U$, $a \in U'$ so that $b - a < x$. Then write $x = b + (x - b)$.

1.34 Consider the set $\{-x \mid x \in S\}$.

1.36 Consider the set $\{x \in Q \mid x^2 < 2\}$.

Section 1.5

1.42 Use the archimedean property with $a = \varepsilon$, $b = 1$.

1.44 Yes, see Exercise 1.26.

Section 1.6

1.45 (c) $|a + b|^2 = (a + b)^2 = a^2 + 2ab + b^2 \le |a|^2 + 2|a||b| + |b|^2 = (|a| + |b|)^2$

1.47 (a) **R, J**   (b) **R** $- \{0\}, \{-1, 1\}$ (c) **R, N** $\cup \{0\}$
(d) **R**, $\{-1, 0, 1\}$ (e) $(0, \infty), (0, \infty)$
(f) $(-\infty, 0) \cup [1, \infty), (-1, 1) \cup (1, 3)$

1.48 Show that the range of f is equivalent to a subset of A and then use the corollary of Theorem 1.11.

1.54 (a) Countable   (b) Uncountable   (c) Countable
(d) Countable   (e) Uncountable   (f) Countable

1.55 For fixed $\alpha \in \mathbf{R} - \mathbf{Q}$, $\{x + \alpha \mid x \in \mathbf{Q}\}$.

Section 2.1

2.1 (a) 0   (b) Does not exist   (c) 2   (d) 0
(e) Does not exist   (f) 1   (g) 0   (h) 0

2.3 $|b_n| \le M$ for every $n = 1, 2, 3, \ldots,$ and $|a_n| \le \varepsilon/M$ for $n > N$. Then
$$|a_n b_n| = |a_n||b_n| \le \frac{\varepsilon}{M}(M) = \varepsilon \quad \text{if } n > N$$

2.5 (a) $a_n = \frac{1}{n}$, $b_n = \sin n$

(b) $a_n = \sin n$, $b_n = \cos\left(n + \frac{\pi}{2}\right)$

(c) $a_n = (-1)^n$, $b_n = \sin \pi(n - \tfrac{1}{2})$

2.7 (a) $a_n = n$, $b_n = \frac{1}{n^2}$ (b) $a_n = n^2$, $b_n = \frac{1}{n}$

(c) $a_n = n$, $b_n = \frac{(-1)^n}{n}$ (d) $a_n = n$, $b_n = \frac{c}{n}$

2.10 Assume $\lim_{n \to \infty} a_n = A$. There is a natural number N such that $|a_n - A| < \varepsilon/2$ whenever $n > N$. Choose n so large that
$$\frac{|a_1 - A| + |a_2 - A| + \cdots + |a_N - A|}{n} < \frac{\varepsilon}{2}$$

and $n > N$. Then

$$|\hat{a}_n - A| = \left|\frac{a_1 + a_2 + \cdots + a_n}{n} - A\right| = \left|\frac{(a_1 - A) + (a_2 - A) + \cdots + (a_n - A)}{n}\right|$$

$$\leq \frac{|a_1 - A| + |a_2 - A| + \cdots + |a_n - A|}{n} = \frac{|a_1 - A| + |a_2 - A| + \cdots + |a_N - A|}{n}$$

$$+ \frac{|a_{N+1} - A| + \cdots + |a_n - A|}{n} < \frac{\varepsilon}{2} + \frac{n - N}{n}\frac{\varepsilon}{2} < \varepsilon$$

If $a_n = (-1)^n$ then $\lim_{n \to \infty} \hat{a}_n = 0$, but $\{a_n\}$ diverges.

Section 2.2

2.13 (a) e (b) e^2 (c) 0 (d) 1

(e) $\sqrt{e}$ (f) $\dfrac{1}{e}$ (g) $\dfrac{1}{e}$ (h) $e^{8/3}$

2.15 (a) $\dfrac{1 + \sqrt{5}}{2}$ (b) 2

2.18 (a) $\{0, 1, -2\}$ (b) $\{0, 1, \sqrt{2}, -1, -\sqrt{2}\}$
(c) $\varnothing$ (d) $\{0\}$ (e) $\{-\tfrac{1}{2}\}$

2.20 (a) $\limsup a_n = 1$, $\liminf a_n = -2$ (b) $\limsup a_n = \sqrt{2}$, $\liminf a_n = -\sqrt{2}$
(c) $\limsup a_n = \infty$, $\liminf a_n = \infty$ (d) $\limsup a_n = 0$, $\liminf a_n = 0$
(e) $\limsup a_n = -\tfrac{1}{2}$, $\liminf a_n = -\tfrac{1}{2}$

2.21 $a_n = n \sin(n\pi/2)$,

2.22 $a_n = r_n$, where $\{r_1, r_2, r_3, \ldots\} = \mathbf{Q}$

2.23 Let $I_n = (0, 1/n)$.

Section 2.3

2.27 (a) $\varnothing$ (b) $\mathbf{R}$ (c) $\{1, -1\}$

(d) $\bigcup_{k \in J} [4k, 4k + 1]$ (e) $\{0\} \cup \bigcup_{n=1}^{\infty} \left[\dfrac{1}{4n + 1}, \dfrac{1}{4n}\right]$

2.31 $\{1, \tfrac{1}{2}, \tfrac{1}{3}, \ldots\}$ and $\{-1, -\tfrac{1}{2}, -\tfrac{1}{3}, \ldots\}$ or $\mathbf{Q}$ and $\mathbf{R} - \mathbf{Q}$

2.33 $A_n = \left\{\dfrac{1}{n}\right\}$ or $A_n = \left[\dfrac{1}{4n + 1}, \dfrac{1}{4n}\right]$

Section 3.1

3.5 Consider $f(x) = -1/x$ and $g(x) = -1$ on $(0, 1)$.

3.7 (a) $\left|\dfrac{\sin x}{1 + x^2}\right| \leq |\sin x| \leq 1$ on $(-\infty, \infty)$

(b) $\left|\dfrac{\sin(1/x)}{x + 2}\right| \leq \dfrac{1}{x + 2} < \dfrac{1}{2}$ on $(0, 2)$

(c) $\left|\dfrac{x^4 - 3x^2 + 2}{2 - \cos x}\right| \le |x^4 - 3x^2 + 2| = |x - 1| \cdot |x + 1| \cdot |x^2 - 2|$

$\le (2\pi - 1)(2\pi + 1)(4\pi^2 - 2)$ on $[0, 2\pi]$

(d) $\left|\dfrac{\sin x}{3x - 2x \sin x}\right| = \left|\dfrac{\sin x}{x}\right| \dfrac{1}{3 - 2 \sin x} \le 1$ on $(0, \infty)$

(e) $\left|\dfrac{\cos x}{x^2 - 2x + 2}\right| = \dfrac{|\cos x|}{(x - 1)^2 + 1} \le |\cos x| \le 1$ on $(-\infty, \infty)$

(f) $\left|\dfrac{5x^2 + 3x + 1}{x^2 - 2}\right| \le |5x^2 + 3x + 1| \le 5x^2 + 3|x| + 1 \le 9$ on $[-1, 1]$

(g) $\left|\dfrac{\sin x}{\sqrt{x}}\right| = \sqrt{x} \left|\dfrac{\sin x}{x}\right| \le \sqrt{x} \le 1$ on $(0, 1]$

$\left|\dfrac{\sin x}{\sqrt{x}}\right| \le |\sin x| \le 1$ on $[1, \infty)$

(h) $\left|\dfrac{1 - \cos x}{x^2}\right| = \left|\dfrac{\sin x}{x}\right|^2 \dfrac{1}{1 + \cos x} \le 1$ on $\left[-\dfrac{\pi}{2}, \dfrac{\pi}{2}\right]$

$\left|\dfrac{1 - \cos x}{x^2}\right| = \dfrac{|1 - \cos x|}{x^2} \le \dfrac{|1 - \cos x|}{(\pi/2)^2} \le \dfrac{8}{\pi^2}$ if $|x| \ge \dfrac{\pi}{2}$

(i) $\left|\dfrac{1 + x^2}{1 + x^3}\right| \le 1 + x^2 \le 2$ on $[0, 1]$

and

$\left|\dfrac{1 + x^2}{1 + x^3}\right| \le 1$ on $[1, \infty]$

(j) $\left|\dfrac{1 - x^2}{1 + x^3}\right| = \left|\dfrac{1 - x}{1 - x + x^2}\right| = \dfrac{1 - x}{1 - x + x^2} \le 1$ on $(-1, 1)$

3.9 If $|f(x)| \le M$ for all $x \in A$ then

$|f(x) + g(x)| \ge |g(x)| - |f(x)| \ge |g(x)| - M$ on A

3.10 Let $f(x) = 1/x$ and $g(x) = x$ on $(0, \infty)$.

3.11 Let $g(x)$ be unbounded on A; consider
(a) $f(x) = 0$ on A (b) $f(x) = 1$ on A

3.12 (a) $\sup f(x) = 4$, $\inf f(x) = -5$ (b) $\sup f(x) = 2$, $\inf f(x) = -1$
(c) $\sup f(x) = 0$, $\inf f(x) = -1$ (d) $\sup f(x) = \infty$, $\inf f(x) = -\infty$

(e) $\sup f(x) = \infty$, $\inf f(x) = 0$ (f) $\sup f(x) = 1$, $\inf f(x) = -\dfrac{2}{3\pi}$

(g) $\sup f(x) = 1$, $\inf f(x) = -1$ (h) $\sup f(x) = \infty$, $\inf f(x) = -\dfrac{4}{9\pi^2}$

3.13 (a) $(f + g)(-x) = f(-x) + g(-x) = f(x) + g(x) = (f + g)(x)$

$(kf)(-x) = kf(-x) = kf(x) = (kf)(x)$

$(f \cdot g)(-x) = f(-x)g(-x) = f(x)g(x) = (f \cdot g)(x)$

(b) $(f+g)(-x) = f(-x) + g(-x) = -f(x) - g(x) = -[f(x) + g(x)] = -(f+g)(x)$

$(kf)(-x) = kf(-x) = k[-f(x)] = -kf(x) = -(kf)(x)$

$(f \cdot g)(-x) = f(-x)g(-x) = [-f(x)][-g(x)] = f(x)g(x) = (f \cdot g)(x)$

(c) $(f \cdot g)(-x) = f(-x)g(-x) = -f(x)g(x) = -(f \cdot g)(x)$

3.15 $f: \mathbf{R} \to \mathbf{R}, f(x) \neq 0$ for all $x \in \mathbf{R}$ and $f(-x) = f(x)$

$$\frac{1}{f}(-x) = \frac{1}{f(-x)} = \frac{1}{f(x)} = \frac{1}{f}(x)$$

If f is odd then necessarily $f(0) = 0$ and so $1/f$ is not defined at $x = 0$.

Section 3.2

3.18 (a) $|(3x - 1) - 5| = 3|x - 2| < \varepsilon \quad$ if $|x - 2| < \frac{\varepsilon}{3}$

(b) $\left|\frac{x^2 + x - 2}{x + 2} - (-3)\right| = |(x - 1) + 3| = |x + 2| < \varepsilon \quad$ if $0 < |x + 2| < \varepsilon$

(c) $\left|\frac{1}{x^2} - 4\right| = \frac{|1 - 4x^2|}{x^2} = \frac{|1 + 2x||1 - 2x|}{x^2}$

$= \frac{2}{x^2}|1 + 2x||\frac{1}{2} - x| < 80|\frac{1}{2} - x| < \varepsilon \quad$ if $|\frac{1}{2} - x| < \min(1/4, \varepsilon/80)$

(d) $|x^3 - 1| = |x - 1||x^2 + x + 1| < 7|x - 1| < \varepsilon \quad$ if $|x - 1| < \min(1, \varepsilon/7)$

(e) $\left|\frac{1}{8x + 1} - 1\right| = \left|\frac{-8x}{8x + 1}\right| = \frac{8}{|8x + 1|}|x| < 16|x| < \varepsilon \quad$ if $|x| < \min(1/16, \varepsilon/16)$

(f) $|\sqrt[3]{x} - 2| = \left|\frac{x - 8}{x^{2/3} + 2x^{1/3} + 4}\right| < \frac{1}{4}|x - 8| < \varepsilon \quad$ if $|x - 8| < \min(1, 4\varepsilon)$

3.20 $\left|\frac{1}{\sqrt{x}} - \frac{1}{\sqrt{a}}\right| = \left|\frac{x - a}{\sqrt{x}\sqrt{a}(\sqrt{x} + \sqrt{a})}\right| < \frac{2(\sqrt{2} - 1)}{a\sqrt{a}}|x - a| < \varepsilon$

$$\text{if } |x - a| < \min\left(\frac{a}{2}, \frac{a\sqrt{a}\,\varepsilon}{2(\sqrt{2} - 1)}\right)$$

3.23 (a) $\frac{1}{2}$ (b) -2 (c) n (d) $\frac{1}{48}$ (e) 0

3.25 Let $f(x) = 1$ if $x > a$ and $f(x) = -1$ if $x < a$.

3.28 $0 \leq |x \sin 1/x| \leq |x|$

3.30 (a) $\frac{1}{2}$ (b) 2 (c) $\frac{1}{2}$ (d) $\sqrt{2}/2$

Section 3.3

3.37 (a) 0 (b) $\sqrt{2}$ (c) -2 (d) $1, 1$, does not exist
(e) $\frac{4}{3}, \frac{2}{3}, 0$ (f) 0 (g) -4 (h) -2

3.38 (a) -3 (b) 3 (c) does not exist (d) -1 (e) 3

3.41 Consider $f(x) = 1/x$ with $a = 0$.

3.49 $|f(x) - L| < 1$ if $x \geq a$; then $L - 1 < f(x) < L + 1$ for $x \in [a, \infty)$.

3.53 $f(x) \geq M$ if $x \geq b$; then $g(x) \geq f(x) \geq M$ if $x \geq \max(a, b)$.

3.54 (a) $(2 + \sin x)^x = 1$ if $x = (3\pi/2) + 2k\pi$, for $k = 1, 2, 3, \ldots$.
(b) $x^2 \operatorname{sgn}(\cos x) = 0$ if $x = -(\pi/2) - 2k\pi$ for $k = 1, 2, 3, \ldots$.
(c) $x^{(1+\sin x)} = 1$ if $x = (3\pi/2) + 2k\pi$ for $k = 1, 2, 3, \ldots$.
(d) $e^x |\operatorname{sgn}(x - [\![x]\!])| = 0$ if $x = 1, 2, 3, \ldots$.

3.55 $f(x) \geq M/\alpha$ if $x \geq b$. Then $f(x) \cdot g(x) \geq (M/\alpha)\alpha = M$ if $x \geq \max(a, b)$.

3.59 $0 < g(x) \leq K$ if $x \geq a$ and $f(x) \geq MK$ if $x \geq b$. Then

$$\frac{f(x)}{g(x)} \geq MK \frac{1}{K} = M \quad \text{if } x \geq \max(a, b)$$

Section 3.4

3.60 (a) Not monotone (b) Strictly increasing (c) Not monotone (d) Not monotone
(e) Strictly decreasing (f) Monotone increasing (g) Monotone decreasing
(h) Monotone (i) Strictly increasing (j) Monotone increasing

3.64 $f(x) = \begin{cases} 1 & \text{if } x \in \mathbf{Q} \cap (a, b) \\ 0 & \text{if } x \in (\mathbf{R} - \mathbf{Q}) \cap (a, b) \end{cases}$

Section 4.1

4.2 (a) Jump discontinuities at $x = 0$ and $x = 3$; discontinuity of second kind at $x = 4$; left-continuous at $x = 0$; right-continuous at $x = 3$ and $x = 4$.
(b) Jump discontinuity at each integer and left continuous at each integer.
(e) Jump discontinuity at $x = 1$; discontinuity of the second kind at $x = 0$; left-continuous at $x = 0$; right-continuous at $x = 1$.
(f) Discontinuity of the second kind at $x \neq 0$; jump discontinuity at $x = 0$; left-continuous at $x = 0$.

4.3 $|\cos x - \cos a| = \left| -2 \sin \frac{x+a}{2} \sin \frac{x-a}{2} \right| \leq |x - a|$

4.5 Let $g = -f$.

4.7 Consider $f = 0$ and $f = 1$.

4.10 If $f(x_0) > 0$ then

$$|\sqrt{f(x)} - \sqrt{f(x_0)}| = \frac{|f(x) - f(x_0)|}{\sqrt{f(x)} + \sqrt{f(x_0)}}$$

If $f(x_0) = 0$ then

$$|\sqrt{f(x)} - \sqrt{f(x_0)}| = \sqrt{f(x) - f(x_0)}$$

4.12 $f\left(\dfrac{\pi}{4}\right) = \dfrac{4 - \pi}{4}$

4.15 Removable discontinuity at 0, continuous everywhere else.

4.16 Show that $f(0) = 0$ and then show that $\lim_{x \to a} f(x) = \lim_{u \to 0} f(u + a)$.

4.19 Let

$$g(x) = \begin{cases} f(x) & \text{if } a \leq x \leq b \\ f(b) & \text{if } x > b \\ f(a) & \text{if } x < a \end{cases}$$

4.21 If either of the limits does not exist, g cannot be continuous at that point. If both limits exist then define

$$g(x) = \begin{cases} f(x) & \text{if } a < x < b \\ \lim_{x \to a^+} f(x) & \text{if } x = a \\ \lim_{x \to b^-} f(x) & \text{if } x = b \end{cases}$$

4.22 Use the result of Exercise 4.21.

Section 4.2

4.26 (a) $M = 4 = f(1)$, $m = 5 = f(4)$ (b) $M = 2 = f(1)$, $m = -1 = f(-2)$
(c) $M = 1$, $m = 0$ (d) $M = 1 = f(0)$, $m = -3$

4.28 See Exercise 4.22. Consider $f(x) = \sin(1/x)$ on $(0, 1)$.

4.32 If $p(x)$ has odd degree there are real numbers a and b such that $p(a) > 0$ and $p(b) < 0$.

4.34 $f(0) > 0$ and $f(1) < 0$.

4.35 $f(\sqrt{2}/2) = 0$, f must be constant.

4.36 There exists $x_1 \in (0, \infty)$ and $x_2 \in (0, \infty)$ such that $f(x_1) < \sqrt{3}/2 < f(x_2)$.

4.38 f is constant on the intervals $[a, x_0]$ and $(x_0, b]$.

4.39 Apply the intermediate-value theorem to $g - f$.

4.41 Let $f(a) = f(b) = 0$, where f assumes each real value exactly twice. Then f has one sign between a and b, and the opposite sign outside $[a, b]$. But this implies that f is unbounded on $[a, b]$ and this contradicts the continuity of f.

4.43 Use Theorem 4.9.

Section 4.3

4.45 (a) $|x^3 - a^3| = |x - a| |x^2 + ax + a^2| \le 3|x - a|$; choose $\delta = \varepsilon/3$.
(b) $(\sqrt{x} - \sqrt{a})^2 = x + a - 2\sqrt{x}\sqrt{a} < x - a$ if $0 < a < x$; choose $\delta = \varepsilon^2$.

4.47 $f(x) = g(x) = x$ on $\mathbf{R}$.

4.49 (a) Yes   (b) No   (c) Yes   (d) No   (e) No   (f) Yes
(g) Yes   (h) No   (i) Yes   (j) Yes

4.51 $f \cdot g$ is continuous on (a, b), and $\lim_{x \to a^+} f(x) \cdot g(x)$ and $\lim_{x \to b^-} f(x) \cdot g(x)$ both exist.

4.52 Boundedness is not necessary; consider $f(x) = g(x) = \sqrt{x}$ on $(0, \infty)$.

4.55 False, $f(x) = \sin x^2$.

4.58 Let $f(x)$ be any function which is continuous on $\mathbf{R}$ but not uniformly continuous on $\mathbf{R}$, for example, $f(x) = x^2$.

4.60 Let $g(x) = f(-x)$ for $x \in [-b, \infty)$; use Theorem 4.15.

4.61 Use Theorem 4.14.

Section 4.4

4.62 For any x_0 and any $\varepsilon > 0$ choose $\delta = \min |r - x_0|$ for rationals $r = (p/q)$ $(r \ne x_0)$ with $1/q \ge \varepsilon$; then for each $x \in N_\delta^*(x_0)$, $|f(x)| < \varepsilon$. Therefore $\lim_{x \to x_0} f(x) = 0$.

4.65 $F = \bigcup_{n=1}^{\infty} F_n$. Define

$$f(x) = \begin{cases} \dfrac{1}{n} & \text{if } x \in F_n \cap \mathbf{Q} \text{ and } x \notin F_j \text{ for } j = 1, \ldots, n-1 \\ -\dfrac{1}{n} & \text{if } x \in F_n - \mathbf{Q} \text{ and } x \notin F_j \text{ for } j = 1, \ldots, n-1 \\ 0 & \text{if } x \notin F \end{cases}$$

Section 5.1

5.1 $f'(x) = 3x|x|$

5.3 $f'(x) = \begin{cases} 2x & x > 2 \\ 4 & x \le 2 \end{cases}$

5.4 $a = 3, \quad b = -2$

5.8 $f(x) = \begin{cases} x^2 & \text{if } x \text{ is rational} \\ 0 & \text{if } x \text{ is irrational} \end{cases}$

5.9 Yes; no

5.10 Differentiable at every noninteger; discontinuous at every integer except at $x = 1$.

5.11 (a) Yes   (b) No

Section 5.2

5.15 Apply the mean-value theorem to an interval $[x_1, x_2]$ as in the proof of Theorem 5.7.

5.17 $\left| \dfrac{f(x) - f(y)}{x - y} \right| \le |x - y| \quad \text{for all } x \ne 0$

and so $f'(x) = 0$.

5.18 $f'(1)$ does not exist.

5.20 $|f(x_1) - f(x_0)| \le M|x_1 - x_0|$ and then pick $\delta = \varepsilon/M$.

5.21 False; consider $f(x) = \sqrt{x}$ on $(0, 1)$.

5.23 Converse is false; consider $f(x) = x^3$ on $(-1, 1)$.

Section 5.3

5.26 Use Theorem 5.14 and its analog for the limit from the left.

5.28 Similar to the proof of Theorem 5.11.

5.30 (a) $\ln a - \ln b$ (b) 5   (c) 1   (d) 0   (e) 0   (f) $-\tfrac{1}{2}$
(g) 0   (h) 1   (i) e^{-6} (j) 0

Section 5.4

5.34

$f'_+(x) = \begin{cases} 1 & x < -1 \\ 2x & -1 \le x < 0 \\ 4 & 0 \le x \le 2 \\ 2x & x > 2 \end{cases} \qquad f'_-(x) = \begin{cases} 1 & x \le -1 \\ 2x & -1 < x \le 0 \\ 4 & 0 < x \le 2 \\ 2x & x > 2 \end{cases}$

f is differentiable at all points except $x = -1$ and $x = 0$.

5.36 All zero.

5.38 True.

Section 6.1

6.2 If $1 \in [x_{i-1}, x_i]$ then $U(P, f) = x_i$, $L(P, f) = x_{i-1}$, $\overline{\int} f = 1$, $\underline{\int} f = 1$.

6.3 (a) $f(x) = \begin{cases} 1 & \text{if } x \in \mathbf{Q} \cap [a, b] \\ 0 & \text{if } x \in (\mathbf{R} - \mathbf{Q}) \cap [a, b] \end{cases}$

(b) $f(x) = \begin{cases} 1 & \text{if } x \in \left[a, \dfrac{a+b}{2}\right] \\ 0 & \text{if } x \in \left(\dfrac{a+b}{2}, b\right] \end{cases}$

(c) $f(x) = \left|x - \dfrac{a+b}{2}\right|$ on $[a, b]$

(d) $f(x) = \begin{cases} 0 & \text{if } x = a, b \\ 1 & \text{if } x \in (a, b) \end{cases}$

6.4 (a) f is continuous on $[0, 1]$, and so f is Riemann-integrable on $[0, 1]$.
(b) f is continuous on $[0, 1]$, and so f is Riemann-integrable on $[0, 1]$.
(c) f is monotone on $[0, 1]$, and so f is Riemann-integrable on $[0, 1]$.
(d) f is not Riemann-integrable on $[0, 1]$ since f is not bounded on $[0, 1]$.
(e) f is continuous on $[0, 1]$ and so f is Riemann-integrable on $[0, 1]$.
(f) For any partition P of $[0, 1]$, $U(P, f) = 1$ and $L(P, f) = x_{n-1} - x_1$. Therefore $\overline{\int} f = 1$ and $\underline{\int} f = 1$, and so f is Riemann-integrable on $[0, 1]$.
(g) $\overline{\int} f = \frac{2}{3}$ and $\underline{\int} f = 0$; therefore f is not Riemann-integrable on $[0, 1]$.
(h) f is monotone on $[0, 1]$, and so f is Riemann-integrable on $[0, 1]$.
(i) f is not Riemann-integrable on $[0, 1]$ since f is not bounded on $[0, 1]$.
(j) Let P be any partition of $[0, 1]$. $L(P, f) = 0$; suppose $(2/\pi) \in [x_{i-1}, x_i]$. If $(2/\pi) \le x \le 1$ then $\sin(1/x) \ge \sin 1$. Hence $U(P, f) \ge (\sin 1)(1 - x_{i-1}) \ge (\sin 1)(1 - 2/\pi)$. It follows that $\underline{\int} f = 0$ and $\overline{\int} f \ge (\sin 1)(1 - 2/\pi) > 0$. Therefore f is not Riemann-integrable.

6.7 Both functions are Riemann-integrable on $[0, 1]$ by Exercise 6.6

6.8 $$L(P_n, f) = \frac{b^2 - a^2}{2} - \frac{(b-a)^2}{2n}$$

$$U(P_n, f) = \frac{b^2 - a^2}{2} + \frac{(b-a)^2}{2n}$$

$$\overline{\int} f = \frac{b^2 - a^2}{2} \quad \underline{\int} f = \frac{b^2 - a^2}{2} \quad \int_a^b f = \frac{b^2 - a^2}{2}$$

6.10 (a) $L(P, f) = ab - a^2$ (b) $\underline{\int} f = ab - a^2$
(c) $U(P_n, f) = \dfrac{b^2 - a^2}{2} + \dfrac{(b-a)^2}{2n}$ (d) $\overline{\int} f = \dfrac{b^2 - a^2}{2}$

6.11 (a) $U(P, f) = 8$ (b) $\overline{\int} f = 8$ (c) $L(P_n, f) = 4 - \dfrac{2}{n}$ (d) $\underline{\int} f = 4$

Section 6.2

6.13 Suppose $\lim_{\|P\| \to 0} S(P, f, \xi)$ exists (and hence equals $\int_a^b f$). Given arbitrary $\varepsilon > 0$ there is a $\delta > 0$ such that $|S(P, f, \xi) - \int_a^b f| < \varepsilon$ whenever $\|P\| < \delta$. Choose P_0 so that $\|P_0\| < \delta$ and choose intermediate points ξ_i such that $S(P_0, f, \xi) = \alpha$. Then $|\alpha - \int_a^b f| < \varepsilon$. By the arbitrariness of $\varepsilon > 0$ it follows that $\int_a^b f = \alpha$.

6.14 (a) $f(x) = x^3$ is integrable on $[a, b]$, and so by Theorem 6.5 $\lim_{\|P\| \to 0} S(P, f, \xi)$ exists. For any partition P of $[a, b]$ choose $\xi_i = [\frac{1}{4}(x_i^3 + x_i^2 x_{i-1} + x_i x_{i-1}^2 + x_{i-1}^3)]^{1/3}$ for

$i = 1, 2, \ldots, n$. Then

$$S(P, f, \xi) = \sum_{i=1}^{n} f(\xi_i) \Delta x_i = \sum_{i=1}^{n} \tfrac{1}{4}(x_i^3 + x_i^2 x_{i-1} + x_i x_{i-1}^2 + x_{i-1}^3) \Delta x_i$$

$$= \sum_{i=1}^{n} \tfrac{1}{4}(x_i^4 - x_{i-1}^4) = \tfrac{1}{4}(x_n^4 - x_0^4) = \tfrac{1}{4}(b^4 - a^4)$$

By Exercise 6.13, $\int_a^b x^3 \, dx = \tfrac{1}{4}(b^4 - a^4)$.

(b) $f(x) = 1/x^2$ is Riemann-integrable on $[a, b]$, where $a > 0$, and so by Theorem 6.5, $\lim_{\|P\| \to 0} S(P, f, \xi)$ exists. For any partition P of $[a, b]$ choose $\xi_i = \sqrt{x_i x_{i-1}}$ for $i = 1, 2, \ldots, n$. Then

$$S(P, f, \xi) = \sum_{i=1}^{n} f(\xi_i) \Delta x_i = \sum_{i=1}^{n} \frac{1}{x_i x_{i-1}} \Delta x_i$$

$$= \sum_{i=1}^{n} \left(\frac{1}{x_{i-1}} - \frac{1}{x_i} \right) = \frac{1}{x_0} - \frac{1}{x_n} = \frac{1}{a} - \frac{1}{b}$$

By Exercise 6.13, $\int_a^b \frac{1}{x^2} \, dx = \frac{1}{a} - \frac{1}{b}$.

(c) $f(x) = 1/\sqrt{x}$ is Riemann-integrable on $[a, b]$, where $a > 0$, and so by Theorem 6.5, $\lim_{\|P\| \to 0} S(P, f, \xi)$ exists. For any partition P of $[a, b]$ choose

$$\xi_i = \left(\frac{\sqrt{x_i} + \sqrt{x_{i-1}}}{2} \right)^2 \quad \text{for } i = 1, 2, \ldots, n$$

Then

$$S(P, f, \xi) = \sum_{i=1}^{n} f(\xi_i) \Delta x_i = \sum_{i=1}^{n} \frac{2}{\sqrt{x_i} + \sqrt{x_{i-1}}} \Delta x_i = \sum_{i=1}^{n} 2(\sqrt{x_i} - \sqrt{x_{i-1}})$$

$$= 2(\sqrt{x_n} - \sqrt{x_0}) = 2(\sqrt{b} - \sqrt{a})$$

By Exercise 6.13, $\int_a^b \frac{1}{\sqrt{x}} \, dx = 2(\sqrt{b} - \sqrt{a})$.

6.16 f is Riemann-integrable on $[a, b]$; use Theorem 6.10.

6.17 $f(x) = \begin{cases} 1 & \text{if } x = a, b \\ 0 & \text{if } x \in (a, b) \end{cases}$

6.18 Consider $g = f$ and use Theorem 6.10.

6.19 Let $f(x) = \begin{cases} 1 & \text{if } x \in \mathbf{Q} \cap [a, b] \\ -1 & \text{if } x \in (\mathbf{R} - \mathbf{Q}) \cap [a, b] \end{cases}$

Section 6.3

6.21 (a) 10   (b) $r[\pi + \sqrt{2} + \ln(1 + \sqrt{2})]$ (c) $-\frac{157}{48}$

6.24 (a) $1/x^2$ is not bounded on $[-1, 1]$.   (b) $1/\sqrt{x}$ is not bounded on $[0, 1]$.

6.26 (a) $\ln \frac{b}{a}$ (b) $\ln \frac{a}{b}$ (c) $\frac{1}{n+1}(b^{n+1} - a^{n+1})$

6.28 Let $g(x) = -1/2x^2$ on $[a, b]$, where $a > 0$. For any partition P of $[a, b]$ choose intermediate points $\xi_i \in (x_{i-1}, x_i)$ according to the mean-value theorem of differential calculus. Then

$$g'(\xi_i) = \frac{g(x_i) - g(x_{i-1})}{x_i - x_{i-1}} \quad \text{for } i = 1, 2, \ldots, n$$

and so

$$f(\xi_i) = \frac{-(1/2x_i^2) + 1/2x_{i-1}^2}{x_i - x_{i-1}} = \frac{x_i + x_{i-1}}{2x_i^2 x_{i-1}^2}$$

Now

$$S(P, f, \xi) = \sum_{i=1}^{n} f(\xi_i) \Delta x_i = \sum_{i=1}^{n} \frac{x_i + x_{i-1}}{2x_i^2 x_{i-1}^2} (x_i - x_{i-1})$$

$$= \sum_{i=1}^{n} \left(\frac{1}{2x_{i-1}^2} - \frac{1}{2x_i^2} \right) = \frac{1}{2x_0^2} - \frac{1}{2x_n^2}$$

$$= \frac{1}{2a^2} - \frac{1}{2b^2} = \frac{b^2 - a^2}{2a^2 b^2}$$

$\lim_{\|P\| \to 0} S(P, f, \xi)$ exists by Theorem 6.5, and so by Exercise 6.13,

$$\int_a^b f = \frac{b^2 - a^2}{2a^2 b^2}$$

Section 6.4

6.29 (a) 1 (b) $\frac{3}{4}$ (c) $1 - \sin 1$ (d) 2
6.30 (a) 1 (b) 0 (c) 0 (d) 2
6.31 Only (b)
6.32 (a) and (b)
6.34 (a) and (c)
6.35 Let $K = \{x_1, x_2, x_3, \ldots\} \subset [a, b]$ and let

$$f(x) = \begin{cases} \dfrac{1}{n} & \text{if } x = x_n \\ 0 & \text{otherwise} \end{cases}$$

$\mathfrak{D}_f = K$ and f is Riemann-integrable by Theorem 6.19 and Exercise 6.33.

6.37 If $K = \mathbf{Q} \cap [a, b]$, then f is not Riemann-integrable on $[a, b]$. On the other hand if $K = \{a + (b - a)n \,|\, n = 1, 2, 3, \ldots\}$ then f is Riemann-integrable on $[a, b]$.

6.39 Not necessarily. Let $f(x) = 0$ for all $x \in [a, b]$ and let $g(x)$ be defined by

$$g(x) = \begin{cases} 1 & \text{if } x \in \mathbf{Q} \cap [a, b] \\ 0 & \text{if } x \in (\mathbf{R} - \mathbf{Q}) \cap [a, b] \end{cases}$$

Then $g(x) = f(x)$ for all $x \in [a, b]$ except for those $x \in E = \mathbf{Q} \cap [a, b]$, but $\mathfrak{D}_g = [a, b]$, and so g is not Riemann-integrable on $[a, b]$.

6.40 Not necessarily. Let

$$f(x) = \begin{cases} \dfrac{1}{x - a} & \text{if } x \in (a, b] \\ 0 & \text{if } x = a \end{cases}$$

For any given $\varepsilon > 0$ (assume $\varepsilon < b - a$) define $g(x)$ by

$$g(x) = \begin{cases} \dfrac{1}{\varepsilon^2}(x-a) & \text{if } x \in [a, a+\varepsilon] \\ \dfrac{1}{x-a} & \text{if } x \in (a+\varepsilon, b] \end{cases}$$

Then $g = f$ outside the interval $(a, a + \varepsilon)$ and g is continuous on $[a, b]$. But f is not Riemann-integrable on $[a, b]$ since it fails to be bounded.

Section 7.1

7.1 1

7.3 $\sum_{k=N}^{\infty} a_k = \sum_{k=1}^{\infty} a_k - S_{N-1}$

7.4 $\sum_{k=1}^{n} ca_k = c \sum_{k=1}^{n} a_k = cS_n$

7.5 $\sum_{k=1}^{n} (a_k + b_k) = \sum_{k=1}^{n} a_k + \sum_{k=1}^{n} b_k$

7.7 (a) Converges (comparison test) (b) Converges (comparison test)
(c) Diverges (integral test) (d) Diverges (limit form of comparison test)
(e) Converges (limit form of comparison test)
(f) Converges (limit form of comparison test)

7.9 If $R < 1$ then for any ρ satisfying $R < \rho < 1$ there is an N such that $\sqrt[k]{a_k} \le \rho$ for all $k > N$. Then $a_k \le \rho^k$ for $k > N$ and so $\sum_{k=1}^{\infty} a_k$ converges by comparison with the geometric series.

If $R > 1$ then $\sqrt[k]{a_k} > 1$ for infinitely many natural numbers k. Then $a_k > 1$ for infinitely many natural numbers k and so $\lim_{k \to \infty} a_k \ne 0$.

7.11 (a) Converges (ratio test) (b) Converges (ratio test)
(c) Converges (ratio test) (d) Converges (sum of two geometric series)

7.12 (a) Converges conditionally (b) Diverges
(c) Converges absolutely (d) Converges absolutely

7.13 No guarantee there is an N such that $\sum_{k=N+1}^{\infty} |a_k| < \varepsilon/2$.

7.14 The sequence of partial sums of a regrouped series is always a subsequence of the sequence of partial sums of the original series.

7.15 (a) For some natural number N, $b_k/a_k \le 1$ whenever $k > N$. Thus $b_k \le a_k$ whenever $k > N$.
(b) For some natural number N, $b_k/a_k \ge 1$ whenever $k > N$. Thus $b_k \ge a_k$ whenever $k > N$.

7.16 (a) $E \le \dfrac{1}{\sqrt{6}}$ (b) $E \le \dfrac{13}{212}$ (c) $E \le \dfrac{16}{15}$

7.17 Use the lemma preceding Theorem 7.3.

Section 7.2

7.21 (a) $S(x) = \begin{cases} 1 & \text{if } x = 0 \\ 0 & \text{if } x \in (0, 1] \end{cases}$

The convergence is not uniform.

(b) $S(x) = \begin{cases} \frac{1}{2} & \text{if } x = \pm 1 \\ 0 & \text{if } x \in (-1, 1) \end{cases}$

The convergence is not uniform.

(c) $S(x) = \begin{cases} 1 & \text{if } x = 1 \\ 0 & \text{if } x \in (-1, 1) \end{cases}$

$\{S_n(-1)\}$ diverges. The convergence is not uniform.
(d) $S(x) = 0$ on $(-1, 1)$. The convergence is not uniform.
(e) $S(x) = 0$ on $[0, 1]$. The convergence is uniform.
(f) $S(0) = S(1) = 0$. $\{S_n(x)\}$ diverges for each $x \in (0, 1)$.

7.23 $|S_n(x) - S(x)| \leq T_n$ for all $x \in I$. If $\lim_{n \to \infty} T_n = 0$ then given any $\varepsilon > 0$ there is a natural number N such that $|T_n| < \varepsilon$ for $n > N$. Then $|S_n(x) - S(x)| < \varepsilon$ for $n > N$ and for all $x \in I$. Hence $\{S_n(x)\}$ converges uniformly to $S(x)$ on I. Conversely, suppose $\{S_n(x)\}$ converges uniformly to $S(x)$ on I. Given any $\varepsilon > 0$ there is a natural number N such that $|S_n(x) - S(x)| < \varepsilon/2$ for $n > N$ and for all $x \in I$. Then $T_n = \sup_{x \in I} |S_n(x) - S(x)| \leq \varepsilon/2 < \varepsilon$ for $n > N$, and so $\lim_{n \to \infty} T_n = 0$.

7.24 (a) For all x such that $|(1 - x)/(1 + x)| < 1$; that is, for all $x > 0$
(b) $-1 \leq x < 1$ (c) $-1 \leq x \leq 1$ (d) for all $x \in \mathbf{R}$ (e) $-1 < x < 1$

7.25 $|a_k \cos kx| \leq |a_k|$ and $|a_k \sin kx| \leq |a_k|$

7.26 $\sum_{k=1}^{\infty} f_k(x)$ converges uniformly on I if and only if $\sum_{k=1}^{\infty} f_k$ converges.

7.27 $|S_n(x) - S(x)| = \left|\frac{1 - x^n}{1 - x} - \frac{1}{1 - x}\right| = \left|\frac{x^n}{1 - x}\right|$; $\sup_{x \in (-1, 1/2)} \left|\frac{x^n}{1 - x}\right| = \frac{1}{2}$

and so by Exercise 7.23, $\{(1 - x^n)/(1 - x)\}$ does not converge uniformly to $1/(1 - x)$ on $(-1, \frac{1}{2})$.

Section 7.3

7.29 Consider $|S_n(x) - S_{n-1}(x)| = |x^n/n!|$. Given $0 < \varepsilon \leq 1$, $|S_n(n) - S_{n-1}(n)| \geq 1 \geq \varepsilon$. By the Cauchy criterion, $\{S_n(x)\}$ does not converge uniformly on $\mathbf{R}$.

7.30 $S_n(0) = 0$ for all n; if $x_0 \in (0, 1]$ choose the natural number N so large that $1/N < x_0$. Then $S_n(x_0) = 0$ for $n \geq N$, and so $\{S_n(x)\}$ converges to $S(x) = 0$ on $[0, 1]$. The convergence is not uniform on $[0, 1]$ since $T_n = \sup_{x \in [0, 1]} |S_n(x) - S(x)| = n$ (see Exercise 7.23).

7.32 (b) Let

$$S_n(x) = \begin{cases} \frac{1}{x} & \text{if } x > \frac{1}{n} \\ 0 & \text{otherwise} \end{cases}$$

Then

$$S(x) = \begin{cases} \frac{1}{x} & \text{if } x > 0 \\ 0 & \text{otherwise} \end{cases}$$

7.33 Conjecture is true.

7.35

$$\lim_{n \to \infty} S'_n(x) = \begin{cases} -1 & \text{if } x = 1 \\ 0 & \text{if } 0 \leq x < 1 \end{cases} \quad \text{where } S'_n(x) = (n + 1)x^{n-1}\left(\frac{n}{n+1} - x\right)$$

SOLUTIONS AND HINTS TO SELECTED EXERCISES

7.36
$$S'_n(x) = \begin{cases} \left(1 + \dfrac{1}{n}\right)x^{1/n} & \text{if } x > 0 \\ 0 & \text{if } x = 0 \\ -\left(1 + \dfrac{1}{n}\right)(-x)^{1/n} & \text{if } x < 0 \end{cases}$$

and

$$\lim_{n \to \infty} S'_n(x) = \begin{cases} 1 & \text{if } x > 0 \\ 0 & \text{if } x = 0 \\ -1 & \text{if } x < 0 \end{cases} = \text{sgn}(x)$$

The convergence of $\{S'_n(x)\}$ to sgn (x) is not uniform since each $S'_n(x)$ is continuous but sgn (x) is not.

7.37 (a) $S(x) = \begin{cases} 1 & \text{if } x = 0 \\ 0 & \text{if } 0 < x \le 1 \end{cases}$

(b) $S(x) = \begin{cases} 1 & \text{if } 0 < x < 1 \\ 0 & \text{if } x = 1 \\ -1 & \text{if } 1 < x < 2 \end{cases}$

(c) $S(x) = \begin{cases} 0 & \text{if } -1 < x < 1 \\ 1 & \text{if } x \in [-2, -1] \cup [1, 2] \end{cases}$

In each of the above cases the sequence of continuous functions converges to a function which is not continuous.

7.40 Theorem 7.13 is not contradicted because the functions S_n are not restricted to a common closed bounded interval $[a, b]$.

Section 7.4

7.41 (a) $R = 1$ (b) $R = 1/e$ (c) $R = \tfrac{1}{3}$ (d) $R = \tfrac{1}{3}$
      (e) $R = 5$ (f) $R = 5$ (g) $R = 0$ (h) $R = \infty$

7.42 (a) $[-1, 5]$ (b) $(-4, -2]$ (c) $(-\infty, \infty)$ (d) $(1, 9)$

7.47 $E \le \tfrac{1}{7!}(0.5)^7 = 0.0000016$

7.48 $E \le 0.25$

Section 8.1

8.1 $|U + V|^2 = (U + V) \cdot (U + V) = |U|^2 + 2U \cdot V + |V|^2$
                $\le |U|^2 + 2|U||V| + |V|^2 = (|U| + |V|)^2$

8.4 (a) $\{(x, y) \mid x^2 + y^2 \le 1\}$ (b) $\{(x, y) \mid x^2 + y^2 \le 1\}$
(c) $\{(x, y) \mid y = \sin 1/x, x \ne 0\} \cup \{(0, y) \mid |y| \le 1\}$ (d) $\mathbf{R}^2$
(e) $\{(1/n, 1/m) \mid n, m \text{ are nonzero integers}\} \cup \{(1/n, 0) \mid n \text{ is a nonzero integer}\} \cup \{(0, 1/m) \mid m \text{ is a nonzero integer}\} \cup \{(0, 0)\}$

8.5 $(x - 1, y - 7, z + 1) = t(1, -6, 1), t \in \mathbf{R}$; $P = (x, y, z)$ is any point on the line.

8.7 $x - 2y + 2z + 3 = 0$

8.8 (a) Open, connected, bounded; boundary $= \{(x, y) \mid 2x^2 + 4y^2 = 1\}$.
      (b) Closed, not connected, unbounded; boundary $= \{(x, y) \mid x^2 - y^2 = 1\}$.

(c) Closed, connected, bounded; boundary = $\{(x, y) | |y| = 1$ and $|x| \leq 1\} \cup \{(x, y) | |x| = 1$ and $|y| < 1\}$
(d) Neither open nor closed, not connected, unbounded; boundary = $\mathbf{R}^2$

8.13 (a) 1   (b) Does not exist   (c) Does not exist
(d) $\frac{1}{2}$ (e) 0   (f) 1   (g) Does not exist

8.16 (a) Continuous at all points (x, y) except those for which $|x| = |y|$; the discontinuity is removable at those points (x, y) where $x = y$ except at $(0, 0)$. Define $f(x, y) = 1/(x + y)$ at these points.

(b) Continuous at all points (x, y) except those on the line $x = 0$. No discontinuities are removable.

(c) Continuous everywhere.

(d) Continuous at all points (x, y) except those for which $x \leq y$. No discontinuities are removable.

Section 8.2

8.20 (a) $f_x(x, y) = \dfrac{y^3 - x^2 y}{(x^2 + y^2)^2}$ $f_y(x, y) = \dfrac{x^3 - xy^2}{(x^2 + y^2)^2}$ where $(x, y) \neq (0, 0)$

(b) $f_x(x, y) = \dfrac{1}{x - y}$ $f_y(x, y) = \dfrac{1}{y - x}$ where $x > y$

(c) $f_x(x, y) = -\dfrac{y}{x^2} \cos \dfrac{y}{x}$ $f_y(x, y) = \dfrac{1}{x} \cos \dfrac{y}{x}$ where $x \neq 0$

8.21 $f_x(0, 0) = f_y(0, 0) = 0$

8.22 (a) Not differentiable at $(0, 0)$ (b) Not continuous at $(0, 0)$
(c) Continuous, but not differentiable at $(0, 0)$ (d) Differentiable at $(0, 0)$

8.26 Consider the line through $(0, 0)$ given by $(x, y) = t(\beta_1, \beta_2)$, where $\beta_1^2 + \beta_2^2 = 1$ and $t \in \mathbf{R}$.

$$f(x, y) = f(t(\beta_1, \beta_2)) = f(t\beta_1, t\beta_2)$$

$$= \frac{t^3 \beta_1^2 \beta_2}{t^4 \beta_1^4 + t^2 \beta_2^2} = \frac{t\beta_1^2 \beta_2}{t^2 \beta_1^4 + \beta_2^2} \to 0 \quad \text{as } t \to 0$$

8.27

$$f_x(x, y) = \begin{cases} \dfrac{x^4 y - y^5 + 4x^2 y^3}{(x^2 + y^2)^2} & \text{if } (x, y) \neq (0, 0) \\ 0 & \text{if } (x, y) = (0, 0) \end{cases}$$

$$f_y(x, y) = \begin{cases} \dfrac{x^5 - 4x^3 y^2 - xy^4}{(x^2 + y^2)^2} & \text{if } (x, y) \neq (0, 0) \\ 0 & \text{if } (x, y) = (0, 0) \end{cases}$$

$f_{xy}(0, 0) = -1$ and $f_{yx}(0, 0) = 1$

8.28 $f \in \mathcal{C}(\mathbf{R}^2); f \in \mathcal{C}^1(\mathbf{R}^2); f \notin \mathcal{C}^2(\mathbf{R}^2)$

8.29 Let $f(x, y) = x^k |x|$; then $f \notin \mathcal{C}^{k+1}(\mathbf{R}^2)$ but $f \in \mathcal{C}^k(\mathbf{R}^2)$.

8.30 Let

$$f(x, y) = \begin{cases} x^2 \sin \dfrac{1}{x} + y^2 \sin \dfrac{1}{y} & \text{if } x \neq 0 \text{ and } y \neq 0 \\ 0 & \text{if } x = 0 \text{ or } y = 0 \end{cases}$$

$f_x(0, 0) = f_y(0, 0) = 0$. $f(x, y)$ is differentiable at $(0, 0)$; but f_x and f_y are discontinuous at $(0, 0)$.

Section 8.3

8.31 $\dfrac{\partial z}{\partial u} = 2e^{2u} \sin 2v$ $\qquad \dfrac{\partial z}{\partial v} = 2e^{2u} \cos 2v$

8.32 $\dfrac{\partial z}{\partial u} = aF_x + bF_y$ $\qquad \dfrac{\partial z}{\partial v} = cF_x + dF_y$

$\dfrac{\partial^2 z}{\partial u^2} = a^2 F_{xx} + 2ab F_{xy} + b^2 F_{yy}$

$\dfrac{\partial^2 z}{\partial u \, \partial v} = ac F_{xx} + bc F_{yx} + ad F_{xy} + bd F_{yy}$

$\dfrac{\partial^2 z}{\partial v^2} = c^2 F_{xx} + 2cd F_{xy} + d^2 F_{yy}$

8.33 (a) $\dfrac{\partial w}{\partial x} = \dfrac{dw}{du} \dfrac{\partial u}{\partial x}$ $\qquad \dfrac{\partial w}{\partial y} = \dfrac{dw}{du} \dfrac{\partial u}{\partial y}$ $\qquad \dfrac{\partial w}{\partial z} = \dfrac{dw}{du} \dfrac{\partial u}{\partial z}$

(b) $f_{xx} + f_{yy} + f_{zz} = 4(x^2 + y^2 + z^2)g''(x^2 + y^2 + z^2) + 6g'(x^2 + y^2 + z^2)$

8.36 $\dfrac{\partial^2 z}{\partial r^2} + \dfrac{1}{r^2} \dfrac{\partial^2 z}{\partial \theta^2} + \dfrac{1}{r} \dfrac{\partial z}{\partial r} = 0$

8.38 (a) $\dfrac{dz}{dt} = \dfrac{\partial z}{\partial x} \dfrac{dx}{dt} + \dfrac{\partial z}{\partial y} \dfrac{dy}{dt} = F_x(x, y)g'(t) + F_y(x, y)h'(t)$

(b) $\dfrac{\partial z}{\partial x} = F_1(x, y, u) + F_3(x, y, u) \dfrac{\partial u}{\partial x}$ $\qquad \dfrac{\partial z}{\partial y} = F_2(x, y, u) + F_3(x, y, u) \dfrac{\partial u}{\partial y}$

(c) $\dfrac{\partial z}{\partial x} = F_1(x, y, u, v) + F_3(x, y, u, v) \dfrac{\partial u}{\partial x} + F_4(x, y, u, v) \dfrac{\partial v}{\partial x}$

$\dfrac{\partial z}{\partial y} = F_2(x, y, u, v) + F_3(x, y, u, v) \dfrac{\partial u}{\partial y} + F_4(x, y, u, v) \dfrac{\partial v}{\partial y}$

(d) $\dfrac{dz}{dt} = F_1(x, y, t)f_1(y, t)g'(t) + F_1(x, y, t)f_2(y, t) + F_2(x, y, t)g'(t) + F_3(x, y, t)$

8.40 (a) $-1 - x - \dfrac{x^2}{2} + \dfrac{(y - \pi)^2}{2} - \dfrac{x^3}{6} + \dfrac{x}{2}(y - \pi)^2$

(b) $(x - 1) + (y - 1) - \dfrac{(x - 1)^2}{2} - \dfrac{(y - 1)^2}{2} + \dfrac{1}{3}(x - 1)^3 + \dfrac{1}{3}(y - 1)^3$

(c) $1 + 2(x - 1) + 2y + 2(x - 1)y + (x - 1)^2 + y^3$

Section 8.4

8.43 (a) $F(x, y, z) = e^{x^2 + y^2 + z^2} - 1$; $F_z(0, 0, 0) = 1$; no solution for z in a neighborhood of $(0, 0)$.

(b) $F(x, y, z) = z + 2\cos(x + y + z) - 2$; $F_z(0, 0, 0) = 1$; a solution $z = g(x, y)$ exists in a neighborhood of $(0, 0)$.

$$\frac{\partial z}{\partial x} = \frac{2\sin(x + y + z)}{1 - 2\sin(x + y + z)} \qquad \frac{\partial z}{\partial y} = \frac{2\sin(x + y + z)}{1 - 2\sin(x + y + z)}$$

(c) $F(x, y, z) = x^2 y^2 z^2 - 1$; $F_z(1, 1, 1) = 2$; a solution $z = g(x, y)$ exists in a neighborhood of $(1, 1)$.

$$\frac{\partial z}{\partial x} = -\frac{z}{x} \qquad \frac{\partial z}{\partial y} = -\frac{z}{y}$$

(d) $F(x, y, z) = x^2 + y^2 - z^2 - 1$; $F(0, 0, 1) \neq 0$; therefore $(0, 0, 1)$ does not lie on the graph of the equation $x^2 + y^2 - z^2 = 1$.

8.44 $\quad \dfrac{\partial x}{\partial y} = -\dfrac{F_y}{F_x} \qquad \dfrac{\partial y}{\partial z} = -\dfrac{F_z}{F_y} \qquad \dfrac{\partial z}{\partial x} = -\dfrac{F_x}{F_z}$

Thus $\dfrac{\partial x}{\partial y}\dfrac{\partial y}{\partial z}\dfrac{\partial z}{\partial x} = -1$.

8.46 $h'(x) = \dfrac{dz}{dx} = f_1(x, y) + f_2(x, y)\dfrac{dy}{dx}$

$$= f_1(x, y) + f_2(x, y)\left(-\frac{g_1(x, y)}{g_2(x, y)}\right)$$

$$= \frac{f_1(x, y)g_2(x, y) - f_2(x, y)g_1(x, y)}{g_2(x, y)}$$

8.51 (a) $\{(x, y) \mid x^2 + y^2 \leq 1 \text{ and } y \geq 0\}$
(b) Region inside the square with vertices at $(0, 0), (1, 1), (0, 2), (-1, 1)$.
(c) Region bounded by the curves (1) $y = 0$, $|x| \leq 1$; (2) $x = 1 - (y/2)^2$, $0 \leq y \leq 2$; (3) $x = (y/2)^2 - 1$, $0 \leq y \leq 2$
(d) $\{(x, y) \mid y = -2x \text{ and } -5 \leq x \leq 2\}$

8.53 $\quad \dfrac{\partial(u, v)}{\partial(x, y)} = \begin{vmatrix} \frac{1}{J}g_v & -\frac{1}{J}f_v \\ -\frac{1}{J}g_u & \frac{1}{J}f_u \end{vmatrix} = \frac{1}{J^2}\begin{vmatrix} g_v & -f_v \\ -g_u & f_u \end{vmatrix} = \frac{1}{J^2}(f_u g_v - f_v g_u) = \frac{1}{J^2}J = \frac{1}{J}$

8.54 (a) $J = e^{2u} \neq 0$; local inverses exist everywhere.
(b) $J = 4(u^2 + v^2) \neq 0$ unless $(u, v) = (0, 0)$; local inverses exist everywhere except at $(0, 0)$.
(c) $J = u - v$; local inverses exist at each point (u, v) where $u \neq v$.
(d) $J = 0$ everywhere; transformation is not one-to-one in a neighborhood of any point.

Section 8.5

8.55 (a) $f(x, y) = -(x^4 + y^4)$; $f(0, 0)$ is a local max
(b) $f(x, y) = x^4 + y^4$; $f(0, 0)$ is a local min
(c) $f(x, y) = x^3 y^3$; $(0, 0, 0)$ is a saddle point
8.56 (a) $f(3, 1) = -6$ is a local min. (b) $f(\tfrac{1}{2}, -\tfrac{1}{2}) = \tfrac{7}{4}$ is a local max.
(c) No local extrema; $(0, 0, 0)$ is a saddle point.
8.57 (a) $q^2 - pr < 0$ and $p < 0$ (b) $q^2 - pr < 0$ and $p > 0$ (c) $q^2 - pr > 0$
8.58 $x = \tfrac{1}{3}w$ and $\theta = \pi/3$

8.59 Highest tempearture is 25; lowest temperature is -25.

8.60 $\dfrac{|d|}{\sqrt{a^2 + b^2 + c^2}}$

8.63 Maximum value is $\left(\sum\limits_{i=1}^{n} c_i^2\right)^{1/2}$

Section 9.1

9.4 $U(P, f) = \sum\limits_{i=1}^{n} \sum\limits_{j=1}^{m} M_{ij} A(R_{ij}) = \sum\limits_{i=1}^{n} \sum\limits_{j=1}^{m} \left(\dfrac{2i}{n} + \dfrac{3j}{m}\right)\dfrac{1}{mn} = \dfrac{n+1}{n} + \dfrac{3(m+1)}{2m}$

and so $\iint_R (2x + 3y) = \frac{5}{2}$.

9.5 (a) If $B_1 \subseteq B$ then $\mathcal{O}(B_1) \le \mathcal{O}(B)$ (b) $\mathcal{O}(B_1 \cup B_2) \le \mathcal{O}(B_1) + \mathcal{O}(B_2)$

9.7 (a) Suppose A is a set with n elements. For any partition P, $\sum_{C1 \cup C3} A(R_{ij}) \le n\|P\|^2$. Then $\mathcal{O}(A) = \inf_P \sum_{C1 \cup C3} A(R_{ij}) \le n\|P\|$. By the arbitrariness of P, $\mathcal{O}(A) = 0$, since $\|P\|$ can be made as small as we like.

(b) Let $B = \{(x, x) | 0 \le x \le 1\}$ (also see Theorem 9.3).

(c) Let x_0 be an interior point of B and choose the rectangle R such that $x_0 \in R \subseteq B$. Then $\mathfrak{J}(B) \ge A(R) > 0$.

9.9 Let $B = \{(x, y) | x^2 + y^2 \le 1\} \cup \{(x, y) | 1 < x^2 + y^2 \le 2$ and $x, y \in \mathbf{Q}\}$. The boundary $\partial B = \{(x, y) | 1 < x^2 + y^2 < 2\}$ does not have zero area, and so B does not have area.

9.10 (a) $\operatorname{gr}(f) \subseteq \partial(\operatorname{gr}(f))$. If $\operatorname{gr}(f)$ has area then $\partial(\operatorname{gr}(f))$ has zero area, and so by Exercise 9.5 $\operatorname{gr}(f)$ has zero area.

(b) Let $\{A_n\}$ be a sequence of pairwise disjoint countable dense subsets of $[a, b]$ and let $\{r_1, r_2, r_3, \ldots\}$ be the set of rationals in $[0, 1]$. Define $f(x) = r_n$ if $x \in A_n$ and $f(x) = 0$ if $x \in \bigcup_{n=1}^{\infty} A_n$. Then $f: [a, b] \to \mathbf{R}$ is bounded and $\partial(\operatorname{gr}(f)) = [a, b] \times [0, 1]$ does not have zero area. Therefore $\operatorname{gr}(f)$ does not have area.

9.11 $f^*(x, y) = 0$ for every point $(x, y) \in R_1 - R$.

9.13 $\iint_B f$ exists by Theorem 9.5. Let R be a rectangle with $B \subseteq R$ and let f^* be the auxiliary function. Since f^* is bounded on R, there is an $M > 0$ such that $|f^*(x, y)| \le M$ for all $(x, y) \in R$. Since B has zero area, there is a partition P of R such that the sum of the areas of those R_{ij} which contain at least one point of B is less than ε. Then

$$0 \le U(P, f^*) = \sum_{i,j} M_{ij} A(R_{ij}) \le \sum_{C3} M A(R_{ij}) < M\varepsilon$$

Since ε was arbitrary, $\overline{\iint} f^* = 0$, and so $\iint_R f^* = 0$. Therefore $\iint_B f = 0$.

9.15 (a), (b), (c), and (f) are integrable functions on $[0, 1] \times [0, 1]$.

9.16 If $f(x_0, y_0) > 0$ then there is rectangle $\hat{R}$ with $(x_0, y_0) \in \hat{R} \subseteq R$ such that $f(x, y) \ge \frac{1}{2} f(x_0, y_0) > 0$ for all $(x, y) \in \hat{R}$. Then $\iint_{\hat{R}} f \ge \frac{1}{2} f(x_0, y_0) A(\hat{R}) > 0$ and so $\iint_R f > 0$.

Section 9.2

9.18 (a) $\frac{2}{3}$ (b) $\frac{1}{3}$ (c) $\frac{1}{6}$

9.19 $\int_0^1 \int_0^1 f(x, y)\,dy\,dx = 1$. For any fixed $y \in [0, 1]$, $y \ne \frac{1}{2}$, $f(x, y) = 1$ if x is rational in $[0, 1]$ and $f(x, y) = 2y \ne 1$ if x is irrational in $[0, 1]$. Thus $\int_0^1 f(x, y)\,dx$ does not exist when $y \ne \frac{1}{2}$, and so $\int_0^1 \int_0^1 f(x, y)\,dx\,dy$ does not exist. $\iint_R f$ does not exist.

9.20 (a) $\frac{5528}{21}$ (b) $\frac{3394}{105}$

9.21 (a) $\int_0^1 \int_y^1 f(x, y)\,dx\,dy$ (b) $\int_0^4 \int_{\sqrt{x}}^2 f(x, y)\,dy\,dx$

(c) $\int_{-2}^{1} \int_{-y}^{\sqrt{2-y}} f(x, y)\,dx\,dy + \int_1^2 \int_{-\sqrt{2-y}}^{\sqrt{2-y}} f(x, y)\,dx\,dy$

9.22 $\frac{1}{8}$

9.23 $V = \int_{-4}^{1} \int_{3x}^{4-x^2} (x^2 + y^2) \, dy \, dx$

9.24 (a) $\frac{1}{4}(e^4 - 1)$ (b) $\dfrac{2}{3\pi}$

9.26 (a) $g'(x) = 2 \tan^{-1} \dfrac{1}{x}$ (b) $g'(x) = \dfrac{\sin x}{x}$

9.28 (a) $g'(x) = e^{-x^4} - \int_0^x 2xy^2 e^{-x^2 y^2} \, dy$ (b) $g'(x) = e^x \sqrt{1 + e^{3x}} - \cos x \sqrt{1 + \sin^3 x}$

9.29 $g'(x) = \int_{u(x)}^{v(x)} f_x(x, y) \, dy + f(x, v) v'(x) - f(x, u) u'(x)$

Section 9.3

9.31 (a) $\iint_{D^*} r^2 \cos 2\theta \, r \, dr \, d\theta = \int_0^\pi \int_0^{\sqrt{2}} r^3 \cos 2\theta \, dr \, d\theta = 0$

where $D^* = \{(r, \theta) \mid 1 \leq r \leq \sqrt{2} \text{ and } 0 \leq \theta \leq \pi\}$

(b) $\iint_{D^*} \dfrac{1}{2} \sin 2\theta \, r \, dr \, d\theta = \int_0^{\pi/2} \int_0^2 \dfrac{r}{2} \sin 2\theta \, dr \, d\theta = 1$

where $D^* = \{(r, \theta) \mid 0 \leq r \leq 2 \text{ and } 0 \leq \theta \leq \pi/2\}$

(c) $\iint_{D^*} e^{-r} r \, dr \, d\theta = \int_{-\pi}^{\pi} \int_0^3 r e^{-r} \, dr \, d\theta = 2\pi \left(1 - \dfrac{4}{e^3}\right)$

9.32 (a) $V = 4 \int_0^2 \int_0^{\sqrt{4-x^2}} (4 - x^2 - y^2) \, dy \, dx = 8\pi$

(b) $V = \int_{-\pi}^{\pi} \int_0^2 (4r - r^3) \, dr \, d\theta = 8\pi$

9.33 Let $x = u - v$, $y = v$. Then $J = 1$ and

$$\int_0^1 \int_0^{1-x} \exp\left(\dfrac{y}{x+y}\right) dy \, dx = \int_0^1 \int_0^u \exp\left(\dfrac{v}{u}\right) dv \, du = \tfrac{1}{2}(e - 1)$$

9.35 1

Section 9.4

9.37 (a) Converges for $p < -1$ (b) Converges for $p > -1$

9.38 (a) First kind, diverges (b) First kind, diverges
(c) First kind, converges (d) Third kind, converges

9.39 Note that $0 \leq f(x) + |f(x)| \leq 2|f(x)|$ and use the comparison test.

9.41 All converge absolutely except (c), which diverges.

9.42 (a) Converges (b) Diverges, Cauchy principal value converges

9.47 (a) $\hat{f}(s) = 1/(s - a)$ for $s \in (a, \infty)$ (b) $\hat{f}(s) = 1/s^2$ for $s \in (0, \infty)$
(c) $\hat{f}(s) = 1/(1 + s^2)$ for $s \in (0, \infty)$

9.48 Note that $|x^n| < e^{nx}$ for all $x > 0$.

Section 10.1

10.2 For all $a, b \in \mathbf{R}$ $(|a| - |b|)^2 \geq 0$, and so $|a|^2 + |b|^2 \geq 2|a||b|$. Then

$(x_1 - y_1)^2 + (x_2 - y_2)^2 \leq |x_1 - y_1|^2 + 2|x_1 - y_1||x_2 - y_2| +$

$$|x_2 - y_2|^2 \leq 2|x_1 - y_1|^2 + 2|x_2 - y_2|^2$$

and
$$\sqrt{(x_1 - y_1)^2 + (x_2 - y_2)^2} \leq |x_1 - y_1| + |x_2 - y_2|$$
$$\leq \sqrt{2}\sqrt{(x_1 - y_1)^2 + (x_2 - y_2)^2}$$
Hence $d(x, y) \leq d'(x, y) \leq \sqrt{2}\, d(x, y)$

10.4 (a) Parts (a) to (c) of Definition 10.1 are clear.
$$d'(x, y) = \sum_{k=1}^{n} |x_k - y_k| \leq \sum_{k=1}^{n} (|x_k - w_k| + |w_k - y_k|)$$
$$= \sum_{k=1}^{n} |x_k - w_k| + \sum_{k=1}^{n} |w_k - y_k| = d'(x, w) + d'(w, y)$$

(b) Parts (a) to (c) of Definition 10.1 are clear.
$$d^*(x, y) = \max_{1 \leq k \leq n} |x_k - y_k| \leq \max_{1 \leq k \leq n} (|x_k - w_k| + |w_k - y_k|)$$
$$\leq \max_{1 \leq k \leq n} |x_k - w_k| + \max_{1 \leq k \leq n} |w_k - y_k|$$
$$= d^*(x, w) + d^*(w, y)$$

10.6 (b), (c), and (e) are metric spaces

10.7 Part (a) of Definition 10.1: $d(x, y) = \frac{1}{2}[(d(x, y) + d(x, y)] > \frac{1}{2}d(x, x) = 0$
Part (b) given.
Part (c) $d(x, x) + d(y, x) \geq d(x, y)$, and so $d(y, x) \geq d(x, y)$.
$d(y, y) + d(x, y) \geq d(y, x)$, and so $d(x, y) \geq d(y, x)$.
Part (d) $d(x, y) \leq d(x, w) + d(y, w) = d(x, w) + d(w, y)$ by part (c).

10.10 Parts (a) to (c) of Definition 10.1 are clear.
$$d^*(f, g) = \sup_{x \in X} |f(x) - g(x)| \leq \sup_{x \in X} (|f(x) - h(x)| + |h(x) - g(x)|)$$
$$\leq \sup_{x \in X} |f(x) - h(x)| + \sup_{x \in X} |h(x) - g(x)| = d^*(f, h) + d^*(h, g)$$

10.12 There exist functions $f, g \in \mathcal{R}\,[0, 1]$ with $f \neq g$ and $\tilde{d}(f, g) = 0$.

Section 10.2

10.13 Let $\eta = d(x_1, x_2) > 0$ and choose $r_1 = r_2 = \frac{1}{2}\eta$

10.16 If $t \notin S_r[x_0]$ then $d(t, x_0) > r$. Let $\eta = d(t, x_0) - r > 0$. If $z \in S_\eta(t)$ then $d(z, t) < \eta$ and
$$d(z, x_0) \geq d(t, x_0) - d(t, z) > d(t, x_0) - \eta = r$$
and so $z \notin S_r[x_0]$. Thus $S_r(t) \cap S_r[x_0] = \emptyset$. It follows that t is not a limit point of $S_r[x_0]$. By the arbitrariness of t, $S_r[x_0]$ contains all its limit points and so is a closed set.

10.18 In $(\mathbf{R}, d)$, the real numbers with the usual metric, the set $\mathbf{Q}$ of rationals is not closed and not open.

10.19 If (S, d) is discrete space and $x_0 \in S$ then the set $\{x_0\}$ is both closed and open.

10.20 In discrete space (S, d) with $x_0 \in S$, $S_1(x_0) = \{x_0\}$, $\overline{S_1(x_0)} = \{x_0\}$ and $S_1[x_0] = S$. If (S, d) is any metric space and $x \in \overline{S_r(x_0)}$ then $d(x, S_r(x_0)) = 0$. Given any $\varepsilon > 0$ there is a $y \in S_r(x_0)$ with $d(x, y) < \varepsilon$. Then $d(x, x_0) \leq d(x, y) + d(y, x_0) < \varepsilon + r$. By the arbitrariness of ε, $d(x, x_0) \leq r$, and so $x \in S_r[x_0]$.

10.22 If $D \subseteq F$, where F is a closed proper subset of S, then D is disjoint from the nonempty open set $S - F$ and so is not dense in S. Conversely, if D is not contained in any closed proper subset of S then D has a point in common with each nonempty open set in S and so is dense in S.

10.24 The set $\{r + \sqrt{2} \mid r \in \mathbf{Q}\}$ is a countable set which is dense in the space $\mathbf{R} - \mathbf{Q}$ of irrationals.

10.28 Let $D = \{x_1, x_2, x_3, \ldots\}$ be a countable dense subset of the metric space S, and let $\mathcal{G} = \{S_{1/n}(x_k) \mid n, k \in \mathbf{N}.\}$. Now, if G is any open subset of S and $x \in G$ there is an $r > 0$ such that $x \in S_r(x) \subseteq G$. Choose $n \in \mathbf{N}$ with $1/n < r/2$. Then there is an $x_k \in S_{1/n}(x)$. Therefore $x \in S_{1/n}(x_k) \subseteq S_{r/2}(x_k) \subseteq S_r(x) \subseteq G$, and so G is the union of all those open spheres $S_{1/n}(x_k)$ which it contains.

10.29 By Exercise 10.28 there is a countable collection of open sets such that each open set G_α ($\alpha \in \Gamma$) can be written as a union of sets in $\mathcal{G}$. Let $\hat{\mathcal{G}}$ be the subset of $\mathcal{G}$ consisting of all those open sets which are contained in some set G_α. Then $\bigcup_{G \in \hat{\mathcal{G}}} G = S$. Since $\hat{\mathcal{G}}$ is countable, if for each $G \in \hat{\mathcal{G}}$ we choose a set G_{α_i} with $G \subseteq G_{\alpha_i}$, we shall have a countable collection of sets $G_{\alpha_1}, G_{\alpha_2}, G_{\alpha_3}, \ldots$ with $\bigcup_{n=1}^{\infty} G_{\alpha_n} = S$.

10.30 Let G be an open set in (S, d) and let $x \in G$. Then there is an open sphere $S_r(x) \subseteq G$. Now

$$\left\{y \in S \mid \rho(x, y) < \frac{r}{k}\right\} = \{y \in S \mid k \cdot \rho(x, y) < r\} \subset \{y \in S \mid d(x, y) < r\} = S_r(x)$$

Since $\{y \in S \mid \rho(x, y) < r/k\}$ is the open sphere of radius r/k centered at x in the space (S, ρ), G is an open set in the metric space (S, ρ).

10.31 By Exercise 10.2 $d(x, y) \leq d'(x, y) \leq \sqrt{2}\, d(x, y)$, and so by Exercise 10.30 every open set in $(\mathbf{R} \times \mathbf{R}, d')$ is open in $(\mathbf{R} \times \mathbf{R}, d)$ and every open set in $(\mathbf{R} \times \mathbf{R}, d)$ is open in $(\mathbf{R} \times \mathbf{R}, d')$.

10.32 Let $S_r^*(x) = \{y \in S_1 \mid d(x, y) < r\}$ be the open sphere of radius r centered at x in the subspace (S_1, d). Then $S_r^*(x) = S_r(x) \cap S_1$. $H \subseteq S_1$ is open in (S_1, d) if and only if H is the union of open spheres $S_r^*(x)$; let G be the union of the open spheres $S_r(x)$. Then $H = G \cap S_1$.

Section 10.3

10.33 If $\lim_{n \to \infty} x_n = x$ and $\lim_{n \to \infty} x_n = x'$ then $d(x, x') \leq d(x, x_n) + d(x_n, x')$ implies that $x = x'$.

10.34 $d(x_n, y_n) \leq d(x_n, x_0) + d(x_0, y_0) + d(y_0, y_n)$

10.36 Given $\varepsilon > 0$ there is an N_1 such that $d(x_n, x_m) < \varepsilon/2$ whenever $m, n > N_1$ and there is an N_2 such that $d(x_{n_k}, x_0) < \varepsilon/2$ whenever $n_k \geq N_2$. Choose $n_k^* > \max(N_1, N_2)$. If $n \geq n_k^*$ then

$$d(x_n, x_0) \leq d(x_n, x_{m_k}^*) + d(x_{n_k}^*, x_0) < \frac{\varepsilon}{2} + \frac{\varepsilon}{2} = \varepsilon$$

and so $\lim_{n \to \infty} x_n = x_0$.

10.38 If $f \in \mathcal{C}[a, b]$, define $g(x) = f[(b - a)x + a]$ for all $x \in [0, 1]$. $g \in \mathcal{C}[0, 1]$, and by the Weierstrass approximation theorem there is a polynomial $p(x)$ such that $|g(x) - p(x)| < \varepsilon$ for all $x \in [0, 1]$. Define $q(x) = p((x - a)/(b - a))$. Then $q(x)$ is a polynomial and $|f(x) - q(x)| < \varepsilon$ for all $x \in [a, b]$. Therefore the set of polynomial functions is dense in $\mathcal{C}[a, b]$.

10.39 Consider the open sphere $S_{1/3}(f)$, where f is the function

$$f(x) = \begin{cases} 0 & \text{if } x \in \mathbf{Q} \cap [a, b] \\ 1 & \text{if } x \in (\mathbf{R} - \mathbf{Q}) \cap [a, b] \end{cases}$$

10.40 Consider the function $f(x) = \sin x$ on $\mathbf{R}$ and note that every nonconstant polynomial function is unbounded on $\mathbf{R}$.

SOLUTIONS AND HINTS TO SELECTED EXERCISES 351

10.42 R with the usual metric is a complete metric space. Consider the sequence of sets $A_n = [n, \infty)$.

10.43 If $\{x\}$ is nowhere dense in (S, d) then by Theorem 10.12, $S - \{x\}$ is dense in (S, d) and so for every $r > 0$, $S_r(x)$ contains a point in $S - \{x\}$. Therefore x is a limit point of S. Conversely, if x is a limit point of S then $\{x\}$ is not an open set in (S, d), and so every open set G contains a nonempty open set H disjoint from $\{x\}$ (see Exercise 10.13). Therefore $\{x\}$ is nowhere dense in (S, d).

10.44 If x is not a limit point of S then there is an $r > 0$ such that $S_r(x)$ is disjoint from $S - \{x\}$. Clearly $S_r(x) = \{x\}$ and so $\{x\}$ is open.

10.45 If (S, d) is a complete metric space and S is countable then by Theorem 10.13 there is an $x \in S$ such that $\{x\}$ is not nowhere dense. By Exercise 10.43 x is not a limit point of S.

10.46 Let (S, d) be discrete space and suppose A is nowhere dense in (S, d). Let $x \in S$ be arbitrary. Since $\{x\}$ is an open set, $A \cap \{x\} = \emptyset$ and so $x \notin \bar{A}$. Therefore $\bar{A} = \emptyset$.

10.47 If F is a closed set in the metric space (S, d) which is not nowhere dense then, by Theorem 10.12, $S - F$ is not dense in (S, d). Hence there is a nonempty open set G with $(S - F) \cap G = \emptyset$. Then $G \subseteq F$ and so F contains a nonempty open set.

10.48 If G is an open set in the metric space (S, d) and $S - G$ is not nowhere dense then, by Exercise 10.47, $S - G$ contains a nonempty open set H. Clearly $H \cap G = \emptyset$, and so G is not dense in (S, d).

10.49 If G_n is an open dense set in (S, d) for $n = 1, 2, 3, \ldots$ then, by Exercise 10.48, $S - G_n$ is nowhere dense. By Theorem 10.13, $S - \bigcup_{n=1}^{\infty} (S - G_n) = \bigcap_{n=1}^{\infty} G_n$ is dense in (S, d).

10.50 Let $F_n = \{x \in S \mid |f(x)| < n \text{ for every } f \in \mathcal{F}\}$. F_n is closed by the continuity of the functions $f \in \mathcal{F}$ and $\bigcup_{n=1}^{\infty} F_n = S$ by hypothesis. By Baire's theorem there is a set F_{n_0} which is not nowhere dense in (S, d), and by Exercise 10.47 there is an open set $G \subseteq F_{n_0}$. Then for every $x \in G, |f(x)| \le n_0$ for all $f \in \mathcal{F}$.

Section 10.4

10.51 $f:(S, d) \to (T, \rho)$. If f is continuous and F is a closed set in T then, by Theorem 10.16, $f^{-1}(T - F)$ is open in S. Hence $f^{-1}(F) = S - f^{-1}(T - F)$ is closed in S. Conversely, suppose $f^{-1}(F)$ is closed in S for each closed set F in T and suppose H is an open set in T. Then $f^{-1}(H) = S - f^{-1}(T - H)$ is open in S and, again by Theorem 10.16, f is a continuous function on S.

10.52 Choose a sequence of points $\{x_n\}$ in A with $\lim_{n \to \infty} x_n = x \in \bar{A}$. Then, by continuity and Theorem 10.15, $f(x_n) \to f(x)$ and $g(x_n) \to g(x)$. Since $f(x_n) = g(x_n)$ for every n, it follows from Exercise 10.33 that $f(x) = g(x)$.

10.53 $i_s:(S, \rho) \to (S, d)$ is continuous, but $i_s:(S, d) \to (S, \rho)$ is not.

10.54 Let $S = [0, 1]$ and let ρ be the discrete metric and d the usual metric on S. Then the identity function $i_s:(S, \rho) \to (S, d)$ is one-to-one, onto, and continuous, but it is not a homeomorphism.

10.55 Let $S = \{1, \frac{1}{2}, \frac{1}{3}, \ldots\}$ with ρ the discrete metric and d the usual metric on S. Then the identity function $i_s:(S, d) \to (S, \rho)$ is a homeomorphism, but the Cauchy sequence $\{1/n\}$ in (S, d) has its image sequence (again $\{1/n\}$) not a Cauchy sequence in (S, ρ).

10.56 For any metric space (S, d) consider the space (S, τ), where τ is defined by $\tau(x, y) = \min(1, d(x, y))$ for all pairs $x, y \in S$.

10.59 $\rho(f(x_n), f(x_m)) = d(x_n, x_m) \to 0$

10.60 Follows from Exercises 10.58 and 10.59.

10.61 $|f(x_1) - f(x_2)| = |d(x_1, x_0) - d(x_2, x_0)| \le d(x_1, x_2)$

10.62 The inverse is uniformly continuous.

10.63 Given $\varepsilon > 0$ choose $\delta = \varepsilon$.

10.64 Let $S = \{1, \frac{1}{2}, \frac{1}{3}, \ldots\}$ and let d_1 be the discrete metric and d_2 the usual metric on S.

10.65 Given $\varepsilon > 0$ choose $\delta = \varepsilon/k$.

10.66 For any $x \in \bar{A}$ let $\{x_n\}$ be a sequence of points in A with $\lim_{n \to \infty} x_n = x$. Define $f(x) = \lim_{n \to \infty} f(x_n)$.

10.68 A is not a closed set in $(\mathbf{R}, d)$.

10.70 Let $\{a_n\}$ be a sequence of distinct elements in A. Then there is a point $a \in S$ and a subsequence $\{a_{n_k}\}$ of $\{a_n\}$ with $a_{n_k} \to a$. a is a limit point of A.

10.71 If (S, d) is bounded, say $d(x, y) \leq M$ for all $x, y \in S$, then any singleton subset of S is an $(M + 1)$-net in S. Conversely if $\bigcup_{i=1}^{n} S_\varepsilon(x_i) = S$, let $M = \max_{1 \leq i, j \leq n} d(x_i, x_j)$. Then for any x, $y \in S$,

$$d(x, y) \leq d(x, x_i) + d(x_i, x_j) + d(x_j, y) < M + 2\varepsilon$$

where $x \in S_\varepsilon(x_i)$ and $y \in S_\varepsilon(x_j)$. It follows that (S, d) is bounded.

10.73 Let H_n be a $(1/n)$-net for $n = 1, 2, 3, \ldots$. Then the set $H = \bigcup_{n=1}^{\infty} H_n$ is a countable dense subset of (S, d).

Section 10.5

10.77 Since $\mathcal{C}[0, 1]$ is a complete metric space, there is a function $f \in \mathcal{C}[0, 1]$ such that the Cauchy sequence $\{f_n\}$ converges to f. By Theorem 10.15, $\lim_{n \to \infty} f(x_n) = f(x_0)$. Let $\varepsilon > 0$ be given. There is a natural number N_1 such that $|f(x_n) - f(x_0)| < \varepsilon/3$ whenever $n > N_1$. Also, there is a natural number N_2 such that

$$d(f_n, f) = \sup_{x \in [0, 1]} |f_n(x) - f(x)| < \frac{\varepsilon}{3} \quad \text{whenever } n \geq N_2$$

Hence $|f_n(x) - f(x)| < \varepsilon/3$ for all $x \in [0, 1]$ provided $n > N_2$. Let $N = \max(N_1, N_2)$ and suppose $n \geq N$. Then

$$|f_n(x_n) - f_n(x_0)| < |f_n(x_n) - f(x_n)| + |f(x_n) - f(x_0)| + |f(x_0) - f_n(x_0)| < \frac{\varepsilon}{3} + \frac{\varepsilon}{3} + \frac{\varepsilon}{3} = \varepsilon$$

10.79 If $f_y(x, y)$ is bounded in the rectangle R then

$$\left| \frac{f(x, y + h) - f(x, y)}{h} \right| \leq M$$

whenever (x, y) and $(x, y + h)$ are in R.

10.80 (b) and (c)

10.81 Every contraction mapping is uniformly continuous on (S, d); given $\varepsilon > 0$ choose $\delta = \varepsilon$.

INDEX

Abel's theorem, 210–211, 215 (Ex. 7.45)
Absolute convergence:
  of an improper integral, 295 (Ex. 9.39)
  of a series, 191
Absolute maximum, 131–132, 231
Absolute minimum, 131–132, 231
Absolute-value function, 29–30
Accumulation point, 53
Alternating harmonic series, 191
Alternating series, 192–193
Alternating series test, 192–193
Analytic function, 212
Antiderivative, 176
Archimedean property, 27
Area, 265
Ascoli's theorem, 324 (Ex. 10.76)

Baire's theorem, 314
Bernoulli's inequality, 26 (Ex. 1.38)
Binomial theorem, 26 (Ex. 1.39)
Boundary, 221
Boundary point, 221
Bounded function:
  at a point, 61–62
  on a set, 59–60

Bounded metric space, 299
Bounded sequence, 40, 45–53
Bounded set, 25, 221, 229
Bolzano-Weierstrass theorem, 49, 231 (Ex. 8.11)
  for sets, 54

Cauchy criterion, 186, 198, 289–290
Cauchy mean-value theorem, 133–134
Cauchy principle value, 289, 295 (Ex. 9.42)
Cauchy-Schwarz inequality, 220
Cauchy sequence, 43, 222, 309
Central difference, 144 (Ex. 5.31)
Chain rule, 129, 239–244
Closed interval, 28
Closed set, 55, 220–221, 304
Closed sphere, 303
Closure, 231 (Ex. 8.3), 305
Cluster point, 51–52
Compact metric space, 318
Compact set, 57
Comparison test, 186, 287
  limit form, 187
Complete metric space, 309
Completeness, 19, 222

Completeness axiom, 27
Conditional convergence:
  of an improper integral, 295 (Ex. 9.40)
  of a series, 191
Connected set, 221, 231 (Ex. 8.12)
Content:
  inner, 265
  outer, 265
Content zero, 179
Continuity, 94–96, 228–229, 315–316
  one-sided, 99–100
  uniform, 110–113
Continuous curve, 219
Continuous extension, 102 (Ex. 4.19–4.21), 117 (Ex. 4.61)
Continuous function:
  on an interval, 100
  at a point, 94–95, 228, 315
  on a set, 100, 316
Continuous nowhere differentiable function, 206–207, 324–326
Contraction mapping, 137, 326
Convergence:
  absolute, 191, 295 (Ex. 9.39)
  conditional, 191, 295 (Ex. 9.40)
  pointwise, 195, 199
  uniform, 196, 199, 289–291
Convergent improper integral, 286–288
Convergent sequence, 38, 308
Convergent series, 184
Countable set, 32
Cover, 55, 318
Cross product, 225–226
Curve, 219

Dedekind cut, 19
Dedekind's theorem, 23
Deleted neighborhood, 55
DeMorgan laws, 308 (Ex. 10.17)
Dense set, 29, 307

Denumerable set, 32
Derivative, 126–127
  directional, 232–234
  one-sided, 144–145
  partial, 233–234
Diameter of a set, 306
Differentiable function, 126, 234–235
Directional derivative, 232–234
Discontinuity, 95–97, 228–229
  first kind, 97
  jump, 96
  removable, 95, 228–229
  second kind, 97
  simple, 97
Discrete metric, 298
Discrete space, 298
Divergent improper integral, 286–288
Divergent sequence, 38
Divergent series, 184
Domain of a function, 3
Dot product, 219–220

Equicontinuous set of functions, 324 (Ex. 10.75)
Equivalence classes, 4
Equivalence relation, 4
ε net, 320
Euclidean metric, 299
Euclidean n-space, 299
Even function, 62
Exterior of a set, 221
Exterior point, 221
Extreme-value theorem, 104
Extremum of a function, 254

Fibonacci sequence, 50–51
Field, 17
  complete archimedean ordered, 27
  complete ordered, 20, 27
  ordered, 18, 26

INDEX 355

Finite-intersection property, 324 (Ex. 10.69)
First category, 124–125, 314
Fixed point, 106, 137, 326–327
Fubini's theorem, 272–274
Function, 2
  bounded, 59–60
  bounded at a point, 61–62
  composite, 3
  continuous, 94–96, 228–229, 315–316
  continuous nowhere differentiable, 206–207, 324–326
  differentiable, 126, 234–235
  domain, 3
  even, 62
  extremum, 254
  graph, 3
  integrable, 151, 262
  inverse, 3
  limit of, 65–71, 226–227
  mean value, 173
  monotone, 89–90
  odd, 62–63
  one-to-one, 3
  onto, 3
  oscillation, 122–123, 178–179
  range, 3
  uniformly continuous, 110–113, 317
Functions:
  equicontinuous, 324 (Ex. 10.75)
  sequence of, 195
  series of, 195, 199
  uniformly bounded, 302 (Ex. 10.11)
Fundamental limit theorem, 72–73
Fundamental theorem of calculus, 174–175

Gamma function, 291, 295 (Ex. 9.46)
Geometric progression, 37

Geometric series, 184–185
Gradient, 236
Graph of a function, 3
Greatest lower bound, 24
Greatest integer function, 30–31

Harmonic series, 185
Heine-Borel theorem, 55–57, 318
Hilbert space, 300
Homeomorphic spaces, 317
Homeomorphism, 316–317

Identity function, 4
Implicit function theorem, 246–249
Improper integral, 286
Indefinite integral theorem, 173–174
Infimum:
  of a function, 60
  of a set, 24
Inner content, 265
Inner product, 219–220
Integers, 11–13
Integrable function, 151, 262
Integral, 149–151, 262, 268–269
  improper, 286
  linearity of, 165
  lower, 150, 262
  order preserving property, 168
  upper, 150, 262
Integral test, 187–188
Integration-by-parts formula, 178 (Ex. 6.27)
Interior, 221
Interior point, 221
Intermediate-value property, 93 (Ex. 3.65), 102 (Ex. 4.23), 108
Intermediate-value theorem, 105–106
Interval, 28
Interval of convergence, 209–210

Inverse:
  global, 252–253
  local, 252, 253 (Ex. 8.54)
Inverse function, 3
Inverse function theorem, 251–252, 253 (Ex. 8.50)
Irrational numbers, 21
Isolated point, 315 (Ex. 10.44), 316
Isometric spaces, 317
Isometry, 317
Iterated integral, 271–272

Jacobian, 249–250, 282–283
Jump discontinuity, 96

Lagrange multipliers, 256–259, 260 (Ex. 8.61, 8.62)
Laplace transform, 294–295, 295 (Ex. 9.47–9.49)
Laplace's equation, 245 (Ex. 8.36)
Law of cosines, 219–220
Law of trichotomy, 9
Least upper bound, 24
Least upper bound property, 25
Lebesque number, 322
Left-continuous, 99–100
Left-hand derivative, 145
Leibniz's rule, 279
Level curves, 222, 231 (Ex. 8.9)
L'Hospital's rule, 138–144
Limit:
  of a function, 65–71
  at infinity, 84–87
  to infinity, 82–84
  one-sided, 77–82
  of a sequence, 38–39, 308–309
Limit inferior, 51
Limit point, 53–55, 221, 304
Limit superior, 51
Line segment, 219
Linear transformation, 165

Linearity of the integral, 165
Lipschitz condition, 136–138, 328, 329 (Ex. 10.79)
Local maximum, 131–132, 236, 254
Local minimum, 131–132, 236, 254
Lower bound, 24
Lower integral, 150, 262
Lower sum, 150, 262

M test, 290–291
Mathematical induction, 7
Maximum:
  absolute, 131–132, 231
  local, 131–132, 236, 254
Mean value of a function, 173
Mean value theorem for integrals, 172–173
Mean value theorem of differential calculus, 134, 246 (Ex. 8.42)
Measure zero, 180
Method of steepest descent, 236, 238 (Ex. 8.23)
Metric, 298
  discrete, 298
  euclidean, 299
  rectangular, 299, 302
  usual, 298–299
Metric space, 297–298
  bounded, 299
  compact, 318
  complete, 309
  first category, 314
  second category, 314
  separable, 307
  sequentially compact, 319–320
  subspace, 300
  totally bounded, 320–321
Minimum:
  absolute, 131–132, 231
  local, 131–132, 236, 254

Monotone function, 89–90
Monotone sequence, 45–47

Natural numbers, 6–7
Neighborhood, 54–55, 220
Nested-interval property, 49
Norm of a partition, 149–150, 263
Nowhere dense set, 123, 313–314

Odd function, 62–63
One-sided continuity, 99–100
One-sided derivative, 144–145
One-sided limit, 77–82
One-to-one correspondence, 3
Open covering, 55, 318
Open interval, 28
Open set, 55, 220, 303
Open sphere, 302
Order preserving property of the integral, 168
Orthogonal vectors, 220
Oscillation, 122–123, 178–179
Outer content, 265

Partial derivative, 233–234
Partition:
  of an interval, 149
  of a rectangle, 261–262
  of a set, 4
  norm, 149–150, 263
  refinement, 153, 262
Peano postulates, 6–7
Pie function, 81–82, 101 (Ex. 4.51)
Plane, 224–225
Pointwise convergence, 195, 199
Polar coordinate transformation, 285
Polygonal line, 221, 232 (Ex. 8.15)
Polygonally connected, 221, 231, (Ex. 8.12)

Power series, 208
p series, 188

Radius of convergence, 209–210
Range of a function, 3
Ratio test, 189
Rational numbers, 16–19, 21
Ray, 19–21
Real numbers, 19–21
Rectangular metric, 299, 302
Refinement of a partition, 153, 262
Region, 221
Relation, 4
Removable discontinuity, 95, 228–229
Riemann-integrable, 151
Riemann integral, 149–151
Riemann sum, 162, 263
Right-continuous, 99–100
Right-hand derivative, 144–145
Rolle's theorem, 133
Root test, 189–190

Saddle point, 254
Second category, 124–125, 314
Separable metric space, 307
Sequence, 37, 308
  bounded, 40, 45
  Cauchy, 43, 222, 309
  convergent, 38, 308
  divergent, 38
  of functions, 195
  limit of, 38–39, 308–309
  monotone, 45–47
Sequentially compact metric space, 319–320
Series, 184–185
  alternating, 192–193
  alternating harmonic, 191
  Cauchy criterion, 186
  convergent, 184

Series (*Cont.*):
  divergent, 184
  of functions, 195, 199
  harmonic, 185
  of numbers, 184
  partial sums, 184
  power, 208
  rearrangement, 192
  regrouping, 194 (Ex. 7.14)
  Taylor, 212, 244–245
  terms, 184
Set, 1–2
  boundary, 221
  bounded, 24, 221, 299
  closed, 55, 220, 304
  closure, 231 (Ex. 8.3)
  compact, 57
  complement, 2
  connected, 221, 231 (Ex. 8.12)
  content zero, 179
  countable, 32
  countably infinite, 32
  dense, 29, 307
  denumerable, 32
  diameter, 306
  empty, 1–2
  exterior, 221
  F_σ, 122
  finite, 32
  first category, 124–125, 314
  infinite, 32
  interior, 221
  measure zero, 180
  nowhere dense, 123, 313–314
  open, 55, 220, 303
  power, 6
  second category, 124–125, 314
  uncountable, 32
  universal, 1
Sets:
  cartesian product of, 2
  intersection of, 2
  union of, 2
Signum function, 30–31

Simple discontinuity, 97
Sphere:
  closed, 303
  open, 302
Squeeze play, 39–40
Steepest descent, 236, 238 (Ex. 8.23)
Subcover, 318
Subsequence, 48–51
Subspace, 300
Supremum:
  of a function, 60
  of a set, 24
Surface, 222–225

Tangent plane, 235
Taylor polynomial, 244–245
Taylor series, 212, 244–245
  remainder, 213–214
Taylor's formula, 244–245
Topology, 308 (Ex. 10.31)
Totally bounded metric space, 320–321
Triangle inequality, 220, 231 (Ex. 8.1), 298

Uniform-boundedness principle, 315 (Ex. 10.50)
Uniform continuity, 110–113, 232 (Ex. 8.18)
Uniform convergence;
  of an improper integral, 289–291
  of a sequence, 196
  of a series, 199
Uniformly bounded set of functions, 302 (Ex. 10.11)
Uniformly continuous function, 317
Upper bound, 24
Upper integral, 150, 262
Upper sum, 150, 262

Usual metric, 298–299

Vector, 217–218
  length, 219
Vector space, 218
Vectors:
  angle between, 220
  orthogonal, 220

Wave equation, 242–244, 246 (Ex. 8.37)
Weierstrass approximation theorem, 311, 314 (Ex. 10.38)
Weierstrass M test, 199–200
Well-ordering principle, 10
Whole number, 16

Zero area, 265